Richard Dronskowski

Computational Chemistry of Solid State Materials

Related Titles

Gasteiger, J., Engel, T. (eds.)

Chemoinformatics

A Textbook

680 pages with 379 figures and 46 tables
2003
Softcover
ISBN 3-527-30681-1

Koch, W., Holthausen, M. C.

A Chemist's Guide to Density Functional Theory

313 pages with 22 figures and 58 tables
2001
Softcover
ISBN 3-527-30372-3

Cramer, C. J.

Essentials of Computational Chemistry

Theories and Models

618 pages
2004
Softcover
ISBN 0-470-09182-7

Lipkowitz, K. B., Larter, R., Cundari, T. R., Boyd, D. B. (eds.)

Reviews in Computational Chemistry

Volume 21

443 pages
2005
Set
ISBN 0-471-68239-X

Richard Dronskowski

Computational Chemistry of Solid State Materials

A Guide for Materials Scientists, Chemists, Physicists and others

WILEY-VCH Verlag GmbH & Co. KGaA

The Author

Professor Dr. Richard Dronskowski
RWTH Aachen
Institute of Inorganic Chemistry
Landoltweg 1
52056 Aachen
Germany

1st Edition 2005
1st Reprint 2007

Library of Congress Card No.: applied for.

British Library Cataloging-in-Publication Data:
A catalogue record for this book is available from the British Library.

Bibliographic information published by the Deutsche Nationalbibliothek
The Deutsche Nationalbibliothek lists this publication in the Deutsche Nationalbibliographie; detailed bibliographic data are available in the Internet at http://dnb.d-nb.de.

Printed in the Federal Republic of Germany

Printed on acid-free paper

Printing betz-druck GmbH, Darmstadt
Binding Litges & Dopf Buchbinderei GmbH, Heppenheim

ISBN-13: 978-3-527-31410-2
ISBN-10: 3-527-31410-5

Contents

Computational Chemistry of Solid State Materials. Richard Dronskowski

ISBN: 3-527-31410-5

Foreword
Materials: the Bridge Between Chemistry and Physics

Every science has its own history. Out of which emerge distinctive ways of thinking, ways of understanding. Reductionism is a comfortable philosophy only for those who choose to avoid the reality of the way new knowledge is created.

So it is interesting when two mature sciences are forced by the facts of nature and a shared subject to confront each other's ways of thinking, both of them productive and yet, and yet... seemingly incommensurate. This is what has happened, is happening, between chemistry and physics; their preeminent and fertile shared ground is the contemporary solid state, with its exciting materials. This book, in its unique way, shapes a way not just to coexistence of chemistry and physics in materials science, but to a productive future. A future shaped by computational techniques (for theory definitely has a major role to play here) that are respectful of both chemistry and physics.

Chemistry always had its own identity, founded in part on analysis and synthesis, in another part on an immense practical exploration of reactions, of chemical change leading to new substances. That distinctly chemical identity persisted as the molecular science became quantitative, and contiguous, so to speak, with physics. Sharing with physics the common ground of atoms, physical forces, and thermodynamics, chemistry remains different, in its emphasis on structure, reactivity and that marvelous construct of the chemical bond.

The contrast is easily seen in the confrontation of chemistry and physics with the solid state. Look, for instance, at the classical triangle of bonding in materials, with its vertices of ionic substances (represented, say, by NaCl), metals (Ni, for instance), and covalent solids (diamond or silicon). In approaching theoretically any real substance, which is most unlikely to be at one or another extreme of this graph, the physics community clearly favored the ionic and metallic entry points. While chemists, addicted to the molecular compounds so easy to synthesize and analyze in the 20th century, found it much easier to analyze the new through the perspective of a directionally bonded covalent solid.

Which is right? Neither. Or both. There is much to be said for the intuition of the chemists for bonding in matter; they've had much time to think about it, a near infinity of examples. And maybe they had to think harder, just because they couldn't rely on the crutch of mathematics. But the fascinating properties of the new materials of our time – superconductivity, magnetism in all its rich manifestations, phase behavior – all of these could only be described with new physics, a physics beyond the chemist's familiar bond.

The meeting ground of the sciences is here and now, in new materials with novel properties. One is unlikely to make such materials and understand them if one is wearing pure chemical or physical blinders. Success is more likely to come from a coupled approach, a chemical understanding of bonding (in many ways deeper, as it is intuitive, than that of the physics community) merged with a deep physical description of what is really different in conductivity, magnetism and collective phenomena.

What is needed is something between a Baedeker and a Rosetta stone, something that introduces one to the two (or is it more?) cultures of chemistry and physics. As they approach an understanding of what electrons are up to in their shared ground of materials science. This book, a description of modern computational approaches to the solid state, provides the passport of a common language for creative excursions in this fertile middle ground.

Ithaca, July 2005 ROALD HOFFMANN

Preface

Since 1986, when I first started playing around in Arndt Simon's chemical laboratory as a graduate student, I have never regretted entering the fascinating field of solid-state chemistry. Indeed, I have always found that this fundamental brand of the chemical sciences and also its somewhat more applied sister subject, materials chemistry, brings us into contact with a large part of the "real world" surrounding us, and a creative solid-state (or materials) chemist is in true command, in the same way as a molecular inorganic chemist, of the whole periodic table when he or she thinks of making new compounds with often unforeseeable but exciting physical properties. It does not come as a surprise that the extraordinarily broad field of solid-state chemistry is a truly interdisciplinary one. Thus, solid-state chemistry borders with solid-state physics, crystallography, quantum theory, metal science and inorganic chemistry, to name but a few; also, it is one of the rock-solid platforms on which the increasingly popular fields of nanoscience and nanomaterials may be built.

Some of the breathtaking technological advances of the 20th, and also the early 21st century, would have been totally impossible without the fundamental research originating within solid-state chemistry. Here I am thinking of insulators with designed properties such as *dielectric ceramics* for data transmission, novel *ionic conductors* for energy storage in hand-held electrical devices, *magnetic intermetallics* and *oxides* for data storage applications, *advanced nitrides* for electro-optical and diverse mechanical purposes, and also *superconductors* for energy transport and communication applications. This list could easily be made longer, and it surely will get longer as long as solid-state chemists and materials researchers are doing their part by creating the new, particularly when they are *not* thinking of applications but are concentrating on that wonderful art of *curiosity-driven research*.

When you want to design and make new things, however, you must be able to understand (or, at least, describe) the existing ones; thus, sooner or later, theory comes into play. Not too surprisingly, the theoretical tools available in solid-state and materials chemistry are as diverse as the solid-state chemistry-related fields because this is where they originate. Thus chemists, physicists, crystallographers, and quantum theorists have all contributed to the strange blend of tools for describing, understanding and – which is now becoming increasingly important – predicting solid-state materials. At the present time, it seems that numerical approaches – which I will bravely summarize using the term *computational chemistry* – have reached a certain maturity which allows usage by the nonspecialist. Of course, computers have become more powerful, too, but this is mostly due to better hardware (solid-state materials) and, to a lesser extent, more user-friendly software.

Despite the ever-increasing importance of (quantum-theoretical) computer programs used in theoretical solid-state and materials chemistry, however, a newcomer will probably have difficulty in seeing the wood for the trees. Thus, I have felt the necessity to briefly present the type of theoretical approaches which might be used to successfully understand existing materials and also to navigate in the search for new solid-state compounds. These approaches purposely include traditional, classical ways of thinking but also quantum-mechanical approaches, because both are justified. It would be foolish to run high-scale quantum-mechanical calculations unnecessarily if an empirical back-of-the-envelope scheme is almost as predictive; also, a large amount of understanding is based upon classical ways of thinking. This may change but I am afraid it will take some time. On the other hand, why should we restrict ourselves to limited empirical methods if a reliable quantum-mechanical alternative is available? There are cases where the predictive power of quantum chemistry is so overwhelming that the experiment no longer has to be performed. Some scientists (like myself) will appreciate this, others will not. As you may have guessed, this book tries to bridge the gap.

As I write, I have imagined an intelligent reader (chemist, physicist, material scientist, crystallographer, etc.) who is already somewhat familiar with solid-state or materials chemistry, at least in terms of structure and also structure determination. However, because of space limitations, crystallographic techniques simply cannot be taught here. Also, some basic knowledge of quantum mechanics would do no harm. The rest of the book, however, is designed to be self-contained. I have tried to address a person working in the laboratory who is trying hard to make sense of his or her new discoveries. What is the best way to describe your new compound in terms of, say, energetics? What can be learned from a structural discussion using radii concepts and volume considerations? What about more general structure rationalizations? Can one approximate the strength of chemical bonding without using quantum chemistry? If quantum chemistry is needed, what are the most important ingredients? What are these band structures, really? What do you learn from densities-of-states? Can chemical bonding in the solid state be quantified?

What are the pluses and minuses of the diverse quantum-chemical routes? Although it is a crystal, are the atoms really standing still? Can solid-state materials be predicted? Do we have quantum-theoretical access to chemical thermodynamics? Can one design new compounds? (The answer is yes). In general, how do you interpret the quantum-chemical result and how do you transfer the message into a language that can be used in the laboratory? In the end, the book should enable the reader to theoretically handle his or her own materials in the sense of correctly describing and understanding the compounds under study. If it would make you (yes, *you*) predict and synthesize new compounds, I would be truly happy. To ease the difficult life of the busy synthetic materials scientist, I have tried to make the book an entertaining, extremely light read, and I keep my fingers crossed that the "pure" theorists will kindly agree. For those who want to drill deeper, there are many superb monographs available, stuffed with lots of mathematics, and the appropriate references are also included in this little book.

This book would not be here if I had not gladly accepted the honorable invitation of Tohoku University (Sendai) for a guest professorship in the summer of 2004. I am especially grateful to my colleagues and students at the Center of Interdisciplinary Research, in particular to Professor Hisanori Yamane and his family, who did a spectacular job in making my visit a most memorable one. I am extending my thanks to Professor Shinichi Kikkawa at Hokkaido University (Sapporo) for his hospitality and for the great time I was able to enjoy. Sapporo is a wonderful place for beer, too.

Also, I would like to express my thanks to my own research group at RWTH Aachen University; without them, most of the things I find important to communicate to the reader would simply be nonexistent. Over the last eight years, it has been a great pleasure to lead a group of experimentalists (challenging theory) and theorists (challenging experiment), and I would like to acknowledge the important contributions of all former and present coworkers. Thank you very much for your scientific and personal input. A good number of improvements to the first version of the manuscript were suggested, after careful reading, by Jörg von Appen, Bernhard Eck, Boniface Fokwa, Andreas Houben, Michael Krings, Marck-Willem Lumey, Paul Müller, and Holger Wolff; whom I also thank. During my stay in Japan, Mona Marquardt did a wonderful job in keeping the group together and taking care of the communication; thank you so much. The manuscript has also profited from the critical remarks of some of my colleagues, namely Peter Blöchl (Clausthal), Lothar Fritsche (Karlsruhe), Karl Jug (Hannover), Gordon J. Miller (Ames), Rainer Pöttgen (Münster), Michael Ruck (Dresden), and Gerhard Raabe (Aachen); thank you for your time. It is fascinating to observe how differently chemists, physicists, experimentalists, and theorists may consider the same subject, and I hope to have come up with a sensible compromise. Birgit Renardy did a nice job in proof-reading the many references; thanks are due to her. At Wiley-VCH, Elke Maase, Linda Bristow, Uschi Schling-Brodersen, and Manfred Köhl provided helpful guidance. I also thank Roger De Souza (Aachen) for consulting his English dictionary on my behalf. If there are scientific or typographical errors left (and I am afraid they are there), I am solely responsible for them.

Finally, my colleagues from the Institute of Inorganic Chemistry at RWTH let me flee from the remarkable teaching burden of our institution for one semester, and I am very grateful to them for having made this possible. Needless to say, a thousand thanks go directly to my family for their inspiration and support. Gabriele was of tremendous help in the final formatting steps.[1]

This book appears at a time where the high reputation of Germany's university system and even its top academic institutions are endangered by the thoughtless words and actions of some of our highly political elite. To comment on the recent embarassing large-scale experiments (with whatever future outcome) or to name their political inventors would give these people more honor than they deserve. Instead, I devote this book, with affection, to the University of Münster and its Chemistry and Physics Divisions because this is the great place where I had the pleasure to study in the first half of the 1980s.

Aachen, June 2005 RICHARD DRONSKOWSKI

1) All text was written on a Linux workstation using my beloved editor vim, and the entire book was processed by means of the glorious typesetting system LaTeX2e and further handled using xdvi, dvips, ghostscript/ghostview, ps2pdf, and pdftk. The figures were generated using gnuplot, wxdragon, xfig, and xmgrace. A toast to fast, reliable, compact and open-source software!

1 Classical Approaches

Study the past if you would define the future.
CONFUCIUS

Anyone who wants to harvest in his lifetime cannot wait for the ab initio *theory of weather.*
HANS-GEORG VON SCHNERING

Sooner or later, every solid-state chemist will realize that there is nasty time-related problem in solid-state chemistry. Because of the low mobility of the atoms in the solid state, solid-state reactions usually take much longer to complete than reactions for molecules in liquids and gases, so that a good amount of patience is the main requirement for the busy solid-state chemist! Even when high-temperature reactions involving ceramic routes and *lots* of energy are pursued, a solid-state chemist will sooner or later find himself *waiting* for nature to complete its job [1]. I suppose that these periods of contemplation are responsible for the fact that, clearly, there is a long tradition of deep thinking about three-dimensional *structures* and *energetics* in solid-state chemistry.

At present, a newcomer to solid-state chemistry might therefore believe that this science must have been a key proponent in challenging quantum mechanics (and quantum chemistry, too) for the solution of solid-state chemical problems. Strangely, this is not at all the case. Let us remind ourselves that the puzzle of chemical bonding was ingeniously clarified in 1927, not for a crystalline solid but for the hydrogen *molecule*. The rapidly emerging scientific discipline, quantum chemistry, also focused on the *molecular* parts of chemistry both because of technical and "political" reasons: first, the most important quantum-chemical workhorse (Hartree–Fock theory) has been particularly resistant to adaptation to the solid state (see Section 2.11.3) and, second, we surely must be aware of the fact that the solid-state chemical community is limited in size such that the number of "customers" for quantum chemists is relatively small. As a sad consequence, the solid-state chemists have been left alone for some

Computational Chemistry of Solid State Materials. Richard Dronskowski

ISBN: 3-527-31410-5

decades when it comes to questions of theoretical understanding, although Bloch's theorem (see Section 2.4) dates back to the year 1929! Go and try to find the "solid state" in introductory textbooks of quantum chemistry; good luck!

Theoretically isolated, the solid-state chemistry community therefore had to find other ways to rationalize their synthetic, structural findings, for better or for worse. Let me just mention the extremely useful (but intrinsically non-quantum-mechanical) Zintl–Klemm concept [2] or, to give another prominent example, the grossly oversimplified notion of trying to understand solid-state compounds *only* by considering their electrostatic (Madelung) energies (see Section 1.2). As odd as it may seem, the latter concept is still being used today – and sometimes it *is* a useful concept – although Hans Hellmann already showed in the early 1930s that quantum systems comprising only potential but no kinetic energy are thermodynamically unstable [3]. Also, as Roald Hoffmann carefully observed in 1988, the dominating philosophy within solid-state chemistry had isolated the community even more by not "seeing" chemical bonds between the atoms [4]. That may be the reason why, even today, a freshman student will probably get the wrong idea, by looking into regular chemistry textbooks, that molecules are held together by covalent bonds, whereas solids are taken care of by ionic forces. What a complete nonsense that really is! Nonetheless, let us start with the ionic notion.

1.1
Ionic Radii and Related Concepts

Amongst solid-state materials, those phases which appear to be salt-like (transparent, brittle, insulating) are probably among the easiest to understand in terms of structure[1] and, in many cases, also electronic structure; here we have abstained, for purposes of simplicity, from likewise transparent molecular crystals or covalently bonded arrays (e.g., diamond). In fact, rock salt (sodium chloride, NaCl) may be regarded as the most fundamental insulator, and its crystal structure (see Figure 1.1(b)) is the one most often found for solid-state materials of general composition AB. What can be said about the [NaCl] type in terms of structure – not necessarily in terms of electronic structure – can easily be generalized to other simple compounds such as alkali halides (e.g., LiF, KBr, RbI), silver halides (e.g., AgF, AgCl, AgBr), alkaline-earth chalcogenides (e.g., MgO, BaS, CaSe), simple nitrides (e.g., CeN, CrN, YbN), and many others.

1) For a number of good reasons, solid-state chemists are structure aficionados. For self-training purposes, it is therefore *highly* recommended to get an overview of the structures and structural principles within inorganic chemistry [5].

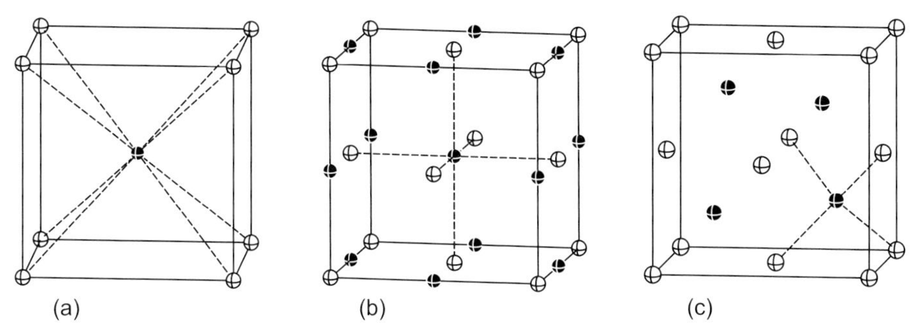

Fig. 1.1 The three fundamental AB structure types with coordination numbers of eight (caesium chloride, (a)), six (sodium chloride, (b)), and four (zinc-blende, (c)). Because the coordinations are identical for atoms A and B, the atomic assignment to A and B may be freely chosen by the reader.

Chemists have long since realized that compositions such as NaCl, MgO, and CeN point towards an ionic description of the chemical bondings, namely because the above formulae suggest formulations such as "Na^+Cl^-", "Mg^{2+}-O^{2-}", and "$Ce^{3+}N^{3-}$" as *plausible*. This is because experimental data show fairly low ionization energies for the metals, many ionic solids form melts composed of ions at higher temperatures, and the atomic charges (or oxidation states) of the nonmetals (−1, −2, and −3 for Cl, O, and N) match our ideas of a noble-gas configuration for the anionic partners; in other words, the compositions seem to be compatible with the *octet rule* of general chemistry. While a quantitative energetic calculus (see Section 1.2) shows this qualitative reasoning to be acceptable *only* for solids and other condensed phases, but *not* for molecules, the idea of a full charge transfer from the metal to the nonmetal is at least good enough to establish the corresponding space partitioning of the crystal structure in terms of *ionic radii*.

A serious, general problem with the determination of any kind of radii is given by the fact that, for obvious reasons, atomic or ionic species do not have a well-defined border; also, the different radii must be a function of the chemical bonding present. For molecular species, however, common sense lets us determine atomic radii quite easily. Since the term *atomic* radius has a somewhat diffuse meaning, one usually speaks of *covalent* radii instead. A few examples may help here. Because the nucleus–nucleus distance in the hydrogen molecule, H–H, is 0.74 Å, the covalent radius of hydrogen *must* be 0.37 Å. Likewise, the covalent radius of oxygen should be half the O–O single-bond distance in H_2O_2 (1.46 Å), that is, 0.73 Å. Note that we can also derive another, shorter bonding radius for oxygen ($\approx$ 0.60 Å), namely from the O=O distance (1.21 Å) in the oxygen molecule, O_2, in which there is a double bond. Trivially, the covalent radius of carbon might be either derived from the C–C distance in diamond or from the one occurring in the ethane molecule, H_3C–CH_3, in both cases yielding 0.77 Å. When combining these covalent radii for the prediction of interatomic distances, one should keep in mind that ionic contributions will have a small impact on the final interatomic distances.

For extended elemental solids, sets of van der Waals radii and also metal radii are simple to determine, at least if the atoms are closely packed. For example, the metal copper crystallizes in its own (Cu) structure type – the face-centered cubic (fcc)[2] crystal structure exhibiting 74% space filling if we assume spherical, nonoverlapping atoms – with a lattice parameter of a = 3.61 Å (see Figure 1.2(b)), and a simple geometrical calculus gives the copper metallic radius as $\frac{a}{4}\sqrt{2}$, that is, 1.28 Å; here we have again assumed that the atomic spheres will *touch* but will not overlap. The same argument holds for the noble gas xenon which also crystallizes in the fcc structure (a = 6.20 Å at a temperature of 40 K) such that the van der Waals radius of Xe is found to be 2.19 Å. Note that this calculation can only be performed because there is only *one* radius that needs to be determined such that the partitioning of space is straightforward. The calculation already becomes a little more complicated for the likewise dense (74% space filling) hexagonal close-packed (hcp) structure (see Figure 1.2(a)) because there are two lattice parameters (a and c) and, therefore, variability in the a/c ratio; thus, a maximum hcp packing density only results for the ideal ratio. The body-centered cubic (bcc) structure (see Figure 1.2(c)) is no longer closely packed (68% space filling) such that a certain arbitrariness is unavoidable for the determination of the metal radius (unless one still assumes touching hard spheres); a useful parameter can be derived, though (see [6]). For intermetallic compounds, things may become quite complicated because the structural hierarchies effectively determine the sizes of the metal radii [7].

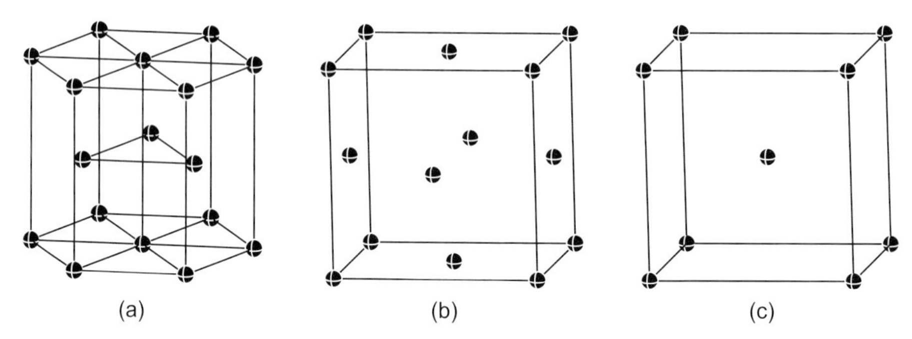

Fig. 1.2 Three fundamental packings of structural chemistry, hexagonal close-packed (hcp, (a)), face-centered cubic (fcc, (b)), and body-centered cubic (bcc, (c)).

2) Our language is rather casual here. The crystal structure of copper is of the cubic close-packed (ccp) type and, because there is only one type of atom, this is equivalent to the face-centered cubic (fcc) structure for this special case. Although not every F-centered crystal structure belonging to the cubic system (e.g., Si or NaCl) can be considered closely packed, the simplified notion ccp = fcc is so ubiquitous in the literature that we also use it in this book.

Coming back to ionic compounds, such as NaCl in Figure 1.1(b), we need to define the sum of cationic and anionic radii to coincide with the interatomic distance; the figure shows that the lattice parameter a (the edge of the cube) is exactly twice the sum of the cationic and anionic radii. It remains to be seen, however, how the interatomic distance between cation and anion can be *partitioned* into the two radii.

A "modern" approach would probably go as follows: Grow a nice NaCl crystal, perform an X-ray diffraction measurement to determine the electron density along the line Na–Cl, search for the minimum density somewhere between the two atoms, and then interpret this value as the crossover from the cation (Na^+) to the anion (Cl^-). Yet, the systems of ionic radii we know of have been generated following other recipes [8], and three approaches seem to be especially worth mentioning.

The first approach from the early 20th century dates back to Victor Moritz Goldschmidt, one of the founders of crystal chemistry [9]. Goldschmidt's system [10] is based on the two standard radii of O^{2-} (1.32 Å) and F^- (1.33 Å) which were derived a little earlier by Wasastjerna [11]; this author simply measured the molar refractions of the alkali halides and set them proportional to the ionic volumes. Goldschmidt then came up with a complete set of ionic radii corresponding to a coordination number (CN) of six. When the coordination polyhedron increases in size, the enlargement of the ionic radius under consideration is usually given by an approximate scaling

$$^{(\mathrm{CN})}r = {}^{(6)}r\left(\frac{(\mathrm{CN})}{6}\right)^{\alpha}, \tag{1.1}$$

in which (CN) designates the new coordination number. The exponent α varies from atom to atom but an operational value lies around $\alpha \approx 1/8$; *any* ionic radius is a function of the coordinating environment but the small α exponent indicates that the variation cannot be excessively large.

At about the same time, another system was proposed by Pauling which is strongly influenced by quantum-chemical ideas [6]. According to Pauling, the ionic radius will enlarge upon filling up the ion's electronic shells and it is therefore proportional to the atomic constant C_n, determined by the main quantum number of the outermost electrons. On the other hand, the ionic radius will shrink as a function of the effective nuclear charge which is the difference between the bare nuclear charge Z and the electronic screening s by the inner core electrons. In short, we simply write

$$r = \frac{C_n}{Z - s}. \tag{1.2}$$

For isoelectronic ions such as Na^+ and F^- which both have the electronic configuration of Ne, [$1s^2 2s^2 2p^6$], the values for C_n must be identical, whatever they are. Also, a quantum-chemical estimate will show that the screening s is the same (4.52) for both ions,[3] and the effective nuclear charges thus arrive at $Z - s = 11 - 4.52 = 6.48$ for Na^+ and at $9 - 4.52 = 4.48$ for F^-. Consequently, the *ratio* of their ionic radii must be 6.48/4.48 = 1.446 such that, for the given Na^+–F^- distance in NaF (2.31 Å, sodium chloride-type), we have $r(Na^+) = 0.95$ Å and $r(F^-) = 1.36$ Å. An important consequence of this procedure is that Pauling's radius for the six-coordinate O^{2-} ion arrives at a larger 1.40 Å if compared to Goldschmidt's 1.32 Å. The complete Pauling set of ionic radii, dubbed *crystal radii* by himself, has been very influential because of the elegance of the arguments and the fame of the author.

Third, a different strategy was pursued by Shannon in the 1970s [14], and he generated a set of self-consistent ionic radii based on crystallographic information, augmented by bond-valence ideas (see Section 1.5). The strength of Shannon's approach surely lies in the large size of the empirical data on which it is based, such that the influence of both the oxidation states of cations/anions and also of the coordination numbers can be sensibly taken into account. This is probably the reason why Shannon's data have replaced – but not superseded – the sets of radii given by Goldschmidt and Pauling. For convenience, we present Shannon's radii in Table 1.1. This tabulation, as well as all others, shows the general trends, that is, that cations are usually smaller than anions, and cationic radii decrease rapidly with overall charge.

3) Quantum-mechanically inspired recipes for the calculation of the screening values s have been proposed by Pauling [12] and Slater [13].

Tab. 1.1 Effective ionic radii according to Shannon [14]. The superscripts indicate the coordination numbers, the electronic state (hs = high-spin, ls = low-spin), and the geometry (sq = square, py = pyramidal).

Ion	$^{(\text{coordination number})}r_{\text{eff}}$ (Å)						
Ac^{3+}	$^{(6)}1.12$						
Ag^{+}	$^{(2)}0.67$	$^{(4)}1.00$	$^{(4,sq)}1.02$	$^{(5)}1.09$	$^{(6)}1.15$	$^{(7)}1.22$	$^{(8)}1.28$
Ag^{2+}	$^{(4,sq)}0.79$	$^{(6)}0.94$					
Ag^{3+}	$^{(4,sq)}0.67$	$^{(6)}0.75$					
Al^{3+}	$^{(4)}0.39$	$^{(5)}0.48$	$^{(6)}0.535$				
Am^{2+}	$^{(7)}1.21$	$^{(8)}1.26$	$^{(9)}1.31$				
Am^{3+}	$^{(6)}0.975$	$^{(8)}1.09$					
Am^{4+}	$^{(6)}0.85$	$^{(8)}0.95$					
As^{3+}	$^{(6)}0.58$						
As^{5+}	$^{(4)}0.335$	$^{(6)}0.46$					
At^{7+}	$^{(6)}0.62$						
Au^{+}	$^{(6)}1.37$						
Au^{3+}	$^{(4,sq)}0.68$	$^{(6)}0.85$					
Au^{5+}	$^{(6)}0.57$						
B^{3+}	$^{(3)}0.01$	$^{(4)}0.11$	$^{(6)}0.27$				
Ba^{2+}	$^{(6)}1.35$	$^{(7)}1.38$	$^{(8)}1.42$	$^{(9)}1.47$	$^{(10)}1.52$	$^{(11)}1.57$	$^{(12)}1.61$
Be^{2+}	$^{(3)}0.16$	$^{(4)}0.27$	$^{(6)}0.45$				
Bi^{3+}	$^{(5)}0.96$	$^{(6)}1.03$	$^{(8)}1.17$				
Bi^{5+}	$^{(6)}0.76$						
Bk^{3+}	$^{(6)}0.96$						
Bk^{4+}	$^{(6)}0.83$	$^{(8)}0.93$					
Br^{-}	$^{(6)}1.96$						
Br^{3+}	$^{(4,sq)}0.59$						
Br^{5+}	$^{(3,py)}0.31$						
Br^{7+}	$^{(4)}0.25$	$^{(6)}0.39$					
C^{4+}	$^{(3)}-0.08$	$^{(4)}0.15$	$^{(6)}0.16$				
Ca^{2+}	$^{(6)}1.00$	$^{(7)}1.06$	$^{(8)}1.12$	$^{(9)}1.18$	$^{(10)}1.23$	$^{(12)}1.34$	
Cd^{2+}	$^{(4)}0.78$	$^{(5)}0.87$	$^{(6)}0.95$	$^{(7)}1.03$	$^{(8)}1.10$	$^{(12)}1.31$	
Ce^{3+}	$^{(6)}1.01$	$^{(7)}1.07$	$^{(8)}1.143$	$^{(9)}1.196$	$^{(10)}1.25$	$^{(12)}1.34$	
Ce^{4+}	$^{(6)}0.87$	$^{(8)}0.97$	$^{(10)}1.07$	$^{(12)}1.14$			
Cf^{3+}	$^{(6)}0.95$						
Cf^{4+}	$^{(6)}0.821$	$^{(8)}0.92$					
Cl^{-}	$^{(6)}1.81$						
Cl^{5+}	$^{(3,py)}0.12$						
Cl^{7+}	$^{(4)}0.08$	$^{(6)}0.27$					
Cm^{3+}	$^{(6)}0.97$						
Cm^{4+}	$^{(6)}0.85$	$^{(8)}0.95$					
Co^{2+}	$^{(4,hs)}0.58$	$^{(5)}0.67$	$^{(6,ls)}0.65$	$^{(6,hs)}0.745$	$^{(8)}0.90$		
Co^{3+}	$^{(6,ls)}0.545$	$^{(6,hs)}0.61$					
Co^{4+}	$^{(4)}0.40$	$^{(6,hs)}0.53$					
Cr^{2+}	$^{(6,ls)}0.73$	$^{(6,hs)}0.80$					
Cr^{3+}	$^{(6)}0.615$						
Cr^{4+}	$^{(4)}0.41$	$^{(6)}0.55$					
Cr^{5+}	$^{(4)}0.345$	$^{(6)}0.49$	$^{(8)}0.57$				
Cr^{6+}	$^{(4)}0.26$	$^{(6)}0.44$					
Cs^{+}	$^{(6)}1.67$	$^{(8)}1.74$	$^{(9)}1.78$	$^{(10)}1.81$	$^{(11)}1.85$	$^{(12)}1.88$	
Cu^{+}	$^{(2)}0.46$	$^{(4)}0.60$	$^{(6)}0.77$				
Cu^{2+}	$^{(4)}0.57$	$^{(4,sq)}0.57$	$^{(5)}0.65$	$^{(6)}0.73$			
Cu^{3+}	$^{(6,ls)}0.54$						
D^{+}	$^{(2)}-0.10$						
Dy^{2+}	$^{(6)}1.07$	$^{(7)}1.13$	$^{(8)}1.19$				

continued on next page

Tab. 1.1 (continued)

Ion	(coordination number) r_{eff} (Å)						
Dy^{3+}	(6)0.912	(7)0.97	(8)1.027	(9)1.083			
Er^{3+}	(6)0.890	(7)0.945	(8)1.004	(9)1.062			
Eu^{2+}	(6)1.17	(7)1.20	(8)1.25	(9)1.30	(10)1.35		
Eu^{3+}	(6)0.947	(7)1.01	(8)1.066	(9)1.120			
F^{-}	(2)1.285	(3)1.30	(4)1.31	(6)1.33			
F^{7+}	(6)0.08						
Fe^{2+}	(4,hs)0.63	(4,sq,hs)0.64	(6,ls)0.61	(6,hs)0.78	(8,hs)0.92		
Fe^{3+}	(4,hs)0.49	(5)0.58	(6,ls)0.55	(6,hs)0.645	(8,hs)0.78		
Fe^{4+}	(6)0.585						
Fe^{6+}	(4)0.25						
Fr^{+}	(6)1.80						
Ga^{3+}	(4)0.47	(5)0.55	(6)0.620				
Gd^{3+}	(6)0.938	(7)1.00	(8)1.053	(9)1.107			
Ge^{2+}	(6)0.73						
Ge^{4+}	(4)0.390	(6)0.530					
H^{+}	(1)−0.38	(2)−0.18					
Hf^{4+}	(4)0.58	(6)0.71	(7)0.76	(8)0.83			
Hg^{+}	(3)0.97	(6)1.19					
Hg^{2+}	(2)0.69	(4)0.96	(6)1.02	(8)1.14			
Ho^{3+}	(6)0.901	(8)1.015	(9)1.072	(10)1.12			
I^{-}	(6)2.20						
I^{5+}	(3,py)0.44	(6)0.95					
I^{7+}	(4)0.42	(6)0.53					
In^{3+}	(4)0.62	(6)0.800	(8)0.92				
Ir^{3+}	(6)0.68						
Ir^{4+}	(6)0.625						
Ir^{5+}	(6)0.57						
K^{+}	(4)1.37	(6)1.38	(7)1.46	(8)1.51	(9)1.55	(10)1.59	(12)1.64
La^{3+}	(6)1.032	(7)1.10	(8)1.160	(9)1.216	(10)1.27	(12)1.36	
Li^{+}	(4)0.590	(6)0.76	(8)0.92				
Lu^{3+}	(6)0.861	(8)0.977	(9)1.032				
Mg^{2+}	(4)0.57	(5)0.66	(6)0.720	(8)0.89			
Mn^{2+}	(4,hs)0.66	(5,hs)0.75	(6,ls)0.67	(6,hs)0.83	(7,hs)0.90	(8)0.96	
Mn^{3+}	(5)0.58	(6,ls)0.58	(6,hs)0.645				
Mn^{4+}	(4)0.39	(6)0.53					
Mn^{5+}	(4)0.33						
Mn^{6+}	(4)0.255						
Mn^{7+}	(4)0.25	(6)0.46					
Mo^{3+}	(6)0.69						
Mo^{4+}	(6)0.650						
Mo^{5+}	(4)0.46	(6)0.61					
Mo^{6+}	(4)0.41	(5)0.50	(6)0.59	(7)0.73			
N^{3-}	(4)1.46						
N^{3+}	(6)0.16						
N^{5+}	(3)−0.104	(6)0.13					
Na^{+}	(4)0.99	(5)1.00	(6)1.02	(7)1.12	(8)1.18	(9)1.24	(12)1.39
Nb^{3+}	(6)0.72						
Nb^{4+}	(6)0.68	(8)0.79					
Nb^{5+}	(4)0.48	(6)0.64	(7)0.69	(8)0.74			
Nd^{2+}	(8)1.29	(9)1.35					
Nd^{3+}	(6)0.983	(8)1.109	(9)1.163	(12)1.27			
Ni^{2+}	(4)0.55	(4,sq)0.49	(5)0.63	(6)0.690			
Ni^{3+}	(6,ls)0.56	(6,hs)0.60					

continued on next page

Tab. 1.1 (continued)

Ion	$^{\text{(coordination number)}}r_{\text{eff}}$ (Å)							
Ni^{4+}	$^{(6,ls)}$0.48							
No^{2+}	$^{(6)}$1.1							
Np^{2+}	$^{(6)}$1.10							
Np^{3+}	$^{(6)}$1.01							
Np^{4+}	$^{(6)}$0.87	$^{(8)}$0.98						
Np^{5+}	$^{(6)}$0.75							
Np^{6+}	$^{(6)}$0.72							
Np^{7+}	$^{(6)}$0.71							
O^{2-}	$^{(2)}$1.35	$^{(3)}$1.36	$^{(4)}$1.38	$^{(6)}$1.40	$^{(8)}$1.42			
OH^{-}	$^{(2)}$1.32	$^{(3)}$1.34	$^{(4)}$1.35	$^{(6)}$1.37				
Os^{4+}	$^{(6)}$0.630							
Os^{5+}	$^{(6)}$0.575							
Os^{6+}	$^{(5)}$0.49	$^{(6)}$0.545						
Os^{7+}	$^{(6)}$0.525							
Os^{8+}	$^{(4)}$0.39							
P^{3+}	$^{(6)}$0.44							
P^{5+}	$^{(4)}$0.17	$^{(5)}$0.29	$^{(6)}$0.38					
Pa^{3+}	$^{(6)}$1.04							
Pa^{4+}	$^{(6)}$0.90	$^{(8)}$1.01						
Pa^{5+}	$^{(6)}$0.78	$^{(8)}$0.91	$^{(9)}$0.95					
Pb^{2+}	$^{(4,py)}$0.98	$^{(6)}$1.19	$^{(7)}$1.23	$^{(8)}$1.29	$^{(9)}$1.35	$^{(10)}$1.40	$^{(11)}$1.45	$^{(12)}$1.49
Pb^{4+}	$^{(4)}$0.65	$^{(5)}$0.73	$^{(6)}$0.775	$^{(8)}$0.94				
Pd^{+}	$^{(2)}$0.59							
Pd^{2+}	$^{(4,sq)}$0.64	$^{(6)}$0.86						
Pd^{3+}	$^{(6)}$0.76							
Pd^{4+}	$^{(6)}$0.615							
Pm^{3+}	$^{(6)}$0.97	$^{(8)}$1.093	$^{(9)}$1.144					
Po^{4+}	$^{(6)}$0.94	$^{(8)}$1.08						
Po^{6+}	$^{(6)}$0.67							
Pr^{3+}	$^{(6)}$0.99	$^{(8)}$1.126	$^{(9)}$1.179					
Pr^{4+}	$^{(6)}$0.85	$^{(8)}$0.96						
Pt^{2+}	$^{(4,sq)}$0.60	$^{(6)}$0.80						
Pt^{4+}	$^{(6)}$0.625							
Pt^{5+}	$^{(6)}$0.57							
Pu^{3+}	$^{(6)}$1.00							
Pu^{4+}	$^{(6)}$0.86	$^{(8)}$0.96						
Pu^{5+}	$^{(6)}$0.74							
Pu^{6+}	$^{(6)}$0.71							
Ra^{2+}	$^{(8)}$1.48	$^{(12)}$1.70						
Rb^{+}	$^{(6)}$1.52	$^{(7)}$1.56	$^{(8)}$1.61	$^{(9)}$1.63	$^{(10)}$1.66	$^{(11)}$1.69	$^{(12)}$1.72	$^{(14)}$1.83
Re^{4+}	$^{(6)}$0.63							
Re^{5+}	$^{(6)}$0.58							
Re^{6+}	$^{(6)}$0.55							
Re^{7+}	$^{(4)}$0.38	$^{(6)}$0.53						
Rh^{3+}	$^{(6)}$0.665							
Rh^{4+}	$^{(6)}$0.60							
Rh^{5+}	$^{(6)}$0.55							
Ru^{3+}	$^{(6)}$0.68							
Ru^{4+}	$^{(6)}$0.620							
Ru^{5+}	$^{(6)}$0.565							
Ru^{7+}	$^{(4)}$0.38							
Ru^{8+}	$^{(4)}$0.36							
S^{2-}	$^{(6)}$1.84							

continued on next page

Tab. 1.1 (continued)

Ion	$^{(\text{coordination number})}r_{\text{eff}}$ (Å)					
S^{4+}	$^{(6)}0.37$					
S^{6+}	$^{(4)}0.12$	$^{(6)}0.29$				
Sb^{3+}	$^{(4,py)}0.76$	$^{(5)}0.80$	$^{(6)}0.76$			
Sb^{5+}	$^{(6)}0.60$					
Sc^{3+}	$^{(6)}0.745$	$^{(8)}0.870$				
Se^{2-}	$^{(6)}1.98$					
Se^{4+}	$^{(6)}0.50$					
Se^{6+}	$^{(4)}0.28$	$^{(6)}0.42$				
Si^{4+}	$^{(4)}0.26$	$^{(6)}0.400$				
Sm^{2+}	$^{(7)}1.22$	$^{(8)}1.27$	$^{(9)}1.32$			
Sm^{3+}	$^{(6)}0.958$	$^{(7)}1.02$	$^{(8)}1.079$	$^{(9)}1.132$	$^{(12)}1.24$	
Sn^{4+}	$^{(4)}0.55$	$^{(5)}0.62$	$^{(6)}0.690$	$^{(7)}0.75$	$^{(8)}0.81$	
Sr^{2+}	$^{(6)}1.18$	$^{(7)}1.21$	$^{(8)}1.26$	$^{(9)}1.31$	$^{(10)}1.36$	$^{(12)}1.44$
Ta^{3+}	$^{(6)}0.72$					
Ta^{4+}	$^{(6)}0.68$					
Ta^{5+}	$^{(6)}0.64$	$^{(7)}0.69$	$^{(8)}0.74$			
Tb^{3+}	$^{(6)}0.923$	$^{(7)}0.98$	$^{(8)}1.040$	$^{(9)}1.095$		
Tb^{4+}	$^{(6)}0.76$	$^{(8)}0.88$				
Tc^{4+}	$^{(6)}0.645$					
Tc^{5+}	$^{(6)}0.60$					
Tc^{7+}	$^{(4)}0.37$	$^{(6)}0.56$				
Te^{2-}	$^{(6)}2.21$					
Te^{4+}	$^{(3)}0.52$	$^{(4)}0.66$	$^{(6)}0.97$			
Te^{6+}	$^{(4)}0.43$	$^{(6)}0.56$				
Th^{4+}	$^{(6)}0.94$	$^{(8)}1.05$	$^{(9)}1.09$	$^{(10)}1.13$	$^{(11)}1.18$	$^{(12)}1.21$
Ti^{2+}	$^{(6)}0.86$					
Ti^{3+}	$^{(6)}0.670$					
Ti^{4+}	$^{(4)}0.42$	$^{(5)}0.51$	$^{(6)}0.605$	$^{(8)}0.74$		
Tl^{+}	$^{(6)}1.50$	$^{(8)}1.59$	$^{(12)}1.70$			
Tl^{3+}	$^{(4)}0.75$	$^{(6)}0.885$	$^{(8)}0.98$			
Tm^{2+}	$^{(6)}1.03$	$^{(7)}1.09$				
Tm^{3+}	$^{(6)}0.880$	$^{(8)}0.994$	$^{(9)}1.052$			
U^{3+}	$^{(6)}1.025$					
U^{4+}	$^{(6)}0.89$	$^{(7)}0.95$	$^{(8)}1.00$	$^{(9)}1.05$	$^{(12)}1.17$	
U^{5+}	$^{(6)}0.76$	$^{(7)}0.84$				
U^{6+}	$^{(2)}0.45$	$^{(4)}0.52$	$^{(6)}0.73$	$^{(7)}0.81$	$^{(8)}0.86$	
V^{2+}	$^{(6)}0.79$					
V^{3+}	$^{(6)}0.640$					
V^{4+}	$^{(5)}0.53$	$^{(6)}0.58$	$^{(8)}0.72$			
V^{5+}	$^{(4)}0.355$	$^{(5)}0.46$	$^{(6)}0.54$			
W^{4+}	$^{(6)}0.66$					
W^{5+}	$^{(6)}0.62$					
W^{6+}	$^{(4)}0.42$	$^{(5)}0.51$	$^{(6)}0.60$			
Xe^{8+}	$^{(4)}0.40$	$^{(6)}0.48$				
Y^{3+}	$^{(6)}0.900$	$^{(7)}0.96$	$^{(8)}1.019$	$^{(9)}1.075$		
Yb^{2+}	$^{(6)}1.02$	$^{(7)}1.08$	$^{(8)}1.14$			
Yb^{3+}	$^{(6)}0.868$	$^{(7)}0.925$	$^{(8)}0.985$	$^{(9)}1.042$		
Zn^{2+}	$^{(4)}0.60$	$^{(5)}0.68$	$^{(6)}0.740$	$^{(8)}0.90$		
Zr^{4+}	$^{(4)}0.59$	$^{(5)}0.66$	$^{(6)}0.72$	$^{(7)}0.78$	$^{(8)}0.84$	$^{(9)}0.89$

A nice feature of Shannon's radii – dubbed *effective* ionic radii by himself – is that the complete set is *additive*, such that experimental cation–anion distances are, in most cases, correctly reproduced. To allow for this, a few ions must have *negative* radii but such an unphysical property does not bother the brave crystal chemist.[4] Second, because of the popularity of Pauling's radii, the size of O^{2-} in six-fold coordination is also fixed to 1.40 Å, such that traditionalists of ionic radii do not have to rethink. Nonetheless, an alternative, likewise additive, set of *crystal* radii – not to be confused with Pauling's crystal radii! – is generated from Shannon's effective radii by subtracting 0.14 Å from all anionic radii and adding this 0.14 Å to all the cationic radii (see below).

Why do we need two sets of Shannon radii? Knowing that LiI also adopts the sodium chloride structure type, we might predict (see again Figure 1.1(b)) the experimental LiI lattice parameter (6.00 Å) either by twice the *effective* ionic radii of Li^+ and I^-, that is, $2 \times (0.76\ \text{Å} + 2.20\ \text{Å}) = 5.92\ \text{Å}$, or by twice their *crystal* radii, namely $2 \times (0.90\ \text{Å} + 2.06\ \text{Å}) = 5.92\ \text{Å}$. If we think in terms of packings of anions, however, a close-packing of I^- anions with $r(I^-) = 2.20\ \text{Å}$ and Li^+ in octahedral holes will come closer to the volume of the unit cell (almost 83%), thus favoring the *effective* ionic radii which are closer to Pauling's radii.

Shannon's *crystal* radii, however, better match the experimentally found electron density distribution mentioned at the beginning. For example, modern X-ray data show that the minimum electron density for LiF ([NaCl] type) corresponds to $r(Li^+) = 0.92\ \text{Å}$ and $r(F^-) = 1.08\ \text{Å}$. This reflects a significantly larger/smaller lithium/fluorine ion than Pauling's (0.60/1.36 Å), Goldschmidt's (0.78/1.33 Å) and also Shannon's effective ionic radii (0.76/1.33 Å), but it is already quite close to Shannon's crystal radii (0.90/1.19 Å). Summarizing, *effective* ionic radii correspond to the idea of closely packed anions but *crystal* radii are closer to the real sizes of anions *and* cations.

The purpose of ionic radii, as presented so far, can be understood as to predict or at least "rationalize" interionic distances and also lattice parameters when the crystal structure type adopted is already known. One is tempted to go somewhat further and try to use ionic radii for the prediction of the crystal structure itself. Let us illustrate this from the simple examples of AB-type structures, depicted in Figure 1.1. For example, the [NaCl] type (b) is usually derived by assuming that the larger anions adopt the elemental fcc structure and the smaller cations just fill up all octahedral sites; indeed, the visual comparison of the fcc type (Figure 1.2(b)) and the [NaCl] type shows that the latter is composed of two fcc structures (by Na^+ and Cl^-) which have been shifted against each other by half the lattice parameter. Despite this elegant

4) The most extreme example is the fate of the H^+ ion which, because the O^{2-} ion is larger than the OH^- ion, is considered to have a strongly negative ionic radius; the reader should not be too concerned about this.

derivation, it seems somewhat strange to assume that both anions and cations should strive for close packing because that is where the anion–anion and cation–cation repulsion is maximized! Neglecting this internal inconsistency for the moment, the AB structure prediction usually proceeds as follows:

No matter what kind of interaction exists between A and B in, for example, the zinc-blende structure type with a coordination number of four (see Figure 1.1(c)), the atoms (or ions) can be expected to minimize their interatomic A–B distance because of energy optimization; so shorter distances should indicate stronger interaction. For a purely electrostatic one (see Section 1.2), this is particularly easy to calculate; the distance between A and B can then be approximated by the sum of the ionic radii of the cation and anion, r_c and r_a.

If we further imagine the cation radius r_c to grow, the A–B distance will also become larger so that the A–B interaction should weaken. Then, however, A might already be so large that six instead of four B partners may coordinate A; in electrostatic terms, the weakening of the A–B interaction is partly overridden by a larger electrostatic field as indicated by the larger Madelung constant for coordination number six (see Section 1.2). Thus, the sodium chloride type will be more favorable than the zinc-blende type for a sufficiently large cation radius or, more generally, for a critical ratio r_c/r_a. The same argument holds true for the change from the sodium chloride to the caesium chloride type (eight-fold coordination) such that, in general, a large radius ratio points towards an enlarged coordination number. This seems logical.

Textbooks of structural chemistry usually go to great lengths to reproduce Goldschmidt's (and also Pauling's) reasoning of how the critical radius ratio between cation and anion can be derived [5]. This elementary calculus will not be repeated here but is left to the reader as an exercise if time allows. Eventually, a critical radius ratio r_c/r_a of less than 0.414 favors four-fold coordination; the zinc-blende type. For larger cations (or smaller anions) which correspond to a radius ratio of between 0.414 and 0.732, six-fold coordination, the [NaCl] type, would be preferred.[5] For even larger ratios beyond 0.732, one should find the [CsCl] type with eight-fold coordination.

Regrettably, the world is not that simple. The corresponding test of the above hypothesis has been pursued by Burdett and coworkers [15] (unfortunately, only many years after the radius-ratio rule had become an essential part of general chemistry), and it is depicted in Figure 1.3.

The maximum coordination number (8) is indeed found for the largest radius ratios, calling for the [CsCl] structure type (open squares). A very large number of structures is of the [NaCl] type with a coordination number of six (shaded circles), and small radius ratios point towards a coordination num-

5) Another choice for six-fold coordination is given by the [NiAs] structure type, with an octahedral environment for Ni and a hexagonal prism for As. This structure type, usually not found for ionic solids, is therefore never considered.

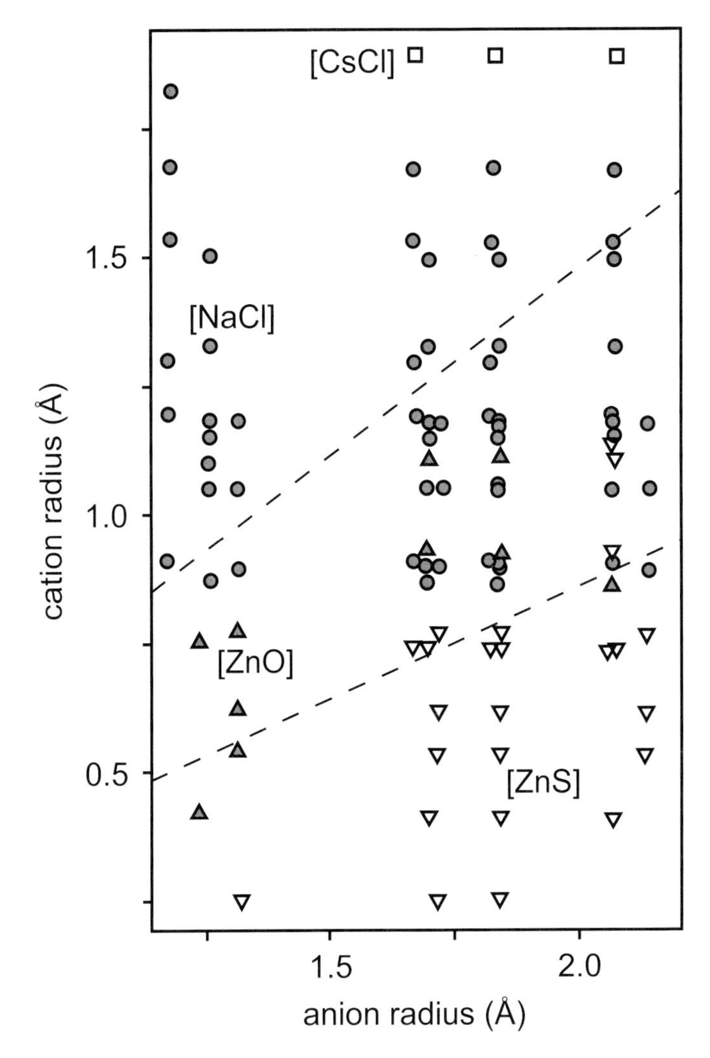

Fig. 1.3 Structure map of AB solids with noble-gas configuration for the ions as a function of cation and anion radii; the experimentally found structures are indicated. All entries above the upper dashed line should crystallize in the [CsCl] type, whereas all under the lower dashed line should adopt the [ZnS] type; the intermediate region is one of the [NaCl] type. For an anion radius of 1.0 Å, the upper and lower lines yield a cation radius of 0.732 and 0.414 Å.

ber of four and the [ZnS] or [ZnO] structure (open and shaded triangles). The disappointing part of the story is that, of the 95 AB-type compounds with supposedly ionic bonding, 37 – close to forty percent! – are incorrectly placed such that the *quantitative* predictive power of the radius ratios is almost imaginary; there must be some variability in these radii upon chemical bonding and the atoms do not behave as billiard balls. Nonetheless, the *qualitative* trend is fine, namely that larger/smaller cations/anions favor large coordination numbers for the cation. Quantitatively, the concept does not serve its purpose, however.

There have been other attempts to derive such structure-sorting maps, with considerable success. One of the most influential strategies is the one by Mooser and Pearson [16] who used the average principal quantum number and electronegativity differences as corresponding indices. Nonetheless, it *is* possible to quantitatively consider atomic size for the derivation of structure-sorting maps [17], but only by a quantum-chemical derivation in the form of *pseudopotential* radii (see Section 2.15.2). The latter express atomic size by the extent of which the outer valence electrons are expelled by the inner core electrons, through the Pauli principle. Thus, the classical idea may be kept but only by transcending it into the realm of quantum chemistry.

Yet another, entirely phenomenological, approach for arriving at structure-sorting maps goes back to the work of Pettifor who used the periodic table of the elements as a starting point [18]. When all atomic numbers are exchanged by Mendeléev numbers M – this effectively corresponds to a total renumbering of the whole periodic table – a structural separation of many binary compounds AB_x may be achieved by using these Mendeléev numbers of the constituent atoms, M^A and M^B, as the axes of the coordinate systems. Why this works – as it often does – is a mystery to me.

1.2 Electrostatics

Amongst the various solid-state materials with either insulating, semiconducting or metallic properties, electric insulators were probably the first to have been studied by theoretical, albeit classical, approaches. Indeed, there is definitely no book on solid-state chemistry which does not, at least qualitatively or unconsciously, mention the "ionic model" of chemical bonding in solids. Let us do that here, too.

A moment's reflection on the possibility of a *gaseous* sodium chloride species based upon experimental energetic data immediately reveals that such a molecular $NaCl_{(g)}$ is unlikely to form if ionic bonding would be involved. The electron loss of Na to form Na^+ will cost an ionization energy I of about 496 kJ/mol, and the capture of the electron by Cl to form Cl^- will gain an electron affinity A of merely −361 kJ/mol such that the system is unstable by 496 kJ/mol − 361 kJ/mol = 135 kJ/mol. Generally, experimental evidence proves that there is no such combination of metals and nonmetals where the electron affinity exceeds the ionization energy; thus, the ionic bonding we always refer to is bound to the condensed matter, period.

This has been known for a long time, and the correct energetic calculus for the formation of solid sodium chloride (we will stay with this example for simplicity) is given by the so-called Born–Haber cycle, depicted in Figure 1.4.

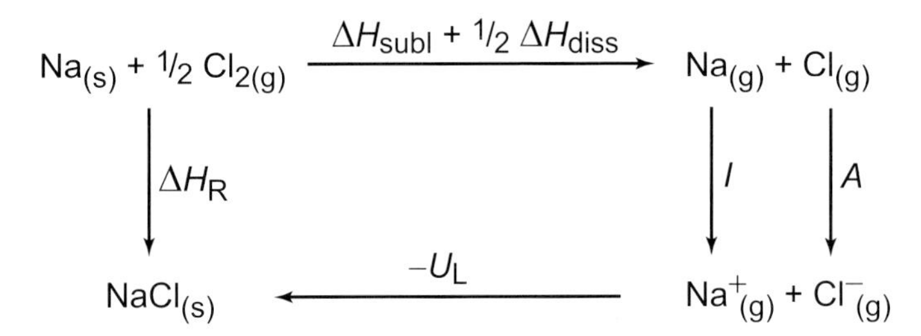

Fig. 1.4 Schematic Born–Haber cycle for the formation of solid NaCl; the energetic data (kJ/mol) are: Na sublimation enthalpy ΔH_{subl} = 100.5; $\frac{1}{2} \times$ Cl_2 dissociation enthalpy ΔH_{diss} = 121.4; Na ionization energy I = 495.7; Cl electron affinity A = −360.5; experimental reaction enthalpy ΔH_R = −411.1.

Summing up all atomic energetic data in the above figure, namely ΔH_{subl}(Na) + $\frac{1}{2} \times \Delta H_{diss}(Cl_2)$ + I(Na) + A(Cl), yields that the formation of gaseous ions needs a total energy of 357.1 kJ/mol (on the right side); on the other hand, the experimental reaction enthalpy is −411.1 kJ/mol (left side) such that, and this is the important result, the ionic solid is more stable than the isolated ions by 357.1 kJ/mol −(−411.1) kJ/mol = 768.2 kJ/mol. It is common practice to call this energy the *lattice energy* U_L of the solid, needed to evaporate the ionic solid into gaseous *ions* – not atoms – and it carries a positive sign by definition.[6] Whether or not the isolated ions exist in the gas phase is another matter (see below).

Because the lattice energy defined as such alludes to structural stability, its theoretical calculation (as opposed to its experimental determination by means of the above Born–Haber cycle) was investigated at an early stage. We might, for example, define the lattice energy to be composed of an electrostatic interaction between the ions which we may approximate by a simple Coulomb expression, augmented by a repulsive part U_{rep} such that the ionic lattice cannot collapse into a singularity, and also account for weak covalent bonding contributions U_{cov}, even weaker van der Waals (dispersive) attractions U_{vdW} and also the zero-point vibration U_{zero}, namely

$$U_L \equiv U_{Coulomb} + U_{rep} + U_{cov} + U_{vdW} + U_{zero}, \qquad (1.3)$$

but it transpires that the three latter terms, being of quantum-mechanical origin and indeed relatively small in size, may be safely ignored for the remain-

6) In the English-speaking world, however, U_L is sometimes found with a negative sign because then one considers the energy which is gained when the ions form the crystal.

ing part of the discussion. This leaves us with an approximate two-particle lattice energy

$$U_{\mathrm{L}} \approx U_{\mathrm{Coulomb}} + U_{\mathrm{rep}} = \frac{1}{4\pi\epsilon_0}\frac{z_1 z_2 e^2}{r} + U_{\mathrm{rep}}, \tag{1.4}$$

where the Coulomb part goes back to the interaction between point charges (+1 and −1 for the case of NaCl) and e is the elementary charge; the repulsion term still needs to be specified. Interestingly, this repulsion term is also of quantum-mechanical origin, and it simply shows that it is impossible to push atoms into each other. The reason, as one might expect, is the exchange interaction between the electrons such that, simply speaking, it is the Pauli principle (see Section 2.9) which forbids the too-close encounter of atoms, when their associated orbitals have already been completely filled; we can always expect such a scenario for the inner shells of any kind of atom. The same reason is responsible for the fact that there is no covalently bonded He_2 molecule (see Section 2.1). One may envisage different ways to describe this quantum effect classically but a simple exponential expression [19],

$$U_{\mathrm{rep}} \approx b e^{-r/\rho}, \tag{1.5}$$

with empirical scaling parameters b and ρ, is a very good approximation since it imitates the exponential decay of the atomic orbitals (see Section 2.2). In practice, a parameter of, say, $\rho = 0.345$ Å does a fine job but ρ may also be

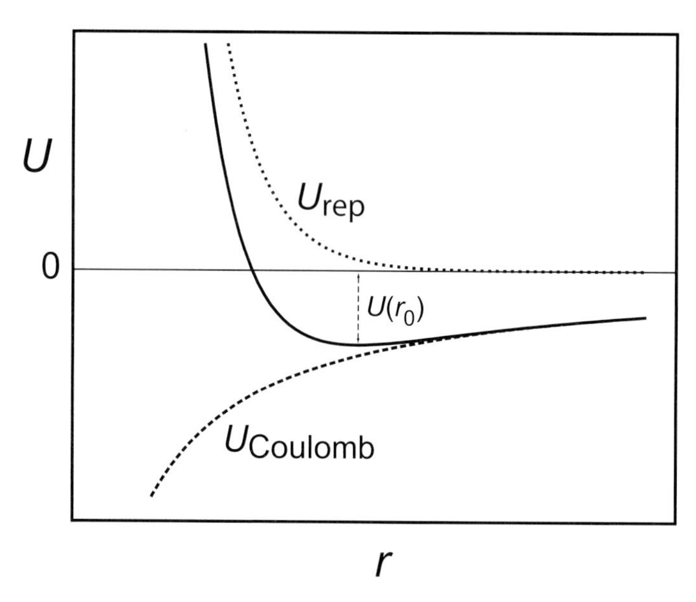

Fig. 1.5 Schematic drawing of Coulomb, repulsive and total energies of a cation–anion pair.

looked upon as a fit parameter, such that the theoretical lattice energy perfectly matches the experimental one, if that is needed.

Thus, a cation–anion *pair* will display a potential energy curve such as the one in Figure 1.5, with an energy minimum at the equilibrium distance r_0. A simple differentiation ($\partial U/\partial r = 0$ at r_0) lets us eliminate the parameter b and yields

$$U_0 = \frac{1}{4\pi\epsilon_0}\frac{z_1 z_2 e^2}{r_0}\left(1 - \frac{\rho}{r_0}\right), \tag{1.6}$$

showing that, for realistic interionic distances around 3 Å or so, the contribution of the repulsive interactions must already weaken the equilibrium energy in the order of 10%. But, as we know from the nonexistence of a gaseous, ionically bonded NaCl molecule, the effect of the infinite lattice still needs to be included into the energetic calculus. For a hypothetical one-dimensional NaCl crystal with a charge distribution that goes as in Scheme 1.1,

----- + − + − + − + − + − + -----

Scheme 1.1

the Coulomb part then simply reads

$$\begin{aligned} U^{\overset{1}{\infty}}_{\text{Coulomb}} &= -\frac{1}{4\pi\epsilon_0}\left(\frac{2e^2}{r} - \frac{2e^2}{2r} + \frac{2e^2}{3r} - \frac{2e^2}{4r} + \ldots\right) \\ &= -\frac{1}{4\pi\epsilon_0}\frac{2e^2}{r}\underbrace{\left(1 - \frac{1}{2} + \frac{1}{3} - \frac{1}{4} + \ldots\right)}_{\ln 2 \approx 0.693} \\ &\approx -\frac{1.386\, e^2}{4\pi\epsilon_0\, r}. \end{aligned} \tag{1.7}$$

Likewise, the repulsive part may be formulated as

$$U^{\overset{1}{\infty}}_{\text{rep}} = 2be^{-r/\rho} + 2be^{-2r/\rho} + 2be^{-3r/\rho} + \ldots \approx 2be^{-r/\rho}, \tag{1.8}$$

where we have taken advantage of the fast decay of the exponential function. Thus, in general, the total energy of the one-dimensional ionic crystal becomes

$$U^{\overset{1}{\infty}}_0 = \frac{1.386\, z_1 z_2 e^2}{4\pi\epsilon_0\, r_0}\left(1 - \frac{\rho}{r_0}\right) \tag{1.9}$$

per ion pair, to be compared with the result of the isolated pair of ions (Equation (1.6)). Obviously, the translational symmetry of the one-dimensional lattice leads to an energetic stabilization by more than 38%, an astonishing result, and the extension of the above strategy into two and three dimensions will increase the size of the pre-factor even more.

According to an oversimplification sometimes made in solid-state chemistry textbooks, the three-dimensional energy summation rapidly converges in real space. This is *not* the case, simply because the Coulomb law extends into infinity, leading to inherent numerical instabilities, such that the question of convergence depends on the *order* of summation. A mathematical trick such as the famous Ewald summation [20] or other, related formulae [21–24], however, can be used to *enforce* convergence, such that there is also a corresponding total energy in three-dimensional space,

$$\overset{3}{U_0^{\infty}} = \frac{\mathcal{M}}{4\pi\epsilon_0} \frac{z_1 z_2 e^2}{r_0} \left(1 - \frac{\rho}{r_0}\right), \tag{1.10}$$

the new pre-factor $\mathcal{M}$ being called the Madelung constant. We eventually arrive at the lattice energy U_L by multiplying U_0 with the (negative) Avogadro number (note that U_L is defined as positive) as given in

$$U_L = -N_A \overset{3}{U_0^{\infty}} = -N_A \frac{\mathcal{M}}{4\pi\epsilon_0} \frac{z_1 z_2 e^2}{r_0} \left(1 - \frac{\rho}{r_0}\right). \tag{1.11}$$

Because the Madelung constant has been computed by a summation over all lattice sites, it adopts characteristic values for all structure types [5,8]. To give a few examples, $\mathcal{M}$ arrives at (dimensionless) values of 1.6381 (zinc-blende-type), 1.7476 (sodium chloride-type), 1.7627 (caesium chloride-type), 5.0388 (fluorite-type), and 25.0312 (corundum-type) and does not scale with (≡ is independent of) the interionic distances. For the case of NaCl, the Madelung constant shows that the three-dimensional lattice surpasses the ionic pair in energy by almost 75%. This is what has made the formation of *solid* NaCl possible, a *collective* stabilization.

When it comes to a comparison between theoretically calculated lattice energies and experimental ones, the NaCl case might again serve as an example. We simply have to insert the experimental Na^+–Cl^- distance from the crystal structure of NaCl (2.82 Å) into Equation (1.11) and arrive at a theoretical U_L value of 756 kJ/mol, not far from the experimental 768 kJ/mol, a mere 1.6% underestimation. The fit can be further improved, for example, by including those corrections which we had discarded at the very beginning (covalent contributions, dispersive interactions, etc.) but one might also optimize the, still fairly qualitative, repulsive interaction and the ρ parameter. This has all been done in the past, but further "corrections" should not obscure the most important oversimplification already included in the Coulomb part.

It is not at all clear which charges have to go into the electrostatic calculus; we may certainly imagine singly-charged Na^+ and Cl^- ions in the gas phase, it is true, but what charge will they have in the solid? Even if the recipe in approximating ionic charges by chemical expectancy charges (oxidation states) does not seem to be too bad for NaCl, how will it perform for other compounds? The dilemma is especially troublesome for the most important class of solid-state materials; oxides. Here, even the *definition* for the lattice energy itself is problematic, at best, because the doubly charged oxide ion, O^{2-}, is *nonexistent* as a free particle. On the contrary, the experimental data show that a free O^{2-} will immediately expel one electron to form O^{-}, and a stable oxide anion is a privilege of the solid state, being stabilized by the Madelung field. Thus, oxide lattice energies relate to a *fictitious reference state* and may therefore be safely considered highly questionable, to say the least.

Nonetheless, the electrostatic traditionalists have solved the Gordian knot largely by ignoring the fundamental difficulties; interestingly, there is a reason for doing so (see below). In addition, the work of Kapustinskii [25] shows that the computational determination of Madelung constants needed for the calculation of theoretical lattice energies is, to a very good approximation, unnecessary. In fact, no structural information is needed! For a given compound of any complexity, the difference between its real Madelung constant and that of the sodium chloride structure (1.7476) is compensated by the difference between the real interionic distances and the sum of the ionic radii for six-fold coordination, that is

$$U_{\mathrm{L}}^{\mathrm{Kapustinskii}} \approx -1202 \frac{\mathrm{kJ}}{\mathrm{mol}} \frac{\nu z_1 z_2\,\text{Å}}{r_\mathrm{c} + r_\mathrm{a}} \left(1 - \frac{0.345\,\text{Å}}{r_\mathrm{c} + r_\mathrm{a}}\right), \tag{1.12}$$

where ν is the number of ions per formula unit. Using this simple, totally empirical, formula for the lattice energy and Pauling's radii, the lattice energy of NaCl ($\nu = 2$, $z_1 = -z_2 = 1$, $r_\mathrm{c} = 0.95$ Å, $r_\mathrm{a} = 1.81$ Å) we obtain is 762 kJ/mol, very close to the experimental 768 kJ/mol. The use of Shannon's radii (r_c = 1.02 Å, r_a = 1.81 Å) leads to a less satisfactory, but still acceptable, 746 kJ/mol.

Also, the lattice energy of MgO by means of Pauling's radii ($\nu = 2$, $z_1 = -z_2 = 2$, $r_\mathrm{c} = 0.65$ Å, $r_\mathrm{a} = 1.40$ Å) yields 3903 kJ/mol, less than 1% below the "experimental" value of 3927 kJ/mol. The latter value has been derived, of course, by bravely ignoring the fact that O^{2-} does not exist, such that there is *no* value which one might denote experimental. For Al_2O_3 ($\nu = 5$, $z_1 = 3$, $z_2 = -2$, $r_\mathrm{c} = 0.50$ Å, $r_\mathrm{a} = 1.40$ Å) we have a theoretical 15 533 kJ/mol, a 2.8% error from the "experimental" 15 110 kJ/mol. Note that such a Kapustinskii-style calculation effectively relates *any* structure to the rock-salt type with six-fold coordination, and the correct trend is the same with whatever kind of tabulation for the octahedral ionic radii (Goldschmidt, Pauling, Shannon). When plotting lattice energies against other physical quantities (e.g., melting temperatures) for selected classes of materials (alkali halides, simple oxides), there are obvious correlations which cannot be refuted.

Finally, electrostatic calculations of this kind have been pushed to their limits (and possibly beyond) by *only* taking the Coulomb interactions into account and neglecting the repulsive ones. Because of these suppressed repulsive energies, bare electrostatic calculations arrive at energy values which are larger than the "real" lattice energies. Surprisingly, these dubbed MAPLE (Madelung part of lattice energy) calculations [26] show, for example, that the Coulomb energies of ternary (or quaternary) phases are often extremely close to the sum of the Coulomb energies of binary phases under the (questionable) assumption that the oxidation states are used as idealized charges. Thus, the MAPLE value for the ternary $MgAl_2O_4$ (a = 8.08 Å; MAPLE = 22 916 kJ/mol) is almost the sum of the MAPLE values of the binary MgO (a = 4.21 Å; 4612 kJ/mol) and the binary Al_2O_3 (a = 4.76 Å; 18 210 kJ/mol) because "$MgAl_2O_4$ = MgO + Al_2O_3". The energy deviation is only 94 kJ/mol, that is, about 0.4% of the sum of the Coulomb energies of the binaries! A huge number of other examples of this kind exist. Note, once and for all, that MAPLE values should not be compared with experimental lattice energies but, instead, with other MAPLE values as demonstrated before; the entire concept does *not* relate to observable quantities!

It seems even more exciting that MAPLE "works" even if there cannot be the slightest doubt that the approximations of ionic charges and also ionic bonding are no longer valid, and are probably entirely false. Let us take a ternary nitride, for example, Ca_6GaN_5 [27]. Even though there is considerable covalent bonding between gallium and nitrogen, and a strongly charged N^{3-} is extremely far from reality (the isolated N^{3-} does not exist either, of course), the Madelung energy of Ca_6GaN_5 (a = 6.28 Å, c = 12.20 Å; MAPLE = 38 075 kJ/mol) is also very close to the sum of twice the MAPLE value of Ca_3N_2 [28] (a = 11.48 Å; 14 106 kJ/mol) and that of GaN (a = 3.18 Å, c = 5.17 Å; 10 555 kJ/mol) because "Ca_6GaN_5 = 2 Ca_3N_2 + GaN". Here the deviation is −692 kJ/mol, an "error" of 1.8%. The lesson learned from the latter example is that, quite obviously, these MAPLE calculations do *not* describe the ionic bonding but simply evaluate whether or not the interatomic distances found in the binary (ternary) crystal structures are reasonable, and the accuracy of the MAPLE concept must result from a fortunate error cancellation both on the ternary (quaternary) side and also on the binary side; the exact sizes of the ionic charges are seemingly unimportant if – *and only if* – they are identical for all binary (ternary) compounds.

Because of this fully empirical, *a posteriori* justification, MAPLE calculations have therefore been (and still are) in routine use in many synthetic solid-state chemical laboratories around the world. For the purpose of a fast check of crystallographically determined structures of new materials, MAPLE remains a simple, inexpensive and plausible approach in the hands of a skilled and experienced solid-state chemist.

1.3 Pauling's Rules

Based on sufficient experimental knowledge of a relatively large number of (mostly ionic) crystal structures clarified in the first three decades of the 20th century, Pauling generalized the crystal-chemical observations in a series of rules touching upon the stability of complex ionic crystals [6], thereby summarizing the contributions of many crystal chemists. As these rules have been quite influential in the (classical) understanding of crystal chemistry and are therefore omnipresent – sometimes unconsciously so – in the minds of many scientists, we will briefly recall them now:

Pauling's **first rule** states that *a coordinated polyhedron of anions is formed about each cation, the cation–anion distance being determined by the radius sum and the ligancy of the cation by the radius ratio.*

As has been shown in the preceding section, it is this distance criterion that has been used in *defining* self-consistent sets of ionic radii of which the tabulation of Shannon is the most complete one. With respect to the radius ratio, the quantitative significance is low, at best (see Section 1.1), and a materials scientist is better off in interpreting this part of the first rule as being mostly qualitative in nature.

The **second rule** postulates that *in a stable ionic structure the valence*[7] *of each anion, with changed sign, is exactly or nearly equal to the sum of the strengths of the electrostatic bonds to it from the adjacent cations.*

This mostly electrostatic, semiquantitative ("or nearly equal to") rule has turned out to be one of the most powerful ones, and the so-called bond-valence (or bond-length–bond-strength) concept is built upon it. For reasons of brevity, we will stop here but cover the latter concept – a *quantitative* one, in fact! – in more detail in Section 1.5.

Pauling's **third rule** states that *the presence of shared edges, and especially of shared faces, in a coordinated structure decreases its stability; this effect is large for cations with large valence and small ligancy.*

Again, this rule is of electrostatic origin and simply indicates that repulsive cation–cation interactions are smaller whenever the degree of linkage between neighboring coordination polyhedra shrinks; this is sketched for the case of connected tetrahedra in Figure 1.6.

An elementary geometrical calculation shows that the positively charged centers of both tetrahedra are separated by $2d$ (= twice the cation–anion distance) for corner-sharing; by a significantly smaller $1.15d$ for edge-sharing; and finally by only $0.67d$ for face-sharing, thereby further destabilizing the latter two alternatives by 74 and 199% in purely electrostatic terms, when com-

7) According to IUPAC, this is the maximum number of univalent atoms (such as H) that may combine with an atom of the element under consideration.

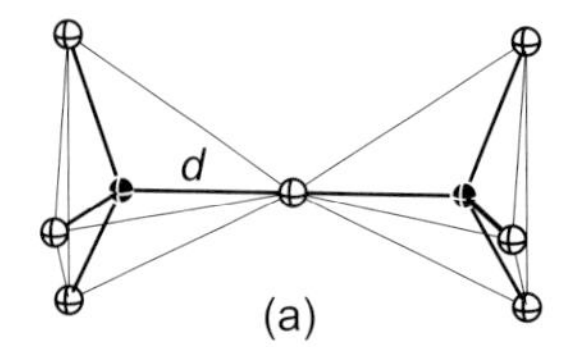

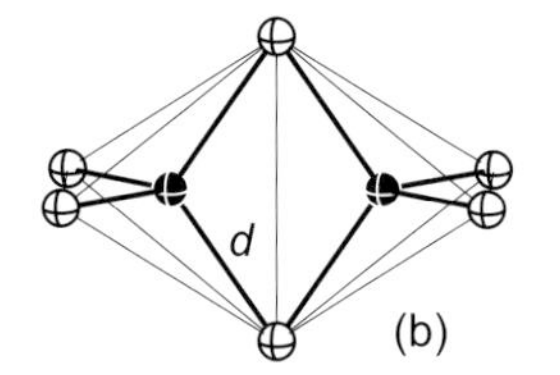

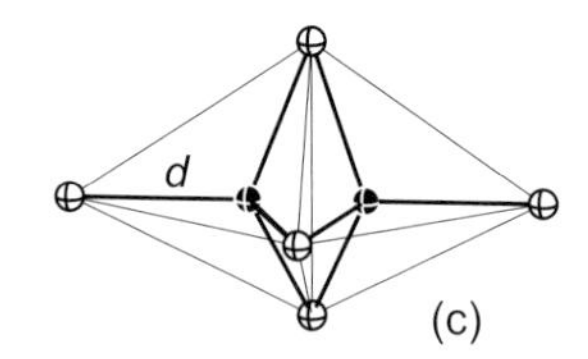

Fig. 1.6 Corner- (a), edge- (b) and face-sharing (c) of regular centered tetrahedra.

pared with the corner-sharing structure (see Section 1.2). Indeed, there is a vast number of oxosilicate structures [29] where corner-sharing – but never edge-sharing – of neighboring SiO_4^{4-} tetrahedra is observed. Only recently has edge-sharing of filled tetrahedra been found in a small number of cases, for example, in nitridosilicate compounds such as $Ba_5Si_2N_6$ and $BaSi_7N_{10}$ [30], and here $Si_2N_6^{12-}$ entities which exhibit edge-sharing may be observed. This astonishing finding goes back to the much more covalent character of the Si–N bond, and the resulting ionic charge on the Si atom must be smaller than in the oxide case. Very similarly, the universe of oxoborate structures being composed of BO_4^{5-} tetrahedra only contains corner-sharing tetrahedra under standard conditions. When enormous pressure is applied upon synthesis, however, neighboring tetrahedra can be squeezed to share an edge. The corresponding example is given by the high-pressure phase $Dy_4B_6O_{15}$ which contains structural fragments that are derived from a $B_2O_6^{6-}$ entity [31]. Nonetheless, the corresponding face-sharing of filled tetrahedra still awaits its discovery, a nice corroboration of Pauling's third rule.

For octahedral structures, the TiO_2 polymorphs usually serve as illustrative examples of Pauling's third rule. In the stable rutile structure, a TiO_6^{8-} octahedron shares two of its edges with neighboring octahedra; in the less stable brookite and anatase polymorphs, however, the number of shared edges increases to three and four [5]. Also, the rutile type is much more common than the other two.

For octahedra, the rule also breaks down whenever covalent bond formation between the central atoms affects the geometrical details. For example, the structural fragment $M_2Cl_9^{3-}$ found for the trivalent cations of Cr, Mo, and W as sketched in Figure 1.7 consists of face-sharing octahedra, and here one would naively expect the M^{3+}–M^{3+} distances to continually increase because of the continuously increasing cationic radii of 0.615 Å (Cr^{3+}), 0.69 Å (Mo^{3+}), and $\approx$ 0.7 Å (W^{3+}) corresponding to six-fold coordination. In reality, however, the M^{3+}–M^{3+} distance *shrinks* from 3.17 Å ($Cs_3Cr_2Cl_9$) to 2.65 Å ($Cs_3Mo_2Cl_9$) to eventually arrive at 2.50 Å ($Cs_3W_2Cl_9$) [32] because a metal–metal bond sets in due to a better *d*–*d* overlap. Clearly, effects like this one cannot be understood in electrostatic terms.

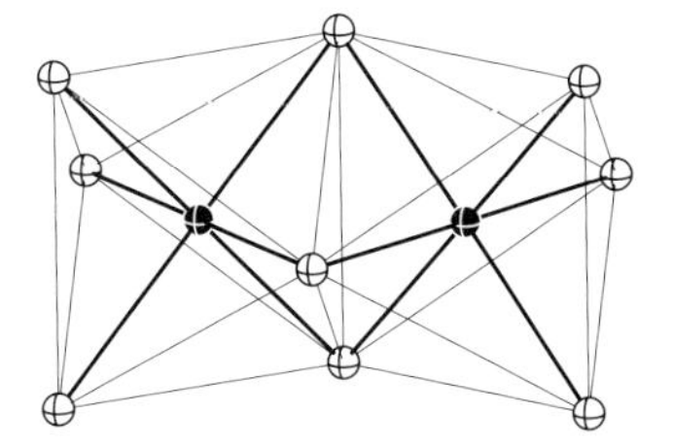

Fig. 1.7 Face-sharing of two centered octahedra.

Pauling's **fourth rule** further exemplifies the preceding one by stating that *in a crystal containing different cations, those with large valences and small coordination number tend not to share polyhedral elements with each other.*

Finally, the **fifth rule** says that *the number of essentially different kinds of constituents in a crystal tends to be small.*

In other words, highly symmetric, simple structures tend to be more stable. We will look at this rule more generally (see Section 1.6) but, for the moment, let us simply sketch a regular and a strongly distorted variant of the sodium chloride crystal structure with identical molar volumes as given in Figure 1.8. In the latter, all ionic positions for Na^+ and Cl^- have been randomly shifted up to a maximum of 1.0 Å. Although it is a bit difficult to recognize, the "idea" behind the two structures (rock-salt type) is indeed identical. To evaluate the situation more quantitatively, we may now adopt a rather simple picture of the chemical bonding – a dominating attractive Madelung field plus a short-

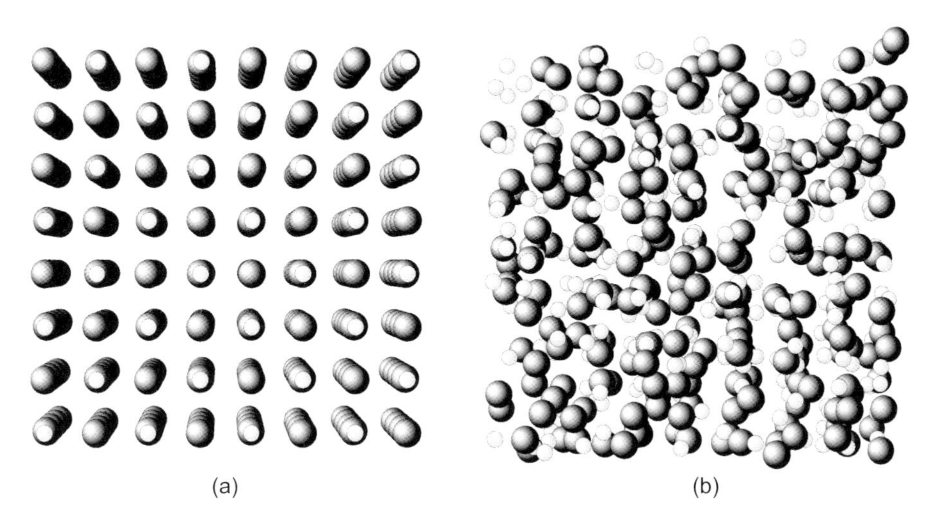

Fig. 1.8 Regular (a) and strongly distorted (b) NaCl structure.

ranged Pauli repulsion (see also Section 1.2) – and will immediately find that the enlarged *repulsive* contributions of the distorted structure have destabilized the latter; it simply costs too much energy if neighboring atoms are pushed together too closely. But even if we were to neglect all short-ranged repulsion energies and concentrate on the *electrostatic* terms only, a numerical evaluation of the Madelung sum of the distorted NaCl variant yields an electrostatic part of the lattice energy of 848 kJ/mol, to be compared with the ideal value of 861 kJ/mol for the regular NaCl structure; thus, *regularity* leads to a lower energy, in particular for the electrostatic interactions with an infinite range. Pauling's fifth rule is in favor of regularity, calling for high-symmetry arrangements of all atoms.

Exceptions from the rule usually show up as quantum-chemical effects which lower the total energies by also lowering the structural symmetries (see Section 3.4). On the other hand, we should add that there is a whole universe of *amorphous* structures which are deliberately excluded from the discussion if we follow Pauling's fifth rule but, admittedly, these refer to thermodynamically metastable states in practically all cases. Indeed, the study of solid-state materials that do *not* exhibit translational invariance albeit chemical, local order – ordinary window glass is an ubiquitous example – has not been particularly excessive when compared with "normal" (that is, crystalline) solid-state materials, simply because the characterization of such matter is much more difficult; for amorphous materials, ordinary X-ray or neutron diffraction loses its enormous analytical power. Thus, our atomistic knowledge of amorphous materials is far from being satisfactory, and Pauling's fifth rule should probably be taken with a pinch of salt.

1.4 Volume Increments

There is an alternative traditional recipe which attempts to describe empirically the crystal chemistry of solid-state materials. In contrast to the preceding ionic-radii concept covered in Section 1.1, implicitly based on hard, nonpenetrating spheres, one might simply take the point of view that the total molar volume V of any solid-state material may be approximated by the sum of individual volume increments v_A which are characteristic for the individual "particles" and, therefore, will contain information about the atom, ion, or ionic groups and also their valence state; thus, we can simply write

$$V = \sum_{A=1}^{M} v_A, \tag{1.13}$$

where the sum should include all atoms per formula unit or per unit cell. In order to do so, the individual increments have to be determined by a rigorous statistical analysis of a large number of crystal structures; trivially, the larger the database, the more reliable the derived volume increments.

A moment's reflection reveals that such an approach, although certainly related in spirit – again, the idea of close-packing pops up – to the concept of ionic radii, does strongly differ in at least two important characteristics. First, since no assumption has been made as regards the spatial shape of the individual entities (atoms, ions, ionic groups), these shapes do *not* have to be spherically symmetric. In fact, they may show any shape to fulfill Equation (1.13) as well as possible. Second, the concept does not rely on the detailed knowledge of the crystal structure such that an important – and possibly too overstated – goal of *volume chemistry* is to predict the volumes of any given material *without* having access to the spatial parameters of the underlying crystal structure. Thus, it probably might not be restricted to crystals, but may also apply to glasses or amorphous solids.

It transpires that the above goal has already been accomplished, partly so, in the early 20th century due to the work of Biltz and coworkers [33]. Based on the available experimental data of pyknometric and X-ray-based densities of inorganic (and also organic!) solid-state materials at that time, a tabulation referring to absolute zero temperature and a standard pressure of 1 atm was proposed; it is given in Table 1.2.

Tab. 1.2 Volume increments according to Biltz [33] including **basis**, regular and *uncertain* values.

Atom	$^{\text{(oxidation state)}}v$ (cm^3/mol)			
Ag	$^{(2+)}$6–7	$^{(+)}$**9**	$^{(0)}$**10.13**	
Al	$^{(3+)}$0–1	$^{(0)}$**9.9**		
Ar	$^{(0)}$**23.4**			
As	$^{(5+)}$**0**	$^{(3+)}$7	$^{(0)}$**13.0**	$^{(3-)}$*40*
Au	$^{(3+)}$*3*	$^{(+)}$10–12	$^{(0)}$**10.12**	
B	$^{(3+)}$**0**	$^{(0)}$**3.25**		
Ba	$^{(2+)}$**16**	$^{(0)}$37		
Be	$^{(2+)}$**0**	$^{(0)}$**4.84**		
Bi	$^{(3+)}$8	$^{(0)}$**21.0**	$^{(3-)}$*60*	
Br	$^{(5+)}$*2*	$^{(0)}$**19.2**	$^{(-)}$**25**	
C	$^{(4+)}$**0**	$^{(0)}$**3.41/5.38**	$^{(4-)}$*20*	
Ca	$^{(2+)}$**6.5**	$^{(0)}$**25.6**		
Cd	$^{(2+)}$**6**	$^{(0)}$**12.7**		
Ce	$^{(4+)}$*≈3*	$^{(3+)}$*7.8*	$^{(0)}$**20.2/20.3**	
Cl	$^{(7+)}$**0**	$^{(5+)}$*1*	$^{(0)}$**16.3**	$^{(-)}$**20**
Co	$^{(3+)}$0	$^{(2+)}$3	$^{(0)}$**6.58**	
Cr	$^{(6+)}$**0**	$^{(3+)}$1	$^{(2+)}$4	$^{(0)}$**7.2**
Cs	$^{(+)}$**26**	$^{(0)}$**65.9**		
Cu	$^{(2+)}$3	$^{(+)}$5	$^{(0)}$**7.05**	
Dy	$^{(3+)}$*6.1*			
Er	$^{(3+)}$*5.7*	$^{(0)}$22.1		
Eu	$^{(3+)}$*6.7*			
F	$^{(0)}$*8.5*	$^{(-)}$9.5		
Fe	$^{(3+)}$1	$^{(2+)}$4	$^{(0)}$**7.05**	
Ga	$^{(3+)}$2	$^{(2+)}$*≈6*	$^{(+)}$11	$^{(0)}$**11.7**
Gd	$^{(3+)}$*6.5*			
Ge	$^{(4+)}$*≈1*	$^{(2+)}$*8*	$^{(0)}$**13.5**	
H	$^{(+)}$**0**	$^{(0)}$**11.4**	$^{(-)}$**11**	
He	$^{(0)}$*≤27*			
Hf	$^{(4+)}$2–3	$^{(0)}$**13.39**		
Hg	$^{(2+)}$**8**	$^{(+)}$**16**	$^{(0)}$**13.75**	
Ho	$^{(3+)}$*5.9*			
I	$^{(7+)}$**0**	$^{(5+)}$6	$^{(0)}$**24.5**	$^{(-)}$**34**
In	$^{(3+)}$*≈4*	$^{(2+)}$*≈11*	$^{(+)}$*≈16*	$^{(0)}$**15.3**
Ir	$^{(4+)}$1	$^{(0)}$**8.46**		
K	$^{(+)}$**16**	$^{(0)}$**43.4**		
Kr	$^{(0)}$**26.3**			
La	$^{(3+)}$8/10	$^{(0)}$**22.0**		
Li	$^{(+)}$1–2	$^{(0)}$**12.56**		
Lu	$^{(3+)}$5			
Mg	$^{(2+)}$2	$^{(0)}$**13.8**		
Mn	$^{(7+)}$**0**	$^{(4+)}$1	$^{(2+)}$5	$^{(0)}$**7.26**

continued on next page

Tab. 1.2 (continued)

Atom	**$^{(\text{oxidation state})}v$ (cm^3/mol)**			
Mo	$^{(6+)}$**0**	$^{(4+)}$1	$^{(2+)}$≈6	$^{(0)}$**9.37**
N	$^{(5+)}$**0**	$^{(3+)}$*5*	$^{(0)}$**13.7**	$^{(3-)}$*19*
NH_4^+	$^{(+)}$**19.5**			
Na	$^{(+)}$**6.5**	$^{(0)}$**22.8**		
Nb	$^{(5+)}$**0**	$^{(4+)}$1	$^{(3+)}$4	$^{(0)}$**10.85**
Nd	$^{(3+)}$*7.4*	$^{(0)}$**20.2**		
Ne	$^{(0)}$**14.0**			
Ni	$^{(2+)}$2	$^{(0)}$**6.55**		
O	$^{(0)}$≈8/**10.9**	$^{(2-)}$**11**		
Os	$^{(8+)}$**0**	$^{(4+)}$1	$^{(0)}$**8.36**	
P	$^{(5+)}$**0**	$^{(3+)}$6	$^{(0)}$**11.4/13.2/15.4**	$^{(3-)}$*35*
Pb	$^{(4+)}$≈4	$^{(2+)}$**12.5**	$^{(0)}$**17.9**	$^{(4-)}$*64*
Pd	$^{(2+)}$4	$^{(0)}$**8.79**		
Pm	$^{(3+)}$*7.2*			
Pr	$^{(4+)}$≈3	$^{(3+)}$*7.6*	$^{(0)}$**20.4**	
Pt	$^{(0)}$**9.04**			
Rb	$^{(+)}$**20**	$^{(0)}$**53.1**		
Re	$^{(7+)}$**0**	$^{(0)}$**8.82**		
Rh	$^{(3+)}$1	$^{(0)}$**8.24**		
Ru	$^{(8+)}$**0**	$^{(4+)}$1	$^{(0)}$**8.09**	
S	$^{(6+)}$**0**	$^{(0)}$**15.0**	$^{(2-)}$29	
Sb	$^{(5+)}$**0**	$^{(3+)}$8	$^{(0)}$**18.1**	$^{(3-)}$*46*
Sc	$^{(3+)}$2	$^{(0)}$*15*		
Se	$^{(6+)}$**0**	$^{(4+)}$*4*	$^{(0)}$**15.8/17.2**	$^{(2-)}$*32*
Si	$^{(4+)}$**0**	$^{(0)}$**12.03**	$^{(4-)}$*36*	
Sm	$^{(3+)}$*6.9*	$^{(2+)}$11	$^{(0)}$19.2	
Sn	$^{(4+)}$≈2	$^{(2+)}$**10**	$^{(0)}$**16.0/20.5**	$^{(4-)}$*62*
Sr	$^{(2+)}$**11**	$^{(0)}$**33.2**		
Ta	$^{(5+)}$**0**	$^{(0)}$**10.9**		
Tb	$^{(4+)}$≈2	$^{(3+)}$*6.3*		
Te	$^{(6+)}$**0**	$^{(4+)}$*5*	$^{(0)}$**20.2**	$^{(2-)}$*40*
Th	$^{(4+)}$6	$^{(0)}$**19.7**		
Ti	$^{(4+)}$0–1	$^{(3+)}$*2*	$^{(0)}$**10.7**	
Tl	$^{(3+)}$≈6	$^{(+)}$**18.5**	$^{(0)}$**16.9**	
Tm	$^{(3+)}$*5.4*			
U	$^{(6+)}$**0**	$^{(4+)}$4	$^{(0)}$**12.2**	
V	$^{(5+)}$**3**	$^{(4+)}$1	$^{(3+)}$2	$^{(0)}$**8.2**
W	$^{(6+)}$**0**	$^{(4+)}$1	$^{(0)}$**9.50**	
Xe	$^{(0)}$**34.9**			
Y	$^{(3+)}$6	$^{(0)}$**20.2**		
Yb	$^{(3+)}$*5.2*	$^{(2+)}$*8*		
Zn	$^{(2+)}$**3**	$^{(0)}$**8.90**		
Zr	$^{(4+)}$*2–3*	$^{(0)}$**13.94**		

In practice, the derivation of these increments was performed in a similar way to Goldschmidt's derivation of ionic radii, namely starting from the alkali halides and thereby deriving basis values, then extending it to other materials. At least two interesting observations can be made from the latter table which, although very dated (1934), is still in use in many solid-state chemical laboratories.

First, anionic increments are larger than cationic increments, and cationic increments may adopt a zero value. The latter finding is typical for cations in their highest oxidation states with zero valence electrons, because these highly charged, tiny cations usually rest in the remaining empty sites of the closely packed anions, with no demand for additional space; in other words, the whole volume of the solid is fully described by the anions.

Second, the spatial requirement of "inert" electron pairs is easily deduced from the table, namely from the entries of In^{+}/In^{3+}, Tl^{+}/Tl^{3+}, Sn^{2+}/Sn^{4+}, Pb^{2+}/Pb^{4+}, and Sb^{3+}/Sb^{5+}, being of the order of 10 cm^3/mol, about the size of the hydrogen anion, H^-. Later studies have shown that the latter anion, a strongly polarizable ion depending on the binding partner, also varies greatly in its incremental volume [34].

Let us illustrate this simple approach by a few examples, but we first add that numerical values for molar volumes (measured in cm^3/mol) are converted into the crystallographer's $Å^3$/f.u. (f.u. = formula unit) by multiplying them by a factor of 1.661 and vice versa. According to the Biltz table, sodium chloride (NaCl) should have a molar volume of 6.5 cm^3/mol (Na^+) + 20 cm^3/mol (Cl^-) = 26.5 cm^3/mol. In a given crystal structure, a single formula unit of NaCl will therefore occupy 44.0 $Å^3$. In fact, the crystal structure of NaCl has a lattice parameter of 5.64 Å and a unit cell volume of 179.4 $Å^3$ with four formula units, which yields an experimental 44.9 $Å^3$ per formula unit and an incremental error of only 2%.[8]

Likewise, this concept helps in identifying correct indexing proposals for the lattice parameters from a number of incorrect ones, a problem every synthetic materials chemists will face sooner or later, when trying to solve complex crystal structures from badly resolved X-ray diffractional data taken on powders. For example, the predicted molar volume of solid Mn_2O_7 is easily computed to be 2×0 cm^3/mol (Mn^{7+}) + 7×11 cm^3/mol (O^{2-}) = 77 cm^3/mol or 127.9 $Å^3$ per formula unit. We reiterate that this calculus implicitly assumes that the highly charged and thus tiny heptavalent Mn cation will not increase the volume requirement of the anionic oxygen substructure which effectively determines the size of the unit cell. Upon crystallizing the molecular material, the X-ray indexing yields, among many other solutions, a monoclinic lattice with a volume of 1055.1 $Å^3$, corresponding to $8.25 \approx 8$ formula units. This is the correct indexing because solid Mn_2O_7 adopts a defect fluorite anti-

8) If you think that the NaCl example comes close to circular reasoning, you are probably right.

type [35] with an fcc packing for the O atoms and two symmetry-independent molecules, each on a general (four-fold) position such that the above volume calculus immediately supports the crystallographic unit cell *before* the structure has been solved. "Fuzzy" information like this is incredibly helpful in the laboratory.

Third, the increment method allows the *qualitative* energetic assessment of new materials. When sodium nitride, Na_3N, was eventually synthesized for the first time [36], its properties pointed towards a metastable compound, and this was backed by quantum-mechanical total-energy calculations. Because the lattice parameter (4.73 Å) of the ReO_3-type sodium nitride corresponds to a molar volume of 63.7 cm^3/mol, one arrives at a huge experimental N^{3-} volume increment of 44.2 cm^3/mol if (three times) the tabulated increment for Na^+ (6.5 cm^3/mol) is subtracted. Thus, the Na_3N-derived nitride increment is more than *twice* the tabulated one (19 cm^3/mol) such that, within metastable Na_3N, the nitride ion is seemingly subject to extreme polarization, a clear sign of low stability.

There has been an attempt, worth discussing, on how to put the concept of volume increments on some quantum-mechanical footing [37] but I personally believe that this inherently classical approach would benefit from an up-to-date statistical analysis of all presently available crystal structure data, combined with a rigorous analysis of their chemical bondings. Given the plethora of digitally stored crystal-structure information, it seems very likely that a lot more interesting information is potentially available through data mining, which could not be extracted more than seven decades ago.

1.5 The Bond-valence Method

The bond-valence or bond-length–bond-strength method is another traditional, but surprisingly lively, approach targeted at better quantifying Pauling's second rule; namely that the valence of a central atom matches the sum of all bond strengths[9] to its coordinating atoms. For ions, this rule locally assures a maximum charge balance whereas, for more covalent structures, it fulfills the octet rule. In order to apply the rule to *asymmetric* coordination polyhedra, the crucial point is to find a reliable correlation between the interatomic distance and the bond strength which results from that particular distance. We should mention that the early ideas of Zachariasen and predecessors [38], touching upon the crystal chemistry of oxidic materials, went into the concept, and

9) It is not easy to specify what chemists really mean by bond *strength*, a nonobservable quantity without a physical unit; which the physicists will kindly allow. To measure the strength of a covalent bond, its dissociation energy or its force constant certainly proves useful.

also Pauling [6] formulated a *molecular* correlation for carbon–carbon bond distances d and their bond strengths s, namely

$$d_{\mathrm{C-C}} = 1.54\,\text{Å} - 0.71\,\text{Å}\,\log s, \tag{1.14}$$

such that the bond strengths of single ($s = 1$), double ($s = 2$) and triple ($s = 3$) bonds yield C–C distances of 1.54, 1.33 and 1.20 Å; organic chemists will surely love this scheme. Strictly speaking, however, the rule of thumb is somewhat questionable since we know of chemical bonds that *weaken* upon shortening. For example, the $^3\Sigma_u^+$ excited state of the C_2 molecule is almost 3% shorter than the more stable ground state; this is covered in more detail in Section 2.3. Also, there are *many* examples of elements/compounds whose more densely packed (i.e., with shorter interatomic distances) allotropes/crystallographic phases are higher in energy than the ground state. For the sake of the empirical concept derived here, we will suppress this counterargument for the moment. In addition, there is hardly anything more appealing than a nicely working concept that is fundamentally flawed (I am, of course, joking here).

One great advantage of the modern bond-valence concept [39], developed by Donnay, Allmann, Brown, Altermatt and many others [40], is that it does not rely on any kind of distinction whatsoever between ionic or covalent bonding; both are *mapped* into the same formalism, and the cation/anion terms simply stand for the less/more electronegative bonding partners, that is, Lewis acids and bases. This is the reason for having replaced "ion" with "atom" when referencing Pauling's second rule at the beginning of this section.

Formally, the bond-valence sum or, simply, the formal valence v_i of a coordinated atom/ion i is equivalent to the sum of all bond-valences (or, bond strengths) s_{ij} to its j coordinating atoms/ions with interatomic/interionic distances r_{ij}, that is,

$$v_i = \sum_j s_{ij} = \sum_j \exp\left(\frac{r_0 - r_{ij}}{B}\right), \tag{1.15}$$

and the standard distance r_0, corresponding to a single bond, is a tabulated constant for a given pair of atom types. In rare cases, r_0 is unknown and needs to be determined. The scaling length B for the monotonically decaying exponential function has been chosen to be identical with 0.37 Å, and this choice has turned out satisfactorily for a plethora of crystal structures. Also, the above formula (there are others [39]) is the most popular one.

In the solid-state chemical community, bond-valence (BV) calculations have been tremendously useful, for example, when it comes to the localization of light atoms that do not show up very easily from X-ray density maps, especially when these include very heavy scatterers. Also, isoelectronic ions such as Al^{3+}/Si^{4+} can be distinguished in alumosilicate structures even when no neutron data are available, simply because of their strongly differing (25%)

bond-valence sums. BV calculations may also be used for routine structural rationalizations such as the following two:

The *tetrahedral* complex ion $InBr_4^-$ with four In^{3+}–Br^- distances of 2.47 Å contains trivalent indium; thus, each bond strength is 3/4 = 0.75, and the r_0 bond-valence parameter for In^{3+}–Br^- must be 2.364 Å. With the help of this parameter, one may predict the In^{3+}–Br^- bond distance within a regular $InBr_6^{3-}$ *octahedron* to be 2.62 Å, provided that it really contains trivalent indium. This is because 2.62 Å corresponds to a bond strength of 0.5 for r_0 = 2.364 Å, and six such bonds give a trivalent state for the central indium atom. In fact, the experimental crystal structure of In_4Br_7 [41] exhibits a corresponding interatomic distance of 2.66 Å, supporting the above calculation. The, slightly too large (0.04 Å), distance found experimentally points towards the necessity for a crystallographic (rigid-body) correction of that octahedron [42], and the 2.66 Å only *seems* to be too large.

Another example: Given that heptavalent manganese in the tetrahedral permanganate ion, MnO_4^-, exhibits four identical Mn^{7+}–O^{2-} bond distances of roughly 1.58 Å, what are the expected distances in molecular dimanganese heptaoxide, Mn_2O_7 = O_3Mn–O–MnO_3? From the permanganate ion (individual bond strength = 7/4 = 1.75), the bond-valence parameter r_0 is found to be 1.79 Å. Thus, Mn_2O_7 should show six Mn=O double bonds of 1.53 Å and two (bridging) Mn–O single bonds of 1.79 Å. In fact, these values were predicted [43] before the crystal structure of this explosive material was determined, yielding bond distances of 1.58 and 1.77 Å [35].

There have been several compilations of bond-valence parameters r_0 (with B fixed at the above 0.37 Å) for a multitude of anion–cation combinations, based on a very large number of well-resolved inorganic crystal structures. In addition, Brese and O'Keeffe have included less common cation–anion combinations, and they have also predicted reference r_0 values for those cases where no experimental data were available. For convenience, we include their data [44] in Tables 1.3 and 1.4, so that they can be used for your own calculations.

Tab. 1.3 Bond-valence r_0 parameters (Å) for oxides, fluorides and chlorides [44].

Cation	O	F	Cl	Cation	O	F	Cl
Ac^{3+}	2.24	2.13	2.63	Hg^{2+}	1.93	1.90	2.25
Ag^{+}	1.805	1.80	2.09	Ho^{3+}	2.023	1.908	2.401
Al^{3+}	1.651	1.545	2.03	I^{5+}	2.00	1.90	2.38
Am^{3+}	2.11	2.00	2.48	I^{7+}	1.93	1.83	2.31
As^{3+}	1.789	1.70	2.16	In^{3+}	1.902	1.79	2.28
As^{5+}	1.767	1.62	2.14	Ir^{5+}	1.916	1.82	2.30
Au^{3+}	1.833	1.81	2.17	K^{+}	2.13	1.99	2.52
B^{3+}	1.371	1.31	1.74	La^{3+}	2.172	2.057	2.545
Ba^{2+}	2.29	2.19	2.69	Li^{+}	1.466	1.360	1.91
Be^{2+}	1.381	1.28	1.76	Lu^{3+}	1.971	1.876	2.361
Bi^{3+}	2.09	1.99	2.48	Mg^{2+}	1.693	1.581	2.08
Bi^{5+}	2.06	1.97	2.44	Mn^{2+}	1.790	1.698	2.13
Bk^{3+}	2.08	1.96	2.46	Mn^{3+}	1.760	1.66	2.14
Br^{7+}	1.81	1.72	2.19	Mn^{4+}	1.753	1.71	2.13
C^{4+}	1.39	1.32	1.76	Mn^{7+}	1.79	1.72	2.17
Ca^{2+}	1.967	1.842	2.37	Mo^{6+}	1.907	1.81	2.28
Cd^{2+}	1.904	1.811	2.23	N^{3+}	1.361	1.37	1.75
Ce^{3+}	2.151	2.036	2.52	N^{5+}	1.432	1.36	1.80
Ce^{4+}	2.028	1.995	2.41	Na^{+}	1.80	1.677	2.15
Cf^{3+}	2.07	1.95	2.45	Nb^{5+}	1.911	1.87	2.27
Cl^{7+}	1.632	1.55	2.00	Nd^{3+}	2.117	2.008	2.492
Cm^{3+}	2.23	2.12	2.62	Ni^{2+}	1.654	1.599	2.02
Co^{2+}	1.692	1.64	2.01	Os^{4+}	1.811	1.72	2.19
Co^{3+}	1.70	1.62	2.05	P^{5+}	1.604	1.521	1.99
Cr^{2+}	1.73	1.67	2.09	Pb^{2+}	2.112	2.03	2.53
Cr^{3+}	1.724	1.64	2.08	Pb^{4+}	2.042	1.94	2.43
Cr^{6+}	1.794	1.74	2.12	Pd^{2+}	1.792	1.74	2.05
Cs^{+}	2.42	2.33	2.79	Pr^{3+}	2.135	2.022	2.50
Cu^{+}	1.593	1.6	1.85	Pt^{2+}	1.768	1.68	2.05
Cu^{2+}	1.679	1.60	2.00	Pt^{4+}	1.879	1.759	2.17
Dy^{3+}	2.036	1.922	2.41	Pu^{3+}	2.11	2.00	2.48
Er^{3+}	2.010	1.906	2.39	Rb^{+}	2.26	2.16	2.65
Eu^{2+}	2.147	2.04	2.53	Re^{7+}	1.97	1.86	2.23
Eu^{3+}	2.076	1.961	2.455	Rh^{3+}	1.791	1.71	2.17
Fe^{2+}	1.734	1.65	2.06	Ru^{4+}	1.834	1.74	2.21
Fe^{3+}	1.759	1.67	2.09	S^{4+}	1.644	1.60	2.02
Ga^{3+}	1.730	1.62	2.07	S^{6+}	1.624	1.56	2.03
Gd^{3+}	2.065	1.95	2.445	Sb^{3+}	1.973	1.90	2.35
Ge^{4+}	1.748	1.66	2.14	Sb^{5+}	1.942	1.80	2.30
H^{+}	0.95	0.92	1.28	Sc^{3+}	1.849	1.76	2.23
Hf^{4+}	1.923	1.85	2.30	Se^{4+}	1.811	1.73	2.22
Hg^{+}	1.90	1.81	2.28	Se^{6+}	1.788	1.69	2.16

continued on next page

Tab. 1.3 (continued)

Cation	O	F	Cl	Cation	O	F	Cl
Si^{4+}	1.624	1.58	2.03	Tl^{3+}	2.003	1.88	2.32
Sm^{3+}	2.088	1.977	2.466	Tm^{3+}	2.000	1.842	2.38
Sn^{2+}	1.984	1.925	2.36	U^{4+}	2.112	2.034	2.48
Sn^{4+}	1.905	1.84	2.28	U^{6+}	2.075	1.966	2.46
Sr^{2+}	2.118	2.019	2.51	V^{3+}	1.743	1.702	2.19
Ta^{5+}	1.920	1.88	2.30	V^{4+}	1.784	1.70	2.16
Tb^{3+}	2.049	1.936	2.427	V^{5+}	1.803	1.71	2.16
Te^{4+}	1.977	1.87	2.37	W^{6+}	1.921	1.83	2.27
Te^{6+}	1.917	1.82	2.30	Y^{3+}	2.014	1.904	2.40
Th^{4+}	2.167	2.07	2.55	Yb^{3+}	1.985	1.875	2.371
Ti^{3+}	1.791	1.723	2.17	Zn^{2+}	1.704	1.62	2.01
Ti^{4+}	1.815	1.76	2.19	Zr^{4+}	1.937	1.854	2.33
Tl^{+}	2.172	2.15	2.56				

Tab. 1.4 Additional bond-valence r_0 parameters (Å) [44].

Atom	Br	I	S	Se	Te	N	P	As	H
Ag	2.22	2.38	2.15	2.26	2.51	1.85	2.22	2.30	1.50
Al	2.20	2.41	2.13	2.27	2.48	1.79	2.24	2.32	1.45
As	2.32	2.54	2.26	2.39	2.61	1.93	2.34	2.41	1.56
Au	2.12	2.34	2.03	2.18	2.41	1.72	2.14	2.22	1.37
B	1.88	2.10	1.82	1.95	2.20	1.47	1.88	1.97	1.14
Ba	2.88	3.13	2.77	2.88	3.08	2.47	2.88	2.96	2.22
Be	1.90	2.10	1.83	1.97	2.21	1.50	1.95	2.00	1.11
Bi	2.62	2.84	2.55	2.72	2.87	2.24	2.63	2.72	1.97
C	1.90	2.12	1.82	1.97	2.21	1.47	1.89	1.99	1.10
Ca	2.49	2.72	2.45	2.56	2.76	2.14	2.55	2.62	1.83
Cd	2.35	2.57	2.29	2.40	2.59	1.96	2.34	2.43	1.66
Ce	2.69	2.92	2.62	2.74	2.92	2.34	2.70	2.78	2.04
Co	2.18	2.37	2.06	2.24	2.46	1.84	2.21	2.28	1.44
Cr	2.26	2.45	2.18	2.29	2.52	1.85	2.27	2.34	1.52
Cs	2.95	3.18	2.89	2.98	3.16	2.53	2.93	3.04	2.44
Cu	1.99	2.16	1.86	2.02	2.27	1.61	1.97	2.08	1.21
Dy	2.56	2.77	2.47	2.61	2.80	2.18	2.57	2.64	1.89
Er	2.54	2.75	2.46	2.59	2.78	2.16	2.55	2.63	1.86
Eu	2.61	2.83	2.53	2.66	2.85	2.24	2.62	2.70	1.95
Fe	2.26	2.47	2.16	2.28	2.53	1.86	2.27	2.35	1.53
Ga	2.24	2.45	2.17	2.30	2.54	1.84	2.26	2.34	1.51
Gd	2.60	2.82	2.53	2.65	2.84	2.22	2.61	2.68	1.93
Ge	2.30	2.50	2.22	2.35	2.56	1.88	2.32	2.43	1.55
H	1.42	1.61	1.35	1.48	1.78	1.03	1.48	1.52	0.74
Hf	2.47	2.68	2.39	2.52	2.72	2.09	2.48	2.56	1.78
Hg	2.40	2.59	2.32	2.47	2.61	2.02	2.42	2.50	1.71
Ho	2.55	2.77	2.48	2.61	2.80	2.18	2.56	2.64	1.88
In	2.41	2.63	2.36	2.47	2.69	2.03	2.43	2.51	1.72
Ir	2.45	2.66	2.38	2.51	2.71	2.06	2.46	2.54	1.76
K	2.66	2.88	2.59	2.72	2.93	2.26	2.64	2.83	2.10
La	2.72	2.93	2.64	2.74	2.94	2.34	2.73	2.80	2.06
Li	2.02	2.22	1.94	2.09	2.30	1.61	2.04	2.11	1.31
Lu	2.50	2.72	2.43	2.56	2.75	2.11	2.51	2.59	1.82
Mg	2.28	2.46	2.18	2.32	2.53	1.85	2.29	2.38	1.53
Mn	2.26	2.49	2.20	2.32	2.55	1.87	2.24	2.36	1.55
Mo	2.43	2.64	2.35	2.49	2.69	2.04	2.44	2.52	1.73
Na	2.33	2.56	2.28	2.41	2.64	1.93	2.36	2.53	1.68
Nb	2.45	2.68	2.37	2.51	2.70	2.06	2.46	2.54	1.75
Nd	2.66	2.87	2.59	2.71	2.89	2.30	2.66	2.74	2.00
Ni	2.16	2.34	2.04	2.14	2.43	1.75	2.17	2.24	1.40
P	2.15	2.40	2.11	2.26	2.44	1.73	2.19	2.25	1.41
Pb	2.64	2.78	2.55	2.67	2.84	2.22	2.64	2.72	1.97
Pd	2.19	2.38	2.10	2.22	2.48	1.81	2.22	2.30	1.47
Pr	2.67	2.89	2.60	2.72	2.90	2.30	2.68	2.75	2.02
Pt	2.18	2.37	2.08	2.19	2.45	1.77	2.19	2.26	1.40
Rb	2.78	3.01	2.70	2.81	3.00	2.37	2.76	2.87	2.26

continued on next page

Tab. 1.4 (continued)

Atom	Br	I	S	Se	Te	N	P	As	H
Re	2.45	2.61	2.37	2.50	2.70	2.06	2.46	2.54	1.75
Rh	2.25	2.48	2.15	2.33	2.55	1.88	2.29	2.37	1.55
Ru	2.26	2.48	2.16	2.33	2.54	1.88	2.29	2.36	1.61
S	2.17	2.36	2.07	2.21	2.45	1.74	2.15	2.25	1.38
Sb	2.50	2.72	2.45	2.57	2.78	2.12	2.52	2.60	1.77
Sc	2.38	2.59	2.32	2.44	2.64	1.98	2.40	2.48	1.68
Se	2.33	2.54	2.25	2.36	2.55	1.93	2.34	2.42	1.54
Si	2.20	2.41	2.13	2.26	2.49	1.77	2.23	2.31	1.47
Sm	2.62	2.84	2.55	2.67	2.86	2.24	2.63	2.70	1.96
Sn	2.55	2.76	2.45	2.59	2.76	2.14	2.45	2.62	1.85
Sr	2.68	2.88	2.59	2.72	2.87	2.23	2.67	2.76	2.01
Ta	2.45	2.66	2.39	2.51	2.70	2.01	2.47	2.55	1.76
Tb	2.58	2.80	2.51	2.63	2.82	2.20	2.59	2.66	1.91
Te	2.53	2.76	2.45	2.53	2.76	2.12	2.52	2.60	1.83
Th	2.71	2.93	2.64	2.76	2.94	2.34	2.73	2.80	2.07
Ti	2.32	2.54	2.24	2.38	2.60	1.93	2.36	2.42	1.61
Tl	2.70	2.91	2.63	2.70	2.93	2.29	2.71	2.79	2.05
Tm	2.53	2.74	2.45	2.58	2.77	2.14	2.53	2.62	1.85
U	2.63	2.84	2.56	2.70	2.86	2.24	2.64	2.72	1.97
V	2.30	2.51	2.23	2.33	2.57	1.86	2.31	2.39	1.58
W	2.45	2.66	2.39	2.51	2.71	2.06	2.46	2.54	1.76
Y	2.55	2.77	2.48	2.61	2.80	2.17	2.57	2.64	1.86
Yb	2.51	2.72	2.43	2.56	2.76	2.12	2.53	2.59	1.82
Zn	2.15	2.36	2.09	2.22	2.45	1.77	2.15	2.24	1.42
Zr	2.48	2.69	2.41	2.53	2.67	2.11	2.52	2.57	1.79

The numerical precision of the bond-valence concept can be further improved if the universality of the scaling length ($B = 0.37$ Å) is abandoned, thus leaving one more parameter for fitting the experimental data. Research by Adams [45] indeed shows that B depends somewhat on the chemical nature of the atoms involved (e.g., ionization energy, electron affinity, polarizability, etc.), and a corresponding set of improved bond-valence parameters has been proposed, taking into account the bond *softness*. The latter idea is taken from the theory of (absolute) chemical hardness by Pearson, later covered in Section 2.14.

1.6 Symmetry Principles

To the unbiased reader, the preceding classical attempts to (numerically) describe and understand solid-state materials probably look like a somewhat ill-fitting patchwork of grossly differing tools. What they all have in common is the tendency to overemphasize and, at the same time, neglect certain aspects of the structural and bonding phenomena. While the method of ionic radii and also the bond-valence concept both reflect a rather local notion, the entire three-dimensional space is taken into account upon electrostatic reasoning and upon partitioning the total volume into individual increments. In some sense, Pauling's rules combine both local and extended ideas, at least in a qualitative way.

Moreover, a moment's reflection reveals that the clue to a full understanding of any given crystal structure must be hidden in the *whole crystal structure* itself because a quantum-mechanical calculation (see the following sections) based on that particular crystal structure yields any desired property from the final wave function, at least in principle; if you know the structure, you know *everything*. This has been assumed from the very beginning of structural research, and one may therefore ask whether the mathematical apparatus used in structural research contains additional and worthwhile information, which it is hard to recognize when using a "naive" structural view. Realizing that the number of known crystal structures is already approaching half a million entries, the question may be rephrased as follows: Is there an ordering scheme from which one may structurally derive certain crystal structures from other, probably more fundamental, structures?

The answer is a clear "yes", and we have already witnessed the relationship between the NaCl structure and the face-centered cubic packing (see Section 1.1) by filling the octahedral holes of the latter. The rigorous mathematical apparatus to fully exploit these structural relationships is given by group theory, in particular the theory of space groups which, unfortunately, cannot be taught here because of space limitations.[10] We merely mention that a *space group* is a set of *symmetry operations* which obey the *group properties* and which are compatible with the *translational symmetry* of the crystal structure [5]. There is a total of 230 types of three-dimensional space groups which have been tabulated [47] and, given the knowledge of the size and symmetry of the crystallographic unit cell as well as all atomic (Wyckoff) positions, the crystal structure is fully defined.

Now, the *qualitative symmetry principle* of crystal chemistry [48] first states that there is a tendency for highly symmetric atomic arrangements. Second, symmetry reductions result from specific atomic properties but they tend to be small. Third, whenever low-symmetry structures are obtained due to phase transformations and solid-state chemical reactions, the initial symmetry is indirectly (macroscopically) preserved by the formation of crystallographic domains.

In *quantitative* terms, the symmetry principle can be based on exact group–subgroup relationships [48, 49], irrespective of the chemical nature of the material. It is also independent of chemical or physical simplifications because of its purely mathematical nature. Starting with a simple crystal structure of high symmetry (called *aristotype* or *basic structure*), more complicated and less symmetric structures (dubbed *hettotypes* or *derivative structures*) are generated either by structural distortions or by atomic substitutions. When doing so, a group-theoretical *family tree*, first proposed by Bärnighausen, emerges in which the initial high-symmetry space group is "thinned out" in terms of symmetry operations. Subgroups may have either lost nontranslational symmetry (then they are called *translationengleich*, and the corresponding unit cell has not changed in size) or they have lost translational symmetry (*klassengleich* subgroups where the unit cell has either become larger or has lost a centering element) [50]. Despite its mathematical and somewhat abstract character, the study of such symmetry principles is highly recommended for the interested reader because it pre-defines rigorous (albeit largely nonchemical) paths for navigating within the many thousands of crystallographic structures.

10) Group theory is a fantastic tool needed for the mathematical analysis of symmetry-related properties of molecules and solids; despite its mathematical nature, the study of (point) group theory is interesting and also exceedingly useful [46].

2
Quantum-chemical Approaches

I can't understand why people are frightened of new ideas. I am frightened of the old ones.
JOHN CAGE

Description is praise.
Imitation is love.
JOHN UPDIKE

Considering the growing popularity of computational – often quantum-chemical – approaches in the fields of solid-state and materials chemistry, one might believe that quantum chemistry must have always played a key role in describing and understanding the various kinds of solid-state compounds. Surprisingly only at first sight, this is far from being true, probably for simple communication problems between the theoretical and experimental scientific communities and probably also because of technical difficulties in applying the most important quantum-chemical workhorse, Hartree–Fock theory (see Section 2.11.3) to the solid state, in particular to metals and metal-rich compounds. On the contrary, it seems that solid-state (materials) chemists have not profited very much from the advent of quantum chemistry, and this can easily be confirmed by looking at quantum-chemistry textbooks which, usually, concentrate on molecules and make a wide arc around solid-state materials [51], with some notable exceptions [4,52]. If we take some major milestones of quantum chemistry into account (1927: Heitler and London solve the H_2^+ problem; 1928: Hartree method; 1930: Hartree–Fock method and Slater's orbital approximation; 1931: Møller–Plesset equations; 1935: Hellmann's pseudopotential approach; 1950: Gaussian basis sets; 1951: Roothaan–Hall equations; 1955: Mulliken's population analysis; 1964/5 Hohenberg–Kohn–Sham approach for density-functional theory), it is only the 1935 and 1964/5 breakthroughs which had a major impact on the solid state (this is a simplification). Also, Bloch's theorem (from 1929) is hardly found in introductory textbooks

Computational Chemistry of Solid State Materials. Richard Dronskowski

ISBN: 3-527-31410-5

of quantum chemistry, but it is an essential part of any textbook on solid-state physics [53].

The changing prospects for quantum-theoretical solid-state chemistry, however, had already begun in the early 1960s when semiempirical methods, such as extended Hückel theory, were invented [54]. These were simple enough (but not too simple) to allow for semiquantitative accuracy, they were not too difficult to understand and, quite importantly, they could be easily transferred to large systems, a prerequisite for the solid state; this fundamental step opened the way towards an understanding of solid-state compounds which went *beyond* the overstrained classical reasonings. Second, as has been alluded to above, density-functional theory was founded and, over the years, further developed by the solid-state physics theorists. Sometimes it seems that the most popular tools of today's quantum solid-state chemistry may be looked upon as illegitimate children of semiempirical/qualitative molecular orbital theory [55] on the one side and density-functional electronic-structure theory on the other. But we will start from the beginning.

2.1 Schrödinger's Equation

As of today, all available experimental (as well as theoretical) knowledge is in agreement that any given system composed of nuclei and electrons – such as atoms, molecules, and also infinite molecules, that is, crystals – can be described in its entirety and, hopefully, also be understood by solving the fundamental quantum-mechanical equation proposed by Schrödinger [56]. Let us neglect, for the moment, time-dependent phenomena and write down the corresponding *stationary*, i.e., time-*independent* Schrödinger equation,

$$\mathcal{H}\Psi = E\Psi, \tag{2.1}$$

in which Ψ is the so-called *wave function* of the system under study, and the energy eigenvalue E is what we are searching for when applying the so-called Hamilton (or Hamiltonian) operator $\mathcal{H}$ onto the wave function; the Hamilton operator itself is just the sum of the kinetic energy $\widehat{T}$ and potential energy $\widehat{V}$ operators, and for a system composed of only one particle it therefore simply reads

$$\mathcal{H} = \widehat{T} + \widehat{V}(\boldsymbol{r}). \tag{2.2}$$

Chemistry is a many-particle science, however, which we may illustrate by explicitly writing down another Hamilton operator for a system containing many nuclei and electrons, i.e., a given molecule or some crystalline material of any complexity; without relativistic corrections, $\mathcal{H}$ can be written:

$$\begin{aligned} \mathcal{H} &= -\sum_{i=1}^{N} \frac{\hbar^2}{2m} \nabla_i^2 - \sum_{A=1}^{M} \frac{\hbar^2}{2M_A} \nabla_A^2 \\ &\quad - \frac{1}{4\pi\epsilon_0} \left(\sum_{i=1}^{N} \sum_{A=1}^{M} \frac{Z_A e^2}{r_{iA}} - \sum_{i=1}^{N} \sum_{j>i}^{N} \frac{e^2}{r_{ij}} - \sum_{A=1}^{M} \sum_{B>A}^{M} \frac{Z_A Z_B e^2}{R_{AB}} \right). \end{aligned} \tag{2.3}$$

Relativistic corrections would take into account the mass increase of the electrons (plus indirect relativistic effects) occurring in heavier atoms (see Section 2.11.1), to be dealt with properly by Dirac's equation, a relativistic theory. But let us concentrate on the above nonrelativistic Hamilton operator and analyze its ingredients. Obviously, $\mathcal{H}$ is composed of five parts; the first two represent the kinetic energies of the N electrons with masses m and the M nuclei with masses M_A (remember that neither electrons nor nuclei can be expected to be standing still); the differential operator

$$\nabla^2 = \nabla \cdot \nabla = \frac{\partial^2}{\partial x^2} + \frac{\partial^2}{\partial y^2} + \frac{\partial^2}{\partial z^2} \tag{2.4}$$

is sometimes also written as $\nabla^2 = \Delta$ and then called Laplace operator; note that the right-hand side of Equation (2.4) refers to cartesian coordinates only.

Next, in Equation (2.3), there is the Coulomb attraction between the electrons (charged $-e$) and the nuclei (charged Z_A) being r_{iA} apart from each other, and this is what brings the system into "motion". In addition, we find the electrostatic repulsion between the electrons (fourth term) and also the electrostatic repulsion between the nuclei (fifth term) separated by R_{AB}; note that we have excluded the double-counting of these interactions (i.e., $j > i$, $B > A$). We must also keep in mind that the electronic coordinates of this nonrelativistic $\mathcal{H}$ only contain spatial and no spin coordinates, and a spin-dependent description is eventually achieved by requiring a certain symmetry property for the many-electron wave function Ψ (see Sections 2.9 and 2.11.3).

The reader will have probably noticed that neither atoms, ions, nor bonds show up in this general Hamilton operator; indeed, there are only electrons and nuclei! In fact, $\mathcal{H}$ does *not* differ (at least in principle) when moving from (molecular) H_2 to HF or from (solid) NaCl to CeN or to MnAl. This is because the chemical understanding of molecules and solid-state materials in terms of covalent, ionic or metallic bonding are *classical* terms – *a posteriori* – which have no counterpart in deriving the Hamiltonian; these notions had been invented many years before quantum mechanics was discovered. It takes a lot of work to re-derive these important chemical concepts later, from the final quantum-mechanical result.

A number of simplifications may now be introduced for $\mathcal{H}$; let us first remind ourselves that we have agreed not to deal quantum-mechanically with time-dependent phenomena. Second, we may assume that the velocities of the nuclei, because of their larger masses, are much smaller than those of the electrons; thus, fixing the nuclear positions while calculating the electronic ones – the famous Born–Oppenheimer approximation [51] – amounts to removing the second entry of $\mathcal{H}$ given in Equation (2.3), thereby deleting the nuclear coordinates in the wave function which hence becomes a function of the electronic coordinates only. Likewise, we then can also drop the fifth entry (the electrostatic interaction energy of the nuclei) from $\mathcal{H}$ because it is constant and therefore immaterial for the calculation of the wave function. This nuclear energy contribution is later included (because this makes the whole calculation easier) by simply adding it to the wave function once it has been found. We also simplify Equation (2.3) by going over to Hartree atomic units,[1] and this leaves us with the electronic Hamilton operator

$$\mathcal{H} = -\sum_{i=1}^{N} \frac{1}{2}\nabla_i^2 - \sum_{i=1}^{N}\sum_{A=1}^{M} \frac{Z_A}{r_{iA}} + \sum_{i=1}^{N}\sum_{j>i}^{N} \frac{1}{r_{ij}}, \tag{2.5}$$

of which the first entry (kinetic energy of the electrons) is fairly easy to access, the second entry (potential of the nuclei acting on the electrons) will help us to later define the electronic-structure *methods*, and the last part (electron–electron interactions) is by far the most difficult to deal with. The conceptional as well as numerical problems arising from the electron–electron interactions are quite challenging but guarantee sustainable business for all manufacturers of supercomputers. We can distinguish, by looking at Equation (2.5), the kind of quantum-chemical route that one follows, namely the degree of sophistication, by the way $\mathcal{H}$ is set up. One usually talks of an *ab initio* or "first principles" method if (almost) no further simplification goes into $\mathcal{H}$ and, in most cases, if all electrons are properly dealt with. An empirical (or semiempirical) method would simplify $\mathcal{H}$ more or less by discarding parts of it or by introducing experimentally available information, thereby speeding up the calculation. A vast number of intermediate methods are also available.[2]

1) This gets rid of the vacuum permittivity $4\pi\epsilon_0$ ($\equiv 1$), the mass of the electron m ($\equiv 1$), the elementary charge e ($\equiv 1$), and Planck's constant/$2\pi = \hbar$ ($\equiv 1$) by introducing the atomic unit length, the Bohr radius a_0 ($\approx$ 0.529 Å). At the same time, the energy of the H atom is found to be -0.5 atomic energy units $\equiv -0.5$ Hartree. One Hartree corresponds to 27.211 eV; this systems is widely used in the quantum-chemistry community. Just to make things more difficult (so it seems), yet another choice of atomic units is preferred in the theoretical solid-state physics community: By dropping the 1/2 in front of the kinetic energy operator ∇^2, the energy directly becomes -13.606 eV, and the new unit is called one Rydberg. Thus, one Hartree equals two Rydbergs, and there are *two* types of atomic unit.

2) According to a world-famous theoretical chemist, there are several dozens of quantum-chemical approaches available, and each of them is superior to all the others.

When it comes to the actual construction of the wave function Ψ for many electrons (therefore called a many-electron or many-body wave function), experience has shown that additional information, touching upon the amount of electron–electron interaction within the system, helps to come up with a good starting point. Within the realm of *molecular* quantum chemistry, two competing approaches (valence-bond and molecular-orbital theory) have been pursued, which we will cover in Section 2.11.3. If one neglects, in a first step, the electron–electron interactions, one may approximate the many-electron wave function Ψ we are looking for by a set of "one-electron" functions ψ_i called molecular orbitals (MOs).[3] We wish the electrons described by these molecular orbitals to be individually independent of each other, and the product of all ψ_i eventually yields Ψ. It is not very straightforward to justify this ansatz because a closer analysis (see Section 2.11.2) still reveals electron–electron interactions to be present and the necessity for an iterative solution of $\mathcal{H}$, but approximating Ψ by a set of ψ_i is indeed a good starting point because the electrons may be considered *statistically* independent of each other. Let us, in a further step, approximate such a molecular orbital ψ as a linear combination of some basic functions resembling those of isolated atoms, and we then arrive at the famous linear combination of atomic orbitals (LCAO) approximation,

$$\psi_i(\boldsymbol{r}) = \sum_{A} \sum_{\substack{\mu=1 \\ \mu \in A}}^{n} c_{\mu i} \phi_\mu(\boldsymbol{r}), \tag{2.6}$$

where the basis functions $\phi_\mu(\boldsymbol{r})$ are atomic orbitals located on the atoms A; the atomic orbitals (AOs) are the solutions of Schrödinger's equation for hydrogen-like atoms (one electron but arbitrary nuclear charge Z_A). There are also other ways to set up ψ_i but this ansatz already implies that the system's molecular orbital(s) cannot be very far from the constituting atomic orbitals ϕ_μ. Very often Equation (2.6) is indeed an excellent strategy since the chemical bonding is a *small* perturbation of the electronic structure of those atoms which are involved in the chemical bonding. We admit that this is a very chemical argument because, in all cases, we can still recognize the atomic nature of the individual atoms even when they have formed a molecule or a crystal. Based upon the above ansatz, the search for the best wave function contained in Schrödinger's equation can be accomplished by following the *variational principle* [51], which states that the *best* wave function will also be characterized by the *lowest* energy, so we will search for the latter. It is given by the *expectation*

3) Note the language subtleties: At least within the field of *ab initio* quantum chemistry, the term *wave function* is almost exclusively used for a many-electron wave function Ψ. A one-electron wave function for atoms and molecules, however, goes by the name (atomic or molecular) *orbital* ψ, not wave function. The exception proves the rule, though.

value of the Hamiltonian acting on the wave function such as in

$$E = \frac{\int \Psi^* \mathcal{H} \Psi d\tau}{\int \Psi^* \Psi d\tau} = \frac{\langle \Psi | \mathcal{H} | \Psi \rangle}{\langle \Psi | \Psi \rangle} \geq E_0. \tag{2.7}$$

Because Ψ is a potentially complex function including an imaginary part, Ψ^* designates the complex conjugate wave function.[4] The compact and quite famous "bracket" notation on the right-hand side of Equation (2.7) bears the name of Dirac, and the "bra" $\langle \Psi |$ and "ket" $| \Psi \rangle$ symbols stand for Ψ^* and Ψ and their integration. Mathematically, an integral such as $\int \Psi^* \Psi d\tau$ has been re-written as a scalar product $\langle \Psi | \Psi \rangle$ within a complex vector space.

We may now get to the lowest energy by systematically optimizing the LCAO mixing coefficients ($c_{\mu i}$ in Equation (2.6)). In addition, we might also optimize the shape of the basis functions (the atomic orbitals ϕ_μ) but that is another strategy which we will not continue here. In the case of a one-electron system, we can simply write ψ instead of the many-electron Ψ (because there is no longer any electron–electron interaction left) and then use Equation (2.6) for the expansion of the molecular orbital in terms of atomic orbitals. This leaves us with

$$\begin{aligned} E_i &= \frac{\sum_A \sum_B \sum_\mu \sum_\nu c_{\mu i}^* c_{\nu i} \langle \phi_\mu | \mathcal{H} | \phi_\nu \rangle}{\sum_A \sum_B \sum_\mu \sum_\nu c_{\mu i}^* c_{\nu i} \langle \phi_\mu | \phi_\nu \rangle} \\ &= \frac{\sum_A \sum_B \sum_\mu \sum_\nu c_{\mu i}^* c_{\nu i} H_{\mu\nu}}{\sum_A \sum_B \sum_\mu \sum_\nu c_{\mu i}^* c_{\nu i} S_{\mu\nu}}, \end{aligned} \tag{2.8}$$

and we have shortened the notation using the elements of the Hamiltonian and overlap matrices. $H_{\mu\mu} = \langle \phi_\mu | \mathcal{H} | \phi_\mu \rangle$ stands for a Coulomb or on-site integral, $H_{\mu\nu} = \langle \phi_\mu | \mathcal{H} | \phi_\nu \rangle$ is an off-site, an interaction, a resonance or a hopping integral, and $S_{\mu\nu} = \langle \phi_\mu | \phi_\nu \rangle$ is an overlap integral. Without further specifying these elements, a differentiation with respect to the coefficients $c_{\mu i}$ is equivalent to applying the variational principle, and this leads to a set of linear equations, namely

$$\begin{array}{ccccccccc} (H_{11} - S_{11}E_i)c_{1i} & + & (H_{12} - S_{12}E_i)c_{2i} & + & \cdots & + & (H_{1n} - S_{1n}E_i)c_{ni} & = 0 \\ (H_{21} - S_{21}E_i)c_{1i} & + & (H_{22} - S_{22}E_i)c_{2i} & + & \cdots & + & (H_{2n} - S_{2n}E_i)c_{ni} & = 0 \\ \vdots & & \vdots & & \vdots & & \vdots & \\ (H_{n1} - S_{n1}E_i)c_{1i} & + & (H_{n2} - S_{n2}E_i)c_{2i} & + & \cdots & + & (H_{nn} - S_{nn}E_i)c_{ni} & = 0. \end{array} \tag{2.9}$$

4) Remember that a complex number z is written as $z = a + ib$ where $i = \sqrt{-1}$ is the imaginary unit. The complex conjugate of z is $z^* = a - ib$.

To solve Equation (2.9), one only needs to set the secular determinant [51] to zero,

$$\begin{vmatrix} H_{11} - S_{11}E_i & H_{12} - S_{12}E_i & \cdots & H_{1n} - S_{1n}E_i \\ H_{21} - S_{21}E_i & H_{22} - S_{22}E_i & \cdots & H_{2n} - S_{2n}E_i \\ \vdots & \vdots & \vdots & \vdots \\ H_{n1} - S_{n1}E_i & H_{n2} - S_{n2}E_i & \cdots & H_{nn} - S_{nn}E_i \end{vmatrix} = 0, \qquad (2.10)$$

thereby yielding first all energies E_i and, in a second step, the mixing coefficients $c_{\mu i}$, also called the eigenvectors of the Hamiltonian. We emphasize that the solution of Schrödinger's equation for a molecule starting with an LCAO expansion always ends with such a matrix calculus. This is why the term "diagonalizing" the matrix is just a synonym for determining the eigenvalues (energies) and eigenvectors; "diagonalizing" the problem means solving it.

It is a beautiful mathematical consequence of the above LCAO approach that, without having to specify the Hamiltonian any further, it is possible to arrive at a general quantum-chemical solution for the hydrogen molecule H_2^+ (or any other two-nuclei one-electron molecule). Starting with an LCAO ansatz for the molecular orbital as in

$$\psi = c_1\phi_1 + c_2\phi_2, \qquad (2.11)$$

in which ϕ is the 1s atomic orbital (see Section 2.2) of hydrogen, the secular determinant is just

$$\begin{vmatrix} H_{11} - E & H_{12} - S_{12}E \\ H_{21} - S_{21}E & H_{22} - E \end{vmatrix} = 0, \qquad (2.12)$$

because the overlap integral $S_{11} \equiv S_{22}$ is unity in the case when the 1s atomic orbital has been properly normalized. The symmetry of H_2^+ also requires that $H_{11} \equiv H_{22}$, $H_{12} \equiv H_{21}$, and also $S_{12} \equiv S_{21}$, such that we have

$$(H_{11} - E)^2 = (H_{12} - S_{12}E)^2. \qquad (2.13)$$

By taking the square-root we find that Equation (2.13) has two solutions, namely

$$E_+ = \frac{H_{11} + H_{12}}{1 + S_{12}} \quad \text{and} \quad E_- = \frac{H_{11} - H_{12}}{1 - S_{12}}, \qquad (2.14)$$

of which one, the bonding solution E_+, is lower in energy while the other, the antibonding solution E_-, is higher in energy. Plugging these two eigenvalues, one after the other, back into the set of linear equations given in Equation (2.9), the LCAO coefficients $c_{\mu i}$ are immediately found, and they determine the composition (and shape) of the molecular orbitals. For H_2^+, the bonding (+) and antibonding (−) molecular orbitals become

$$\psi_+ = \frac{\phi_1 + \phi_2}{\sqrt{2(1 + S_{12})}} \quad \text{and} \quad \psi_- = \frac{\phi_1 - \phi_2}{\sqrt{2(1 - S_{12})}}, \qquad (2.15)$$

where the normalization constant in the denominator assures that both may accomodate a total of two electrons each.

Let us first recall that H_{11} is the energy of the 1s atomic orbital and H_{12} is the interaction energy between the two neighboring, overlapping orbitals with an overlap integral of S_{12}. Equation (2.14) makes it clear that the energetic destabilization of the antibonding molecular orbital (E_- solution) is always larger than the stabilization of the bonding molecular orbital (E_+ solution) because a positive overlap integral S_{12} leads to a smaller denominator, and thus a larger energy. For completeness, we mention that there exists an extremely compact quantum-chemical method, the so-called Hückel approximation (or hopping model in the physics community), in which the general result based on the above energy and overlap integrals is *further* simplified. Setting all overlap integrals between neighboring atomic orbitals (or atoms) to zero, that is, $S_{12} \equiv 0$, but keeping the on-site S_{11} as unity, the Hamiltonian integrals are parameterized as $H_{11} \equiv \alpha$ (atomic eigenvalue) and as $H_{12} \equiv \beta$ (interaction energy), such that the two eigenvalues for H_2^+ become

$$E_+ = \alpha + \beta \qquad \text{and} \qquad E_- = \alpha - \beta. \tag{2.16}$$

A corresponding sketch of the energies of the general H_2^+ solution and that of the Hückel model of H_2^+ is given in Figure 2.1.

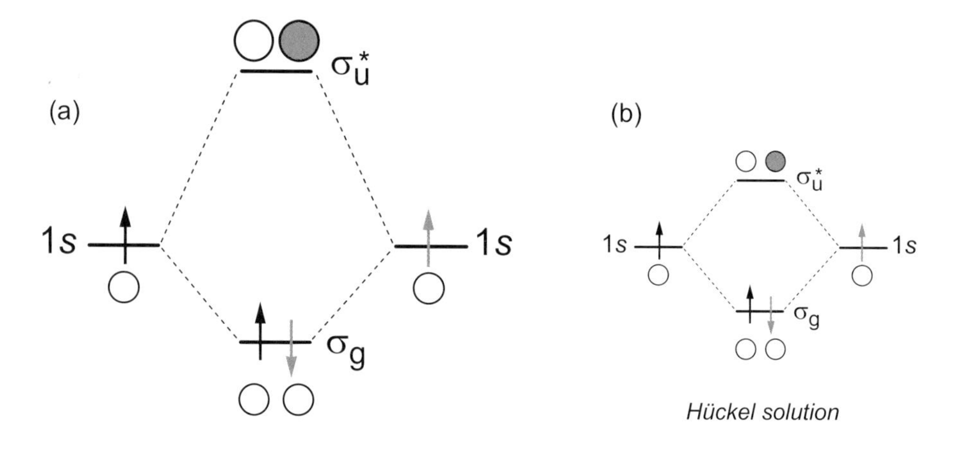

Fig. 2.1 Schematic drawing of the general solution (a) of the H_2^+ (and H_2) problem; sketch (b) shows the solution of the simplistic Hückel method which falsely arrives at energetically symmetric bonding and antibonding solutions. A second electron (H_2 case) is given in grey.

The low-lying bonding solution is characterized by the in-phase combination of the atomic orbitals ($\phi_1 + \phi_2$), and it is labeled as the σ_g molecular orbital of point group $D_{\infty h}$ (note that the linear H_2^+ molecule belongs to this point group). Because the orbital–orbital overlap coincides with the molecular axis, the interaction is called σ type, and the "g" (German *gerade* $\equiv$ even) indicates that a mathematical inversion of the left atomic orbital generates the right atomic orbital *without* a change in their plus/minus signs [46]. On the other hand, the AOs are out-of-phase ($\phi_1 = -\phi_2$) in the antibonding solution, dubbed σ_u^* where the "u" label (German *ungerade* $\equiv$ odd) reflects the plus/minus relationship of the left/right AOs upon inversion, and the asterisk designates an antibonding interaction. In Figure 2.1, the different orbital shadings symbolize the plus/minus signs, and the sizes of the atomic orbitals within the molecular orbitals symbolize the sizes of the mixing coefficients. When the overlap has been deliberately neglected (as in the Hückel model), the bonding/antibonding solutions are energetically symmetric with respect to zero energy. This has also been indicated (Figure 2.1(b)) for that model. Although entirely false, this Hückel weakness is tolerated because the simplistic model has mathematical advantages; many Hückel models can be solved solely with paper and pencil.

Figure 2.1 also indicates where a second electron, which appears for the case of molecular H_2, should go. It seems obvious because it is energetically advantageous to place it, with antiparallel spin, in the low-lying bonding molecular orbital (MO) of σ_g type. By doing so, we have obeyed Pauli's principle in that the two electrons must differ in at least one quantum number but, at the same time, we have implicitly neglected the Coulomb repulsion between the two electrons, later covered in detail in Section 2.9. The important issue of Coulomb repulsion between the electrons is related to the question of *how* to combine the two molecular orbitals ψ_i for the first and the second electron, such that they eventually give the correct two-electron wave function Ψ. Nonetheless, the simple sketch also helps us to understand, semiquantitatively, why there can be no covalently bonded He_2 molecule. For such a species, the high-lying antibonding MO would also be filled with two electrons, and because this particular MO is energetically more destabilizing than the low-lying MO is stabilizing, He_2 must instantaneously explode into two He atoms.

The above recipe for mathematically constructing molecular orbitals is not restricted to covalently bonded molecules, of course. For example, an electronic perturbation due to partial charge transfer because of some ionicity will occur for the LiH^+ (or LiH) molecule, and its valence MOs are shown in Figure 2.2. These MOs are made up from the 2*s* atomic orbital of Li and the 1*s* atomic orbital of H, and the higher electronegativity (energetically deeper-lying 1*s* AO) of hydrogen makes this atom mix more strongly into the bonding MO. As a result, the shared electron between Li and H is more a part of the H atom than of the Li; this is why chemists will intuitively call it lithium *hydride* but not hydrogen lithide. The high-lying MO, however, possesses more

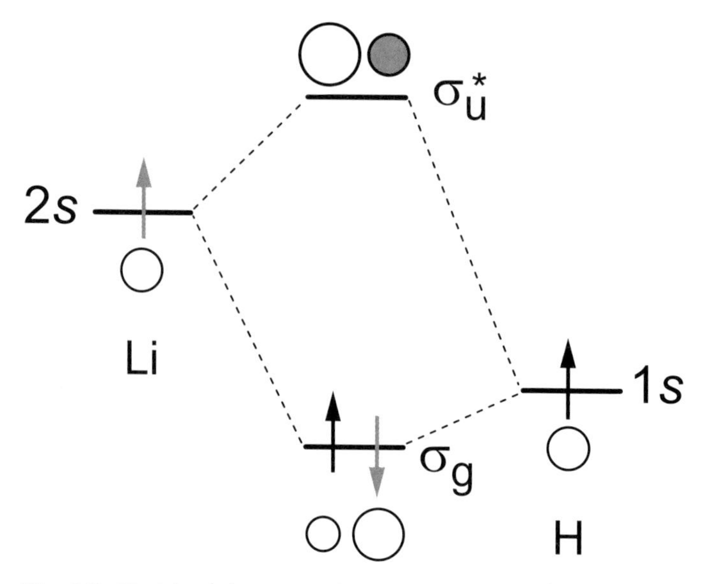

Fig. 2.2 Sketch of the general solution of the LiH^+ (and LiH) problem; see also text. The second electron for the neutral molecule is given in grey.

Li character because the contribution of the Li 2*s* atomic orbital is larger here (see relative sizes of atomic orbitals).

As has been indicated before, these general results for H_2^+, LiH^+ and many other molecules may be derived *without* having to specify the Hamiltonian in detail. When numerical results are needed, however, the variational strategy becomes quite technical, and the quantum-chemical history of the H_2^+ molecule serves as a fascinating example [57]. At present, the computational techniques to deal with molecules of (almost) any complexity have become very diverse and powerful, too [58], but only after having solved those fundamental difficulties arising in systems with more than one electron, later covered in Section 2.9.

2.2 Basis Sets for Molecules

For a given electronic Hamiltonian (that is, a proper description of the electron–nucleus and electron–electron interactions), the LCAO ansatz may deliver the molecular orbitals ψ_i and the many-electron wave function Ψ provided that there is a set of useful basis functions ϕ, for example, atomic orbitals. At this point, it is probably the right time to review briefly what type of atomic orbitals are mostly used within *molecular* quantum chemistry. By doing so, we come to know their general shape and also sense why there are differences in technically dealing with molecules on the one hand and solids on the other, at least when it comes to nonempirical methods.

In general, an atomic orbital – a solution of Schrödinger's equation for a one-electron *atom* of a given atomic number Z – may be formulated as a product of a radius- (r) and an angle-dependent (θ, φ) component such as in

$$\phi_{nlm}(\mathbf{r}) \equiv \phi_{nlm}(r, \vartheta, \varphi) \equiv R_n(r) Y_{lm}(\vartheta, \varphi), \tag{2.17}$$

depending on the main quantum number n, the angular-momentum quantum number l, and the magnetic quantum number m of the individual atom; here, $Y_{lm}(\vartheta, \varphi)$ denote the so-called spherical harmonics which give atomic orbitals their characteristic shapes (s, p, d, f) and which have been tabulated in mathematical monographs and pictorially sketched in most textbooks of general chemistry. Since the s orbitals are spherically symmetric, Figures 2.1 and 2.2 contained spherical sketches of these atomic orbitals.

It is possible, of course, to simply take the numerical solution of the radial Schrödinger equation for the particular atom under question and use it as the radial part of the atomic orbital. It is also possible, and mathematically advantageous, to *approximate* the exponential decay of the radial part of the atomic orbital $R_n(r)$ in the outer region. One particularly successful approach goes back to Slater's recipe [13],

$$R_n(r, \zeta) = \frac{(2\zeta)^{n+1/2}}{\sqrt{(2n)!}} r^{n-1} \exp(-\zeta r), \tag{2.18}$$

in which the ζ orbital exponent is proportional to the difference between nuclear charge (Z) and screening (s) and inversely proportional to the effective principal quantum number (n^*), that is,

$$\zeta = \frac{Z - s}{n^*}. \tag{2.19}$$

As the reader may have noticed, this is just the same expression that determines the sizes of Pauling's ionic radii (see Section 1.1); small orbitals (and thus atoms) will be characterized by large ζ exponents. Sets of Slater exponents have been tabulated for various atoms, and there are several different recipes for determining ζ values at the outset. We mention, however, that any orbital exponent needs modification whenever an atom's effective potential within a molecule deviates substantially from the original atomic potential, because then the ζ orbital exponent must reflect the new potentials. This is a general (and also serious) problem with fixed basis sets.

Nonetheless, Slater-type orbitals (STOs, in short) are a very reasonable approximation for the true atomic orbitals; each atomic orbital is replaced by a Slater orbital. If a higher accuracy is needed, one may also combine two (or even more) STOs for a better approximation; this is called a double-ζ (or higher) expansion of the basis set. One needs to remember, however, that STOs do not possess nodes like a real atomic orbital, such that they are totally useless in the core region; also, the specific kind of exponential function involved makes STOs slow in computing since integrals over STOs need considerable computational effort in the case of electron-repulsion integrals, and even for digital computers [58]. This is tolerable only if a small number of STO-based integrals needs to be calculated.

Because of the latter computational difficulties, another choice of basis sets has been preferred, mainly by the *ab initio* quantum-chemistry community, and this basis set has actually given an important computer program its name. According to Boys [59], a Gaussian-type orbital (GTO) may also serve as a basis function, and the radial-dependent part scales as

$$R_n(r) \sim \exp(-\zeta r^2). \tag{2.20}$$

We have suppressed the normalization which depends on the style of the GTO (there are more than one) and the way the angle-dependent part is treated. The advantage of the GTO approach lies in the numerical evaluation of integrals which is *much* easier than with STOs. This is because the product of two GTOs at two different positions can be rewritten as a GTO, being centered between these two positions. On the other hand, the shape of GTOs is only a poor approximation for the true radial function in the important *outer* region where GTOs simply decay too fast. Thus, it has proved useful to linearly combine a number of different GTOs (with fixed individual ζ parameters, so-called *primitive* Gaussians) to form a so-called *contracted* GTO, such that it will imitate the shape of an STO:

$$\exp(\zeta r) \approx c_1 \exp(\zeta_1 r^2) + c_2 \exp(\zeta_2 r^2) + c_3 \exp(\zeta_2 r^3) \ldots \tag{2.21}$$

By doing so, the correct radial behavior of an STO is expressed by a computationally preferable combination of GTOs. Using the jargon of the community, three superimposed GTOs which are used to fit one STO, yield an STO-3G basis set; this is a so-called minimum basis set which still lacks much in the way of computational accuracy. For comparison, the shapes of the different basis functions are schematically depicted in Figure 2.3.

Also, other subtleties (such as additional polarization functions) play a major role in constructing a good basis set; in fact, there is a wealth of literature available on how to choose this particular kind of fixed basis set [58, 60], but we will not go into the details of these molecule-related technical questions.

As has been mentioned already, Gaussian-type basis sets have dominated (and still do) the numerical quantum chemistry of molecules, and for good

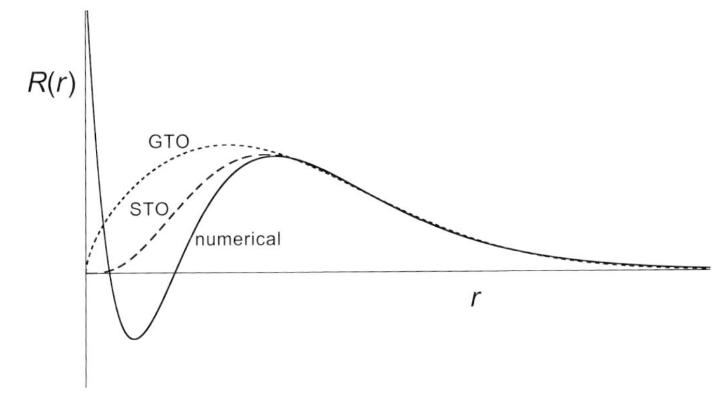

Fig. 2.3 Schematic drawing of a $3s$ numerical radial wave function (full line) of a Na atom, its Slater-type approximation (STO, dashed line), and a combination of three Gaussian-type functions (GTO, dotted line).

reason. Within the molecular *ab initio* field, there is a huge number of integrals that must be computed, and that is why most approaches built upon the molecular quantum-chemistry workhorse, Hartree–Fock theory (see Section 2.11.3), are also based on GTOs [61]. Within solid-state materials, however, the Hartree–Fock method has been less successful, and there is also a more natural type of basis function, simply because of the *translational* symmetry of these systems.

2.3
Three Myths of Chemical Bonding

We are soon to leave the molecular regime and devote ourselves fully to the solid state where Schrödinger's equation also needs to be solved after constructing – or expanding – the system's wave function in terms of a set of known basis functions, but not necessarily atomic orbitals. The strategy is identical, namely finding all the electronic eigenvalues E_i and filling them with electrons such that chemical bonding will result. Although there are more complications, especially when dealing with the electron–electron interactions (see Section 2.9) which also determine the filling of energy levels by the electrons, a maximum of two electrons may enter each orbital, just as for an atomic orbital. The more electrons that are filled into bonding orbitals, the lower the total energy and the stronger the chemical bonding. Before becoming more technical, let us take a short break and remind ourselves of three *misconceptions* of chemical bonding [62].

One sometimes reads that chemical bonding in molecules is due to *spin-pairing* of formerly unpaired electrons. This oversimplification probably goes back to the simple idea that the hydrogen molecule is formed when two hydrogen atoms come together, and then they share their formerly unpaired electrons with antiparallel spins (see sketch in Figure 2.1). While it is correct that the two unpaired electrons of the isolated H atoms are "paired" (antiparallel orientation within the same orbital) upon bond formation in the H–H single bond, a moment's reflection reveals that the pairing of spins is a *side effect* accompanying bond formation in many, but not all, cases. In the H_2 molecule, the second electron must also go into the singly occupied σ_g orbital simply because the latter is lower in energy than the upper σ_u^* orbital. The pairing of spins is a consequence of Pauli's principle because the two electrons *must* have antiparallel spins; nonetheless, let us not forget that the two electrons *repel* each other! Despite this electron–electron repulsion, the concentration of electronic charge in the bonding σ_g molecular orbital *between* the protons leads to a shielding of the two protons from each other and a decrease in the molecule's potential energy. Here is another example:

In its ground state, the paramagnetic O_2 molecule exhibits an electron configuration of $(1\sigma_g)^2$ $(1\sigma_u^*)^2$ $(2\sigma_g)^2$ $(2\sigma_u^*)^2$ $(3\sigma_g)^2$ $(1\pi_u)^4$ $(1\pi_g^*)^2$, and the two energetically highest electrons are found in the highest occupied molecular orbital (HOMO) which is O–O antibonding in character. In its excited singlet $^1\Delta_g$ state,[5] the two electrons still reside within the antibonding and doubly generate HOMO of $1\pi_g^*$ type, but now they are *spin-paired*. Because singlet O_2 is an excited species, it is *less* stable and *less* strongly bonded than the normal ground-state $^3\Sigma_g^-$ triplet O_2 by about 0.95 eV [63]. For synthetic reasons, chemists commonly generate spin-paired, diamagnetic, reactive $^1O_2$ by photochemically activating spin-unpaired, paramagnetic $^3O_2$. Upon spin-pairing, the oxygen molecule becomes less strongly bonded. There is no energetic gain by pairing the spins, as is schematically depicted in Figure 2.4.

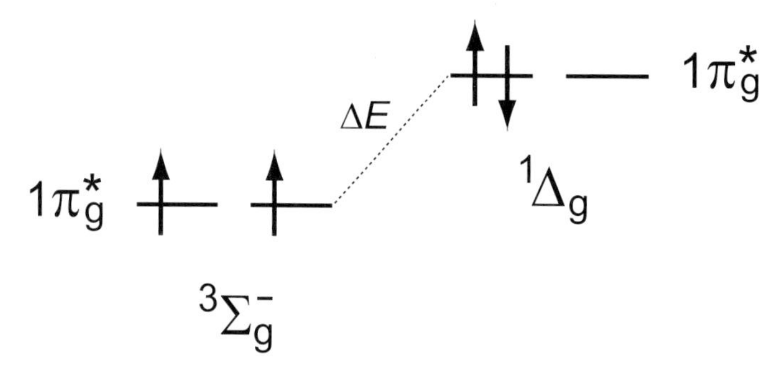

Fig. 2.4 Schematic representation of the electron occupations in the $1\pi_g^*$ antibonding HOMO in the triplet (left) and singlet (right) O_2 molecule.

5) This term symbol and many others are to be found in textbooks on molecular quantum chemistry [51].

Another myth touches upon the relationship between bond length and bond strength; at first sight, shorter bonds are stronger bonds. This rule of thumb (Pauling's bond-length–bond-strength method [6] covered in Section 1.5) has been successfully applied to many (mostly organic) molecules, and it is indeed true that most C=C double bonds, to give a prominent example, are shorter and stronger than C–C single bonds. A typical C–C bond dissociation energy is of the order of 345 kJ/mol whereas the value for a C=C bond is 615 kJ/mol [63]. Nonetheless, the C_2 molecule itself shows that the above rule of thumb is principally incorrect. In its $^1\Sigma_g^+$ ground state, C_2 contains a covalent bond of 1.2425 Å but upon excitation into its $^3\Sigma_u^+$ state the bond length *shortens* by almost 0.035 Å although the excited molecule is less stable (less strongly bonded) by about 1.1 eV (or approx. 100 kJ/mol) [64]; this bond weakens upon shortening. The energy–distance curve for C_2 is depicted in Figure 2.5.

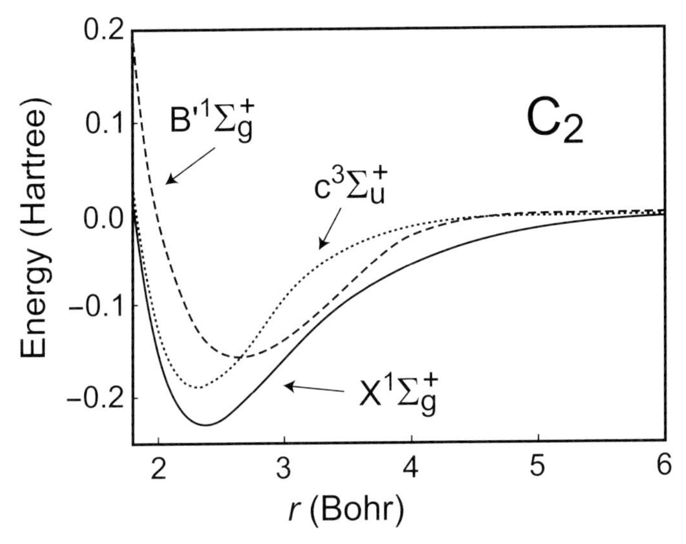

Fig. 2.5 The total-energy–bond-distance curve for various electronic states of the C_2 molecule [64]. Note that the ground state of C_2 is a singlet (opposed to O_2) because the electron configuration is $(1\sigma_g)^2$ $(1\sigma_u^*)^2$ $(2\sigma_g)^2$ $(2\sigma_u^*)^2$ $(1\pi_u)^4$ with a doubly degenerate, fully occupied HOMO; the LUMO is of $3\sigma_g$ type.

There are also many other molecular examples [65] showing that, although there is a correlation between bond length and bond strength, it cannot be generalized as a law. The problem has certainly been known for decades within the solid-state field. There are numerous metastable crystallographic phases (elements, simple compounds) with higher densities and shorter interatomic distances than in their ground-state structures, for example, high-pressure phases which do not spontaneously decay under standard conditions. The cubic phase of silicon nitride, Si_3N_4, is an excellent example because this high-pressure refractory material is more dense but less stable than the α- or β-Si_3N_4 phases found under normal conditions [66].

Third, there is the questionable idea of chemical bonding, still occasionally to be found, which erroneously relates chemical bond formation with a decrease in kinetic energy T. This flaw stems from a misinterpretation of the pioneering quantum-chemical investigation of the H_2^+ molecule and is due to a comparison of non-self-consistent solutions of the Schrödinger equation. Let us explain this in detail. When the 1*s* atomic orbitals of the two H atoms forming the σ_g molecular orbital are not allowed to adjust to the doubled nuclear charge because of fixed orbital exponents ($\zeta = 1.000$, see Section 2.2), the lowered total energy E of H_2^+ compared with H + H^+ is seemingly due to a decrease in the kinetic energy T (see Figure 2.6(a)). Note, however, that the virial theorem is violated ($V \neq -2T$) although it *must* hold for this eigenstate. The error source is given by the $\zeta = 1.000$ atomic orbital which would be fine for a single H atom but not for the H_2^+ molecule. For the latter, the exponent no longer corresponds to the nuclear potential in the molecule, the aforementioned problem of a fixed basis set (Section 2.2).

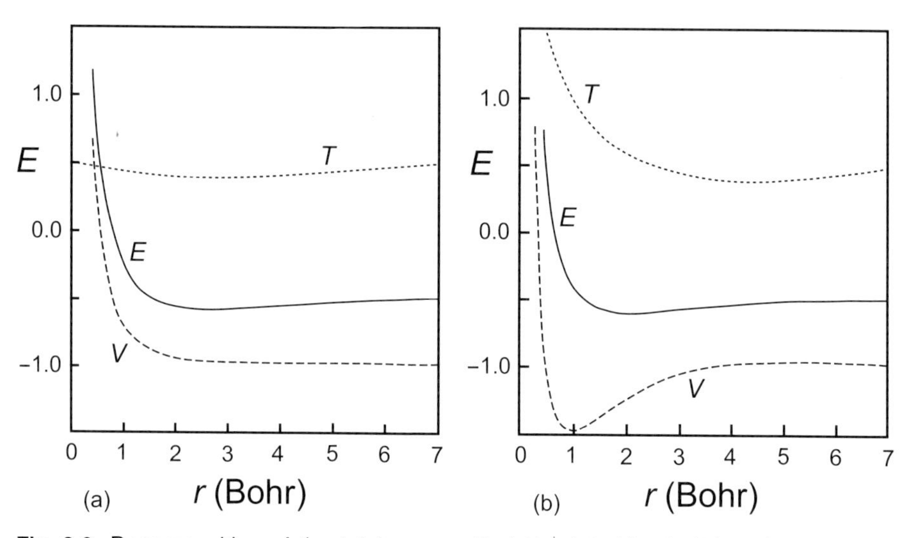

Fig. 2.6 Decomposition of the total energy E of H_2^+ into kinetic (T) and potential energy (V) for the non-self-consistent solution with fixed atomic orbitals (a) and for the self-consistent solution with optimized atomic orbitals (b). Note that the energy decreases are quite small, in particular for the non-self-consistent solution.

When this first approximate wave function is made fully self-consistent by contracting atomic orbitals ($\zeta = 1.228$) which eventually fit the doubled nuclear charge, the total energy decreases by an additional amount, in harmony with the variational principle, and the kinetic energy *increases* strongly; only at interatomic distances which are much longer than the equilibrium (bonding) distances, is a small decrease in the kinetic energy found. At the equilibrium distance, the lowered potential energy overrides the loss in kinetic energy and drives chemical bond formation. For this self-consistent solution (see Figure 2.6(b)), $V = -2T$ holds [62,67].

Summarizing, chemical bonding, following the thermochemical definition, results from a lowered *total* energy which reflects a lowered potential energy through an increased nuclear potential which is acting on the electrons. Neither spin pairing, nor bond shortening, nor a decrease in the kinetic energy is a valid measure of bond formation [68].

2.4 Bloch's Theorem

If compared with a molecular problem, the solution of Schrödinger's equation for any extended solid-state material must appear breathtakingly tedious, at first sight. Since we no longer have just a couple of nuclei and several dozens of electrons but – let us think here of a macroscopic crystal – something on the order of a *mole* of them, the computational problem is almost infinitely more demanding and it will thus overwhelm any computer resources one might hope to possess. This is because the secular determinant of the problem (Equation (2.10)), arrived at by an LCAO procedure (Equation (2.6)), will also be of the order of *one mole* $\times$ *one mole* such that, naively, one has to diagonalize ($\equiv$ solve) incredibly large matrices. Yet, upon exploiting the translational symmetry properties of the crystal in question, the solution is straightforward due to the celebrated theorem of Felix Bloch [69].[6]

Within *molecular* quantum chemistry, the mathematical search for molecular orbitals can also be greatly simplified by taking the molecular symmetry into account. Recall that molecular symmetry, expressed by point groups [46], depends on the symmetry elements of finite objects, namely mirror planes, inversions, rotations, and rotary reflections (or improper rotations). It turns out that all molecular orbitals thus eventually found – or even constructed – can be characterized due to their symmetry properties, and they will "transform" (i.e., behave under symmetry operations) according to some irreducible representation of that particular point group. In the H_2^+ problem, the labels σ_g and σ_u^* were the correct labels for the two symmetry-distinct molecular orbitals within point group $D_{\infty h}$. With respect to the inversion symmetry of the H_2^+ system, one MO is symmetrically even (σ_g), the other is odd (σ_u^*). This is covered in great detail in all textbooks of group theory [46] and molecular

6) Legend has it that Richard P. Feynman, around 1964/5, accidentally derived the theorem during a lecture and was asked by a student whether this was Bloch's theorem; Feynman replied by saying: "I don't think that something so *trivial* can have somebody's name" [70].

quantum chemistry – in particular qualitative molecular orbital theory [55] – so that we do not have to go into the details here.

When it comes to crystals, it is clear that the system under study is translationally invariant in all three spatial directions, and Bloch's theorem *utilizes* the translational symmetry to generate the crystal's wave function, composed of crystal orbitals which are also called electronic bands. We therefore imagine an idealized solid-state material whose electronic potential V possesses the periodicity of the lattice, expressed by a lattice vector $\boldsymbol{T}$, that is

$$V(\boldsymbol{r} + \boldsymbol{T}) \equiv V(\boldsymbol{r}), \tag{2.22}$$

simply because the electronic situation is identical within each crystallographic unit cell. This is already a simplification of the real world, neglecting impurities (false atoms), defects (missing atoms), and also time-dependent phenomena (vibrations of atoms) but it will serve perfectly well for the remaining discussion. Historically, Bloch's theorem was found while searching for an explanation of how the electrons in the crystal can pass the atoms in a metal without any perturbation, that is, without bouncing against the lattice of the nuclei. It turns out that the wave function of the electrons comes close to a plane wave but exhibits a periodic modulation. The mathematics is done by a Fourier transformation.

In modern language, Bloch's theorem can be expressed as follows. For a given wave function $\psi(\boldsymbol{k}, \boldsymbol{r})$ which fulfills Schrödinger's equation, there exists a vector $\boldsymbol{k}$ such that translation by a lattice vector $\boldsymbol{T}$ is equivalent to multiplication by a phase factor:

$$\psi(\boldsymbol{k}, \boldsymbol{r} + \boldsymbol{T}) = e^{ikT} \psi(\boldsymbol{k}, \boldsymbol{r}) \tag{2.23}$$

There are a couple of important consequences of this theorem. First, an extended wave function (or electronic band or crystal orbital) which is a function of a lattice vector $\boldsymbol{T}$ is given on the left-hand side of Equation (2.23). It is generated from the crystal orbital of one particular site in the crystal (right part, $\boldsymbol{r}$ coordinates), so we will *define* the latter as being located within the crystallographic unit cell. If that is so, we only have to know the extended material's crystal orbital $\psi(\boldsymbol{k}, \boldsymbol{r})$ *at this particular point* $\boldsymbol{r}$ (namely, inside the unit cell), and then we know it for the *entire crystal* $\boldsymbol{r} + \boldsymbol{T}$, provided that the material *is* a crystal, i.e., it is translationally invariant. So, if there is a periodic potential $V(\boldsymbol{r} + \boldsymbol{T})$ – if there is a unit cell that is repeated over and over again – then there will be a band. Second, the price paid for this simplification is given by the newly introduced quantum number $\boldsymbol{k}$ which labels which irreducible representation of the translation group $\psi(\boldsymbol{k}, \boldsymbol{r})$ transfers as; in other words, the wave function thus generated is *symmetry-adapted* to the infinite system because of $\boldsymbol{k}$. At the same time, the secular determinant belonging to the translationally invariant system has become "$\boldsymbol{k}$-blocked", and those functions with different $\boldsymbol{k}$ automatically belong to different irreducible representations of the translation group. Fortunately, $\boldsymbol{k}$ is restricted in the values it may adopt.

I trust that people trained in crystallography will immediately realize that the mathematics appearing in Bloch's theorem, namely the above-mentioned Fourier transformation between something local, $\psi(\boldsymbol{k},\boldsymbol{r})$, and something entirely extended, $\psi(\boldsymbol{k},\boldsymbol{r}+\boldsymbol{T})$, bears a striking resemblance to crystallographic computing; in fact, it is basically the same mathematical process. We briefly recall that, within crystal-structure analysis, Bragg's equation,

$$2d_{hkl}\sin\vartheta = n\lambda, \tag{2.24}$$

indicates constructive interference of the radiation (X-rays, electrons, neutrons) with some wave length λ if a given d spacing – corresponding to the lattice plane characterized by the Miller indices hkl – and the angle of diffraction ϑ, match each other. Then, and only then, is the reflection intensity I_{hkl} proportional to the square of the absolute value of the structure factor (or structure amplitude) F_{hkl}, and the latter is given by

$$F_{hkl} = \sum_j f_j \exp\{2\pi i(hx_j + ky_j + lz_j)\}. \tag{2.25}$$

Here, we have dubbed f_j as the atomic scattering factor of atom j, and its atomic position in the unit cell is given by the real-space coordinates x_j, y_j, and z_j; the sum runs over all atoms in the unit cell. Thus, the structure factor F_{hkl} (in reciprocal space because it depends on the reciprocal Miller indices) results as the Fourier-transform of a real-space infinite object (the entire crystal), schematically sketched in Figure 2.7.

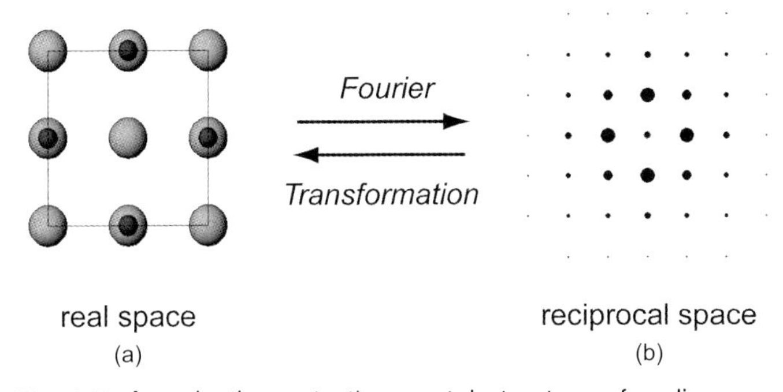

Fig. 2.7 A projection onto the crystal structure of sodium chloride (a) along a lattice vector and the corresponding X-ray diffraction diagram (b) along $[hk0]$, related to each other by a Fourier transformation.

The sketch also illustrates the fact that crystal structure determination does not determine the structure of the macroscopic crystal, but only the structure of the crystallographic unit cell that is repeated over and over again; the price paid for this is that we have to collect the intensities at $h \times k \times l$ grid points, objects of reciprocal space.

Since the mathematics is the same, we might want to rewrite Bloch's theorem such that it resembles the definition of the structure factor as closely as possible, and we can do this by *expanding* the extended, delocalized, $\boldsymbol{k}$-dependent crystal orbital $\psi(\boldsymbol{k}, \boldsymbol{r})$ over a series of localized atomic orbitals ϕ_j which rest at some atomic positions $\boldsymbol{r}_j$ inside the crystallographic unit cell. The orbitals might belong to different atoms but this does not necessarily have to be the case. Thus, a Bloch expansion reads

$$\psi(\boldsymbol{k}, \boldsymbol{r}) = \sum_j \phi_j \exp\{i\boldsymbol{k}\boldsymbol{r}_j\} \equiv \sum_j \exp\{i\boldsymbol{k}\boldsymbol{r}_j\}\phi_j. \tag{2.26}$$

Quite obviously, the atomic orbital ϕ_j has formally replaced the atomic scattering factor f_j, and the quantum number $\boldsymbol{k}$ takes the role of the former Miller indices h, k, and l. Note also that there is a small difference in the argument of the exponential function (the "2π" only showing up in Equation (2.25)) but this is due to different definitions of reciprocal space, probably a relic of some missing communication between crystallographers and quantum theorists some time ago. We might also add that Equation (2.26) is the natural analogue of the molecular LCAO ansatz (Equation (2.6)) for an extended system. We still have atomic orbitals on the far right side of Equation (2.26) which, weighted by mixing coefficients which are now *identical* with the Bloch exponential factor, generate an extended wave function, a crystal orbital; a satisfying solution.

2.5
Reciprocal Space and the k Quantum Number

Looking at the exponential arguments in Equations (2.23) or (2.26), it is clear that the quantum number $\boldsymbol{k}$ must have the dimension of an *inverse* length; it is a resident of reciprocal space. Reciprocal space is an ingenious invention in order to simplify life with infinite objects such as crystals, so let us introduce it now, rather belatedly [53].

Any position in real space is given by the real-space vector $\boldsymbol{R}$ which is a linear combination of the three basic vectors $\boldsymbol{a}_1$, $\boldsymbol{a}_2$, and $\boldsymbol{a}_3$, namely

$$\boldsymbol{R} = n_1\boldsymbol{a}_1 + n_2\boldsymbol{a}_2 + n_3\boldsymbol{a}_3 = \boldsymbol{a}. \tag{2.27}$$

Likewise, any given reciprocal lattice vector $\boldsymbol{K}$ is constructed from the reciprocal basic vectors $\boldsymbol{g}_1$, $\boldsymbol{g}_2$, and $\boldsymbol{g}_3$ according to

$$\boldsymbol{K} = m_1\boldsymbol{g}_1 + m_2\boldsymbol{g}_2 + m_3\boldsymbol{g}_3 = \boldsymbol{g}, \tag{2.28}$$

and real-space and reciprocal-space vectors are connected by the fundamental equation

$$\exp\{i\boldsymbol{K}\boldsymbol{R}\} = 1. \tag{2.29}$$

The latter definition (which would include an additional 2π in the exponential argument if written down by a crystallographer) is equivalent to the orthogonality relations of the basis vectors which are

$$\begin{aligned} \boldsymbol{g}_1 &= \frac{2\pi}{V}(\boldsymbol{a}_2 \times \boldsymbol{a}_3) && \perp \boldsymbol{a}_2, \boldsymbol{a}_3 \\ \boldsymbol{g}_2 &= \frac{2\pi}{V}(\boldsymbol{a}_3 \times \boldsymbol{a}_1) && \perp \boldsymbol{a}_1, \boldsymbol{a}_3 \\ \boldsymbol{g}_3 &= \frac{2\pi}{V}(\boldsymbol{a}_1 \times \boldsymbol{a}_2) && \perp \boldsymbol{a}_1, \boldsymbol{a}_2, \end{aligned} \tag{2.30}$$

and the volume $V = \boldsymbol{a}_1(\boldsymbol{a}_2 \times \boldsymbol{a}_3)$. For the simple case of a one-dimensional infinite system with $\boldsymbol{R} = \boldsymbol{a}_1$, we then simply have $\boldsymbol{g}_1 = \frac{2\pi}{V}(\boldsymbol{a}_2 \times \boldsymbol{a}_3) = 2\pi/\boldsymbol{a}_1$, as expected, because $2\pi/\boldsymbol{a}_1$ is the reciprocal vector of $\boldsymbol{a}_1$.

Coming back to the quantum number $\boldsymbol{k}$, we will soon demonstrate that it is restricted to a small interval, the so-called Brillouin zone, and, for the case of free electrons (the so-called Sommerfeld model) it provides information about the energies E of these electrons and their crystal momentum $\boldsymbol{p}$; using Rydberg atomic units (see Section 2.1) this is

$$E = k^2 \qquad \text{and} \qquad \boldsymbol{p} = \boldsymbol{k}. \tag{2.31}$$

The important term *electronic band structure* is identical with the course of the energy of an extended wave function as a function of $\boldsymbol{k}$, and we seek for $E(\psi(\boldsymbol{k}, \boldsymbol{r}))$, the crystal's equivalent to a molecular orbital diagram. As stated before, the $\boldsymbol{k}$-dependent wave function $\psi(\boldsymbol{k}, \boldsymbol{r})$ is called a crystal orbital, and there may be *many* one-electron wave functions per $\boldsymbol{k}$, just as there may be several molecular orbitals per molecule. Due to the existence of these periodic wave functions, there results stationary states in which the electrons are travelling from atom to atom; the Bloch theorem thereby explains why the periodic potential is compatible with the fact that the conduction electrons do *not* bounce against the ionic cores.

The values of the new quantum number $\boldsymbol{k}$ may be easily deduced from periodic boundary conditions (or, Born–von Karman conditions) [53]; recall that the unit cell is periodically repeated over and over again. For a one-dimensional case, for example, we may imagine a macroscopic crystal of length L that consists of an astronomically large number N of unit cells with lattice parameter a. If we now let the wave function at the left end be identical with the right end (that is, gluing the string of unit cells together to form a circle), the application of Bloch's theorem would yield

$$\psi(k, 0) \equiv \psi(k, L) = e^{ikL}\psi(k, 0) \tag{2.32}$$

for any value of k. Note that this one-dimensional crystal has lost its surface (the left and right end), and this would also be true for the three-dimensional case in which the left/right, top/down, bottom/front faces of the macroscopic crystal are glued together (which is hard to imagine geometrically, of course, but mathematically it is acceptable). In terms of physics (and chemistry, too), "throwing away" the surface is a *severe* qualitative intervention, but only small in magnitude: For a macroscopic crystal of approximately 1 mm^3 in volume, the number of bulk atoms is about one million times larger than those of the surface atoms.

To make Equation (2.32) valid for all k, the exponential argument (kL) must be a multiple of 2π, and using the definition of N this is equivalent to

$$-\frac{\pi}{a} \leq k \leq \frac{\pi}{a} \qquad \text{or, in three dimensions,} \qquad -\frac{\pi}{\boldsymbol{a}} \leq \boldsymbol{k} \leq \frac{\pi}{\boldsymbol{a}}. \tag{2.33}$$

This is usually shortened to $0 \leq |\boldsymbol{k}| \leq \frac{\pi}{a}$, and this part of the reciprocal space is called the Brillouin zone; the (first) Brillouin zone may be looked upon as a unit cell in reciprocal space. Its construction goes as follows. One first chooses one reciprocal lattice point, then connects it with all nearest points and bisects all vectors by planes; the enclosed volume is called the (first) Brillouin zone.[7] The theorem by Kramers [53] assures that the energies of wave functions at either negative or positive $\boldsymbol{k}$ values are the same.

Why are the values of $\boldsymbol{k}$ restricted to a small interval? This is another consequence of infinity (the unit cells are repeated over and over again) and the Fourier transformation which forces $\boldsymbol{k}$ to lie in the (first) Brillouin zone. In comparison, the main quantum number of a finite object (atom) is not restricted, and it may adopt any integer value ($n = 1, 2, 3 \ldots$). To actually calculate a band structure (crystallography: calculate the structure factors from the atomic spatial coordinates), the course of the crystal orbitals as a function of $\boldsymbol{k}$ (crystallography: course of the structure factors as a function of the diffraction angle) is analyzed by letting $\boldsymbol{k}$ adopt all possible values in the Brillouin zone (crystallography: use all hkl-indexed reflections for the intensity calculations). If we are interested in, say, the total energy, all $\boldsymbol{k}$ values used for the calculation of $E(\psi(\boldsymbol{k}, \boldsymbol{r}))$ should lie on a fine three-dimensional mesh within the Brillouin zone. Sometimes, however, it is useful to study the course of a band along special directions in reciprocal space, and this bears striking resemblance with taking crystallographic *axis photographs* of a crystal; along these high-symmetry directions, a lot can be learned about the symmetry of the structure. Likewise, $E(\psi(\boldsymbol{k}, \boldsymbol{r}))$ plots along high-symmetry lines are also much easier to understand.

7) Higher Brillouin zones result from extending the construction vectors to second-nearest lattice points (or lattice points even beyond) but $\boldsymbol{k}$ points of these higher zones can be mathematically back-transformed into the first zone.

The calculations can be further simplified by restricting them to the irreducible Brillouin zones, that is, the irreducible wedge of reciprocal space from which the whole Brillouin zone can be obtained by applying all symmetry operations.[8] This is also known from X-ray (or neutron) crystallography because the number of independent reflections that needs to be collected is a function of the symmetry of the crystal. The latter depends on the space group, and the corresponding symmetry-related information has all been tabulated [71] and also published on the World Wide Web by the Bilbao Crystallographic Server. For illustration, we offer two such Brillouin zones in Figure 2.8 and mention

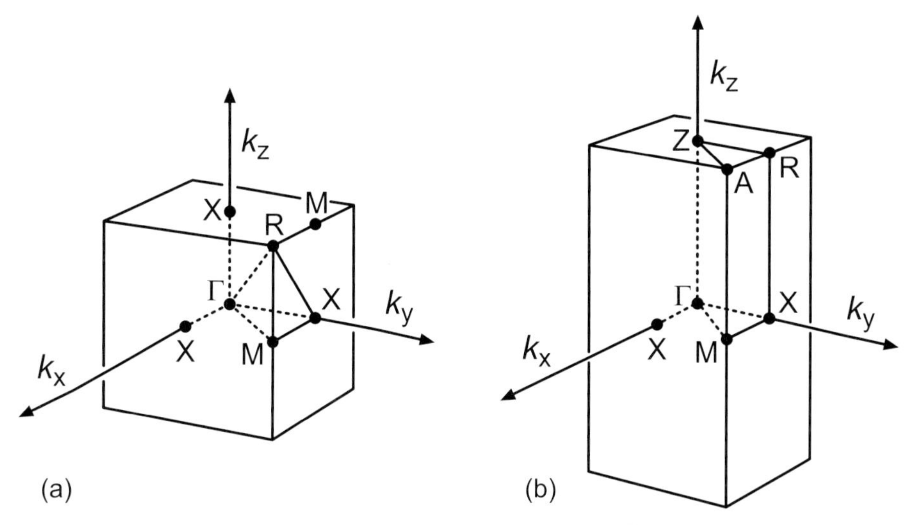

Fig. 2.8 The Brillouin zones for the cubic space group $Pm\bar{3}m$ (a) and the tetragonal space group $I4/mmm$ (b) with their special $\boldsymbol{k}$ points; because both systems are rectangular, $g_1 \equiv g_x$, $g_2 \equiv g_y$, $g_3 \equiv g_z$ holds, and these vectors coincide with the corresponding $\boldsymbol{k}$ vectors $\boldsymbol{k}_x$, $\boldsymbol{k}_y$, and $\boldsymbol{k}_z$.

that so-called *special points* of high symmetry have been given specific names which are, again, a function of the space group. In general, the zone center is called **Γ**, and any other point is expressed as a fraction of the reciprocal vectors, that is, multiples of $2\pi/\boldsymbol{a}$. Thus, the **X** point in Figure 2.8(a) (identical with the **Y** or **Z** points because the system is cubic) is characterized by the coordinates $(\frac{1}{2},0,0)$, $(0,\frac{1}{2},0)$, and $(0,0,\frac{1}{2})$. The **M** and **R** points are defined as $(\frac{1}{2},\frac{1}{2},0)$ and as $(\frac{1}{2},\frac{1}{2},\frac{1}{2})$. In Figure 2.8(b), the tetragonal system gives rise to the new special points **Z** $(0,0,\frac{1}{2})$, **R** $(0,\frac{1}{2},\frac{1}{2})$, and **A** $(\frac{1}{2},\frac{1}{2},\frac{1}{2})$. Fortunately, these mathematical and group-theoretical data do not have to be memorized because they are part of most program packages for the calculation of the electronic structures of extended materials.

8) Thus, the irreducible Brillouin zone in reciprocal space is the equivalent of the asymmetric unit in real space.

How many $\boldsymbol{k}$ points are needed to calculate the total energy of a crystal? Clearly, the larger the number of $\boldsymbol{k}$ points, the better the quality of the calculation. Because of the definition of real and reciprocal space, a small unit cell (real space) corresponds to a large Brillouin zone (reciprocal space) such that a large number of $\boldsymbol{k}$ points is needed for a smooth representation of reciprocal space. Consequently, a huge unit cell is equivalent to a tiny Brillouin zone and a very small number of $\boldsymbol{k}$ points. Keeping in mind that the size of the unit cell (real space) also determines the size of the basis set (that is, the number of atomic orbitals) for the construction of the Bloch functions, an important quality criterion of any given band-structure calculation is that the product between the number of atomic orbitals in the unit cell (real space) and the number of $\boldsymbol{k}$ points (reciprocal space) exceeds a certain number; the latter depends on the computer resources which are available to the theorist (and the amount of money he or she is willing to pay).

2.6 Band Structures

In the era of digital computers, the calculation of band structures (as well as crystallographic computing) is hardly ever done by hand; if manually calculating realistic (not model-like) molecular orbitals is tedious, however, trying the same with $\boldsymbol{k}$-dependent orbitals ($\equiv$ electronic bands) must be even more painful. A few model calculations of these bands will nonetheless help us to understand their behavior, that is, their so-called bandwidth and also their topology [4,72].

2.6.1 One-dimensional Systems

We can retain all general principles by first switching to one-dimensional (instead of three-dimensional) space, and we will also use the simplest system one can imagine, a one-dimensional crystal of equally spaced hydrogen atoms:

----H----H----H----H----H----H----H----H----H----H----

Scheme 2.1

Let us also imagine that each of the translationally invariant H atoms in Scheme 2.1 possesses an (invisible) subscript, ranging from 1 to ∞, which characterizes its position in the chain in terms of the translational vector T. Each translation is a multiple of the interatomic distance a, namely $T = n \cdot a$, and each H atom comes with a single 1s atomic orbital, designated φ_n. The Bloch sum (another expression for the electronic band) goes over all atomic orbitals,

$$\psi(k) = \sum_{n=1}^{N} e^{ikna}\phi_n \qquad \text{with} \qquad 0 \leq |k| \leq \frac{\pi}{a}, \tag{2.34}$$

and because k is a scalar variable, $\psi(k)$ is easy to evaluate; for reasons of brevity, from now on we will shorten $\psi(k, r)$ to $\psi(k)$. At the zone center $\mathbf{\Gamma}$ with $k = 0$, the exponential function is equal to unity such that the crystal orbital is

$$\psi(0) = \sum_{n=1}^{N} \phi_n = \phi_1 + \phi_2 + \phi_3 + \phi_4 + \phi_5 \ldots, \tag{2.35}$$

and the same holds for all other k values of $2\pi/a$, $4\pi/a$ and so on; those points outside the first Brillouin zone are energetically identical with the one inside. For $k = \pi/a$ (and also $3\pi/a, 5\pi/a \ldots$) at the zone edge $\mathbf{X}$, the crystal orbital is

$$\psi\left(\frac{\pi}{a}\right) = \sum_{n=1}^{N} e^{i\pi n}\phi_n = -\phi_1 + \phi_2 - \phi_3 + \phi_4 - \phi_5 \ldots, \tag{2.36}$$

with alternating plus/minus signs for the atomic orbitals. In the middle of the first Brillouin zone, at $k = \pi/2a$ (and also at $5\pi/2a, 9\pi/2a \ldots$), we find

$$\psi\left(\frac{\pi}{2a}\right) = \sum_{n=1}^{N} e^{i\pi n/2}\phi_n = \sum_{n=1}^{N} i^n \phi_n = i\phi_1 - \phi_2 - i\phi_3 + \phi_4 + i\phi_5 \ldots, \tag{2.37}$$

a *complex* wave function, just like the one for the corresponding negative k value of $-\pi/2a$ which is

$$\psi\left(-\frac{\pi}{2a}\right) = \sum_{n=1}^{N} e^{-i\pi n/2}\phi_n = \sum_{n=1}^{N} (-i)^n \phi_n = -i\phi_1 - \phi_2 + i\phi_3 + \phi_4 - i\phi_5 \ldots \tag{2.38}$$

Because $\psi(\pi/2a)$ and $\psi(-\pi/2a)$ have the same energy (recall that this is what the Kramers theorem assures), one can make linear combinations of them, one by addition,

$$\psi_+ \equiv \psi\left(\frac{\pi}{2a}\right) + \psi\left(-\frac{\pi}{2a}\right) \sim -\phi_2 + \phi_4 - \phi_6 + \phi_8 - \phi_{10} \ldots, \tag{2.39}$$

and one by subtraction,

$$\psi_- \equiv \psi\left(\frac{\pi}{2a}\right) - \psi\left(-\frac{\pi}{2a}\right) \sim \phi_1 - \phi_3 + \phi_5 - \phi_7 + \phi_9 \ldots, \tag{2.40}$$

which helps in giving these wave functions a chemical meaning because they are *real* and, therefore, easy to sketch. If we now use some explicit Hamiltonian, the energies of these crystal orbitals can be exactly evaluated according to

$$E(k) = \frac{\langle \psi(k) | \mathcal{H} | \psi(k) \rangle}{\langle \psi(k) | \psi(k) \rangle}, \tag{2.41}$$

and we will simply draw the results by a popular semiempirical method (extended Hückel theory, see Section 2.11.1) to illustrate the course of the energies. It will become clear that the energetic results can be easily estimated by a *visual* interpretation, at least semiquantitatively, and often they do not have to be calculated at all.

Figure 2.9 shows the band structure of the above chain of H atoms within the first Brillouin zone. Because there is only one atomic orbital per unit cell, there can be only one band. For reasons of brevity, we have nevertheless plotted *three* bands in this figure, each for a differently spaced hydrogen chain. Note that only positive k values are usually given, and the energy of the band, $E(\psi(k))$, consists of the course of k quantized values for each $\psi(k)$; although the values that k may adopt are restricted, the number of k values is astronomically large, namely the number of unit cells per crystal. Because of this, $E(\psi(k))$ appears graphically as a continuous line.

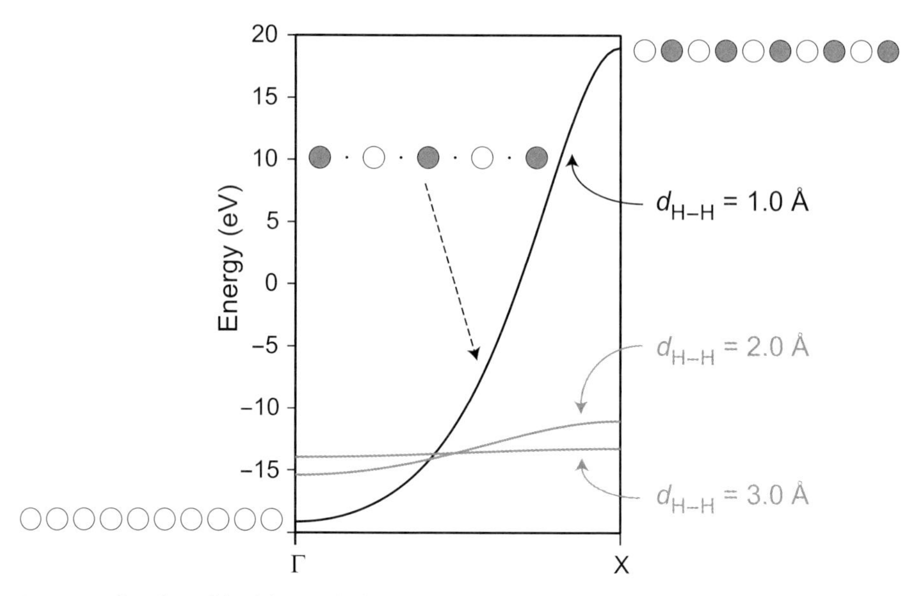

Fig. 2.9 Semiempirical (extended Hückel theory) band structure $E(\psi(k))$ for a one-dimensional chain of hydrogen atoms with H–H distances of 3 Å and 2 Å (in grey) and 1 Å (in black). The shape of the extended wave function has been iconized for three different k points, $\mathbf{\Gamma}$ (left), $k = \pi/2a$ (middle), and $\mathbf{X}$ (right).

The orbital icons characterizing $\psi(0)$, left, $\psi(\pi/2a)$, middle, and $\psi(\pi/a)$, right, in the figure allow a simple interpretation in terms of the interatomic overlap. The band achieves its lowest energy for $k = 0$, the $\mathbf{\Gamma}$ point. This strongest stabilization results from the in-phase, bonding combination of all atomic orbitals, and it is just the one-dimensional analogue of the lowest-energy σ_g wave function of the H_2^+ molecule (see Section 2.1). At $k = \pi/a$, the

X point, the one-electron wave function is high in energy because all atomic orbitals are out-of-phase, and the interaction is ultimately antibonding, similar to the high-energy σ_u^* wave function of H_2^+, reflected by a maximum number of nodes (changing sign of ψ) in going from atom to atom. Because we have included interatomic overlap in the calculation, the energetic destabilization clearly exceeds the energetic stabilization, just as for H_2^+; the upper and lower parts of the band are *not* fully symmetric in terms of energy. In the middle of the zone, at $k = \pi/2a$, the crystal orbital is essentially nonbonding (and medium in energy) because there is a zero atomic orbital coefficient on every second atom. This is all very straightforward.

The principal course does not change when the interatomic distance is altered; for larger H–H distances, however, the interactions become smaller and the *dispersion* of the band, that is, the energetic difference between the highest-energy/lowest-energy $\psi(k)$, decreases. When the interatomic distance shortens, the dispersion increases. This is also equivalent to the H_2^+ molecule although, as a numerical calculation will show, the dispersion is twice as large if compared with the molecular case.

As a rule of thumb, the band's dispersion is proportional to the strength of the atom–atom interaction between the unit cells, the latter being proportional to the overlap integral. Roughly speaking, the overlap integral $S_{\mu\nu}$ increases with decreasing interatomic distance, and $S_{\mu\nu}$ depends on the kind of orbitals involved. Very approximately, S decreases in the order of s–s > p–p > d–d > f–f interactions because of the atomic orbitals becoming more and more contracted upon going from s to p to d to f orbitals. Chemists will remember that in the rare-earth elements, the f orbitals are so contracted that they very much resemble the original atomic orbitals even when the atom is part of a chemical compound.[9]

When it comes to the course and dispersion of more complicated bands, this is easily illustrated by two other one-dimensional examples. Note that the above Bloch formula for the construction of $\psi(k)$ at some k value did *not* depend on the orbital involved; the plus/minus sign changes only resulted from the exponential pre-factor. Since Bloch's theorem just depends on *some* solution of the Schrödinger equation, and this may be another atomic orbital or, equally well, a molecular orbital, let us first assume, in Scheme 2.2, a one-dimensional chain of, say, nitrogen atoms where each N carries a set of one 2s

----N----N----N----N----N----N----N----N----N----N----

Scheme 2.2

and three 2p orbitals ($2p_x, 2p_y, 2p_z$); the interatomic distance will be 2.0 Å.

9) Similar phenomena are found, although to a lesser extent, for the transition metals (with d orbitals) in their compounds.

In the second example (Scheme 2.3), let us consider a one-dimensional chain of hydrogen *molecules*, the molecular axis being perpendicular to the chain direction, and here the orbitals entering the Bloch sum will be the molecu-

```
      H        H        H        H        H        H
----- | ------ | ------ | ------ | ------ | ------ | -----
      H        H        H        H        H        H
```

Scheme 2.3

lar orbitals σ_g and σ_u^*. The vertical intramolecular H–H distance is chosen as 1.0 Å whereas the horizontal intermolecular H–H distance (the length of the unit cell) will be 1.5 Å. Figures 2.10 and 2.12 show the corresponding band structures.

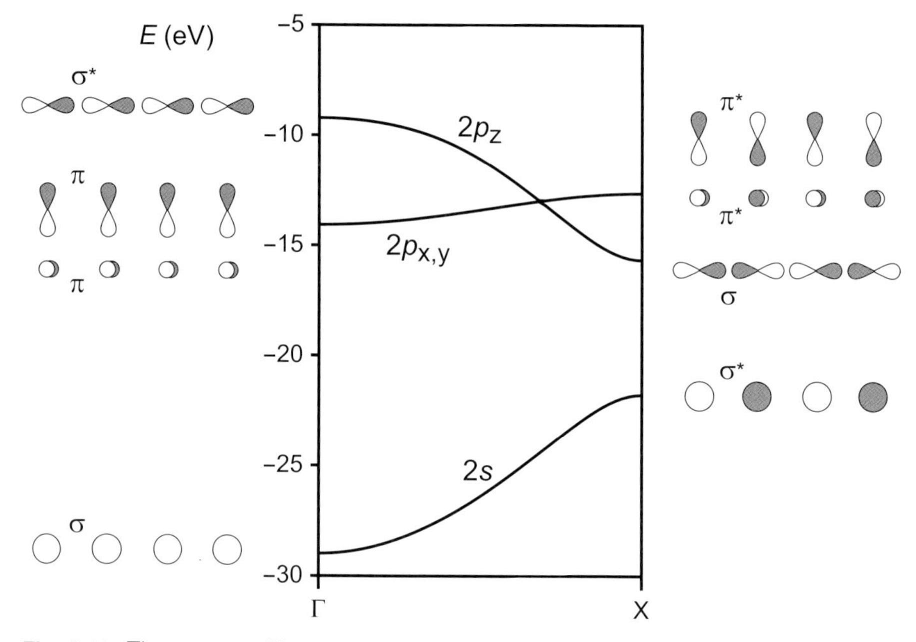

Fig. 2.10 The same as Figure 2.9, but for a one-dimensional chain of nitrogen atoms spaced at 2.0 Å. The labels indicate the types of chemical interaction in terms of the symmetry properties.

If we choose z as the chain direction of the nitrogen atoms, the orbital icons help us to understand why the energetic course of the degenerate $2p_x$ and $2p_y$ functions resembles, as a function of k, that of the $2s$ band lying below. Both $2p$ bands appear as in-phase, π-type bonding combinations at the zone center **Γ** and out-of-phase, π^*-type antibonding ones at the zone edge **X**. Because of the smaller π-type overlap, however, the dispersion is also smaller for the p orbitals. The topology of the $2p_z$ function, on the other hand, is reversed, for obvious reasons, and its dispersion is rather large compared with those of the $2p_{x,y}$ orbitals since the $2p_z$–$2p_z$ overlap is σ- and σ^*-like. Obviously, the topology depends on the orbital symmetry. As has been indicated already, the course of the 2*s* band of nitrogen is similar to that of the 1*s* band of hydrogen. Because of the larger nuclear potential of N, the 2*s* band is much lower in energy.

For this particular system, however, we need to think about its symmetry properties in more detail to fully understand the band structure. The only symmetry element upon going from **Γ** to **X** is the horizontal mirror plane σ coinciding with the infinite chain; with respect to this σ, the 2*s* band is totally symmetric because no sign change occurs for the 2*s* atomic orbitals generating this band upon reflection (see Figure 2.11). Also, the $2p_z$ band – but not the $2p_x$

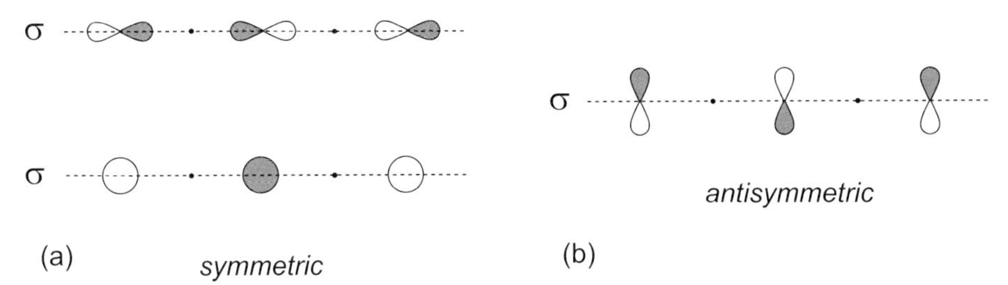

Fig. 2.11 Symmetry properties of the 2*s* and $2p_z$ (a) and $2p_x$ (b) bands belonging to the one-dimensional chain of nitrogen atoms with respect to the horizontal mirror plane σ; the bands are plotted in the middle of the zone for $k = \pi/2a$.

or $2p_y$ bands – is totally symmetric with respect to σ such that the 2*s* and $2p_z$ bands have *identical* symmetry properties and belong to the same irreducible representation.[10] Thus, the 2*s* and $2p_z$ bands *interact* throughout the entire zone (except at **Γ** and **X**) and the labels "2*s*" and "$2p_z$" contained in Figure 2.10 are *simplified* notions and should more correctly read "mostly 2*s*" and "mostly $2p_z$". Without the *s*–*p* mixing occurring for these two bands, the $2p_z$ band would be positioned slightly lower and the 2*s* band slightly higher in energy. The amount of *s*–*p* mixing is inversely proportional to the energy difference of the 2*s* and 2*p* orbitals, and for a very strong mixing the resulting bands will look as if they had repelled each other, the "avoided crossing" scenario [52] which we will meet in a moment.

The band structure of the chain of vertical hydrogen molecules (Figure 2.12) is particularly interesting. The band arising from the σ_g molecular orbital is lower in energy than the one from the σ_u^* level but both destabilize (increase in energy) upon going from **Γ** to **X** because the formerly bonding intermolecular (horizontal) interaction at **Γ** changes to an antibonding interaction at **X**. The bandwidth is larger for the energetically less favorable σ_u^* band because,

10) As noted on page 42 and alluded to on page 61, the study of group theory is very helpful.

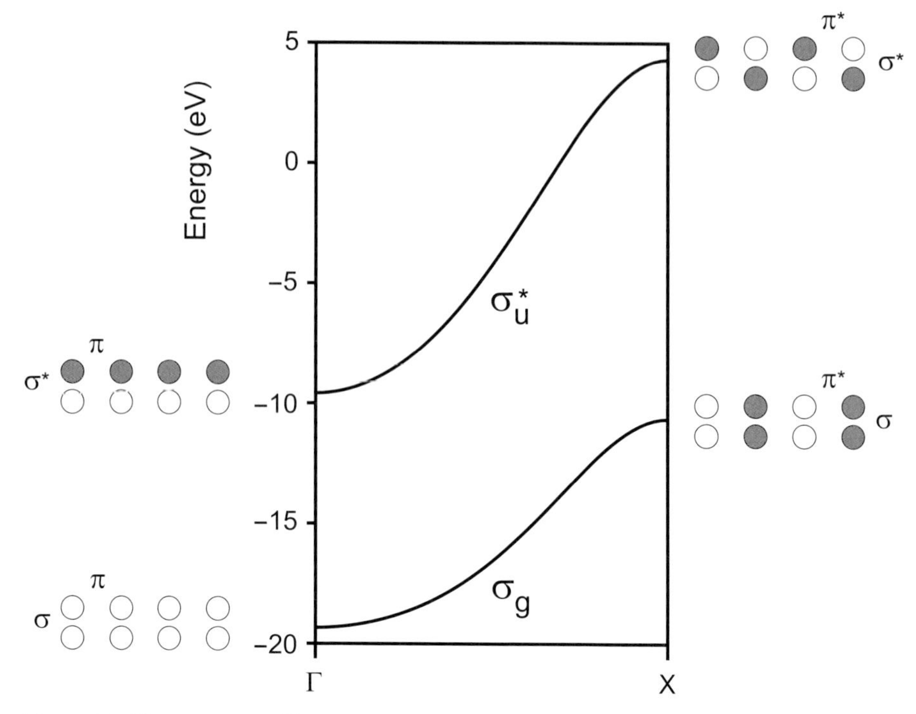

Fig. 2.12 The same as Figure 2.9, but for a one-dimensional chain of hydrogen molecules, their molecular axes being perpendicular to the chain direction. The vertical intramolecular H–H distance is 1.0 Å while the horizontal intermolecular distance is 1.5 Å.

again, overlap is included in the calculation. Inside the unit cell (that is, along the vertical intramolecular interaction), the bonding/antibonding character of σ_g/σ_u^* does not change.

Another nice thing about this band structure is that both bands do *not* interact throughout the entire Brillouin zone, and that is why we may discuss the two bands separately: The σ_g and σ_u^* bands have different symmetry properties for any k. With respect to the above-mentioned mirror plane σ coinciding with the infinite chain, the σ_g band is totally symmetric (see orbital icons) because no sign change occurs for the atomic orbitals generating σ_g upon reflection. The opposite is true for the σ_u^* band which is totally antisymmetric to σ, and the orbital icons of the upper/lower H atoms exhibit different plus/minus signs everywhere. Thus, both bands belong to different point-group representations and cannot mix, in principle.

2.6.2
Structural Distortions

This situation changes completely if we now turn the hydrogen molecules to make them coincide with the chain direction (Scheme 2.4). At the beginning,

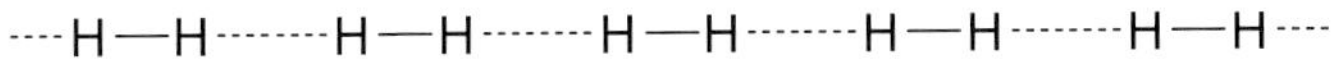

····H—H·········H—H·········H—H·········H—H·········H—H····

Scheme 2.4

we let 0.8 Å be the intramolecular H–H distance, whereas the intermolecular distance is a larger 1.2 Å. The computational result is found in Figure 2.13(a).

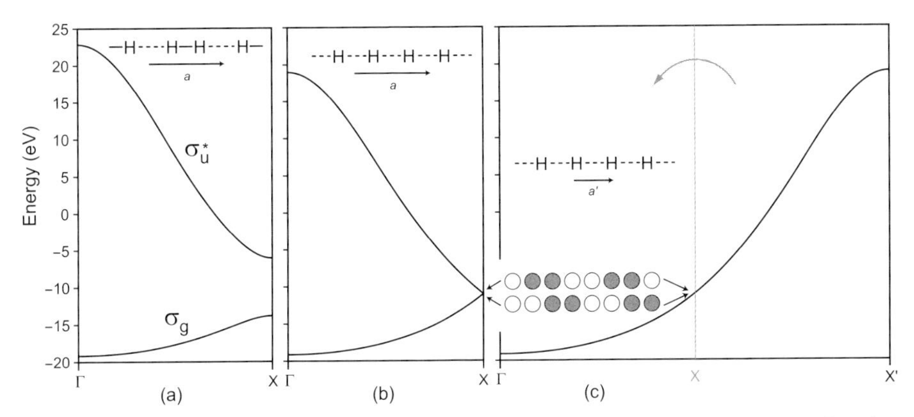

Fig. 2.13 As in Figure 2.12, but now for one-dimensional chains of hydrogen molecules coinciding with the chain direction and intra-/intermolecular H–H distances of 0.8/1.2 Å (a), 1.0/1.0 Å (b), and, again, 1.0/1.0 Å but now with a unit cell that has been reduced in size by 50% (c).

The low-lying σ_g band behaves like an s band and goes upwards; the high-lying σ_u^* is similar in shape, and thus in topology, to the $2p_z$ orbital of the chain of nitrogen atoms. Note that the energy gap between the two crystal orbitals at the **X** point is only a function of the ratio between intra- and intermolecular H–H distances, and the band gap vanishes if these are identical (Figure 2.13(b)). The degeneracy of the crystal orbital at the zone edge then simply reflects that the model under consideration, with identical H–H distances throughout the chain, is just the original chain of H atoms, and its band structure is depicted in Figure 2.13(c). Since we no longer have two atoms in the unit cell but only one, the new lattice parameter is only half as large, and the Brillouin zone ($0 \leq k \leq \pi/a$) has doubled in size. Alternatively, one might think of the middle band structure as the result of a "backfolding" process around the grey vertical line cutting the large Brillouin in two halves. If the upper part of the band in (c) is turned around by 180°, then (b) will be regained.

The degenerate crystal orbitals at the **X** point have also been iconized for Figure 2.13(b) and (c), and the σ_g and σ_u^* bands differ only by a *phase shift* of a single interatomic distance; this is the reason why they are degenerate at **X**. We also note that both bands now have *identical* symmetry properties with respect to the σ mirror plane coinciding with the infinite chain; both bands are totally symmetric and therefore interact with each other. Thus, the smallest collective vibration of the atoms of this chain will immediately raise the upper σ_u^* band and lower the lower σ_g band in energy. Similar consequences, but smaller in magnitude, will affect those parts of the two bands lying deeper in the Brillouin zone further away from **X**. Generally, there is an instability for a one-dimensional system with a half-filled band to undergo such a distortion, first formulated by Peierls [73], opening up a band gap at the highest filled level, the Fermi level. Thus, a system with a band structure such as in Figure 2.13(b) will change into one that corresponds to the band structure in Figure 2.13(a).

For chemists, it is probably not a big surprise that a one-dimensional chain of hydrogen atoms does not exist and that it will immediately decompose into isolated H_2 molecules. The Peierls distortion has important consequences for one-dimensional systems, such as polyacetylene with C–C bond-length alterations (instead of equal C–C distances) [74], infinite molecules with platinum–platinum bonding such as Krogmann's salts $K_2[Pt(CN)_4]X_{0.3} \cdot 3H_2O$ with X = Cl or Br [75], or other one-dimensional systems [76], and it also affects three-dimensional systems, in particular elemental structures (see Section 3.4). From a group-theoretical point of view, Peierls distortions are characterized by a loss of *translational* symmetry; in the above example, the nonequidistant chain of H atoms is less symmetrical (in terms of translational symmetry) than the equidistant one.

The corresponding analogue for molecules in which *nontranslational* symmetry is lost has been known much longer, and it bears the name Jahn–Teller distortion. It asserts that nonlinear molecules with an orbitally degenerate electronic state will undergo a symmetry-lowering distortion to remove this degeneracy [77]; because of its importance, we will briefly discuss it. Although strictly valid only for the instability of degenerate electronic *states*, the present usage is less strict and includes instabilities associated with the asymmetric electronic occupation of degenerate molecular *orbitals* [52,78]. For example, square planar cyclobutadiene, C_4H_4, exhibits all important characteristics of a Jahn–Teller (JT) instability, although a more rigorous group-theoretical analysis designates it a pseudo-JT effect [62]. Its Hückel π molecular orbitals are given in Figure 2.14.

With two electrons in the doubly degenerate, highest occupied molecular orbitals (Figure 2.14(a)), the e_g set (ψ_2 and ψ_3) is half-filled. According to the JT theorem, there exists at least one normal vibrational mode which will

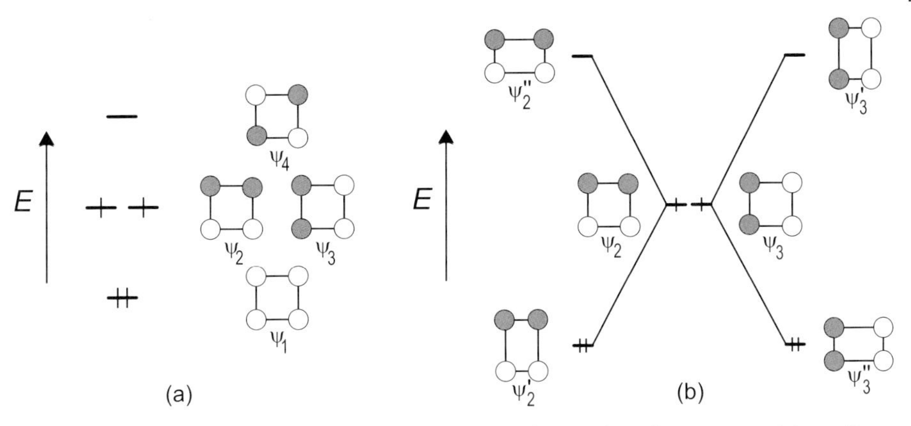

Fig. 2.14 The four Hückel π molecular orbitals and energies of square cyclobutadiene (a) and the consequences of a Jahn–Teller distortion on the two frontier orbitals ψ_2 and ψ_3 (b). The fillings of the molecular orbitals are indicated.

destroy the degeneracy and lower the electronic energy, just as for the one-dimensional H chain. In the present case, a deformation altering the C–C distances and lowering the symmetry from square (D_{4h}) to rectangular (D_{2h}) geometry is the appropriate one (Figure 2.14(b)). The loss of degeneracy in the occupied upmost (frontier) orbitals stabilizes the wave functions ψ_2' and ψ_3'' if the distortion strengthens the bonding interactions and weakens the antibonding interactions. A destabilization occurs for those orbitals whose bonding interactions decrease and antibonding interactions increase (ψ_2'' and ψ_3'), resulting in a net stabilization of the system. It is also clear that the loss of degeneracy is not a fully sufficient criterion for a distortion; the correct electron count is also essential. In this case, a π electron count of four, i.e., a half-filled e_g set, leads to the strongest stabilization. If the system contains two or six π electrons, the same deformation would destabilize the system.[11]

2.6.3 Higher Dimensions

Let us now return to infinite systems. As expected, band structures and their calculation (at least by hand) become increasingly challenging upon going to more complex systems and higher dimensions. For an illustration, we finally work on the band structure of a two-dimensional system, namely a square lattice of iron atoms as given in Scheme 2.5.

The two-dimensional plane is defined by the rectangular lattice vectors x and y, and the z direction is perpendicular to the page; the corresponding

11) These arguments form the core of Hückel's rule for the relative stability (and aromatic character) of cyclic systems as a function of the electron count, a fortunate blend of elementary quantum chemistry and simple topology [55].

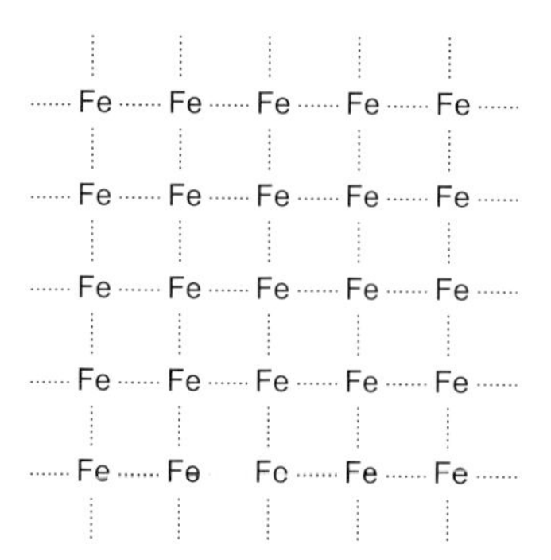

Scheme 2.5

Brillouin zone will also be two-dimensional. A moment's reflection yields that it must be identical with the two-dimensional slice of the primitive cubic Brillouin zone shown in Figure 2.8(a), including the special points **Γ**, **X**, and **M** (the special point **Y** is equivalent to **X** due to the quadratic symmetry). If the Fe–Fe distance defining the size of the unit cell is given as a = 2.2 Å, $\mathbf{k}$ will adopt all values in the reciprocal plane defined by the $k_{x,y}$ coordinates (0,0), $(0,\pi/a)$, $(\pi/a,0)$, and $(\pi/a,\pi/a)$. It would be awesome to draw the course of all possible $E(\psi(k))$ along a random walk within this plane, thus we had better look at them along special high-symmetry lines, that is, going from **Γ** to **X**, then from **X** to **M**, and finally back from **M** to **Γ**; other routes could have been chosen, too. As has been stressed before, the reason for travelling between the high-symmetry points is that the band structure is easier to *understand* than doing otherwise.

The band structure of the two-dimensional crystal of Fe atoms is shown in Figure 2.15. In order to comprehend the course of the $3d$ atomic functions of Fe in detail, we will also sketch the orbital icons at the special $\mathbf{k}$ points **Γ**, **X**, and **M**, keeping in mind that the atomic orbitals comprising the Bloch function change sign at the appropriate edge of the Brillouin zone; the result is given in Figure 2.16 which also shows a sketch of the two-dimensional Brillouin zone.

In the band structure (Figure 2.15), a total of five bands is visible because there is one Fe atom per unit cell which has five $3d$ atomic orbitals; other orbitals ($4s$- or $4p$-type) have been deliberately excluded. All bands change their energies, spaghetti-like, while the $\mathbf{k}$ quantum number changes. The reciprocal-space distances of **Γ–X** and **X–M** are exactly π/a whereas the **M–Γ** distance is larger by a factor of $\sqrt{2}$ because the latter is the diagonal of the two-dimensional Brillouin zone (Figure 2.16). Nonetheless, we plot them all in the band structure (somewhat sloppy, indeed) just as if they were equidistant in reciprocal space. Note also that, for example, the directions **Γ–X** and **X–M** are *perpendicular* to each other (see again Figure 2.16) but, since we are

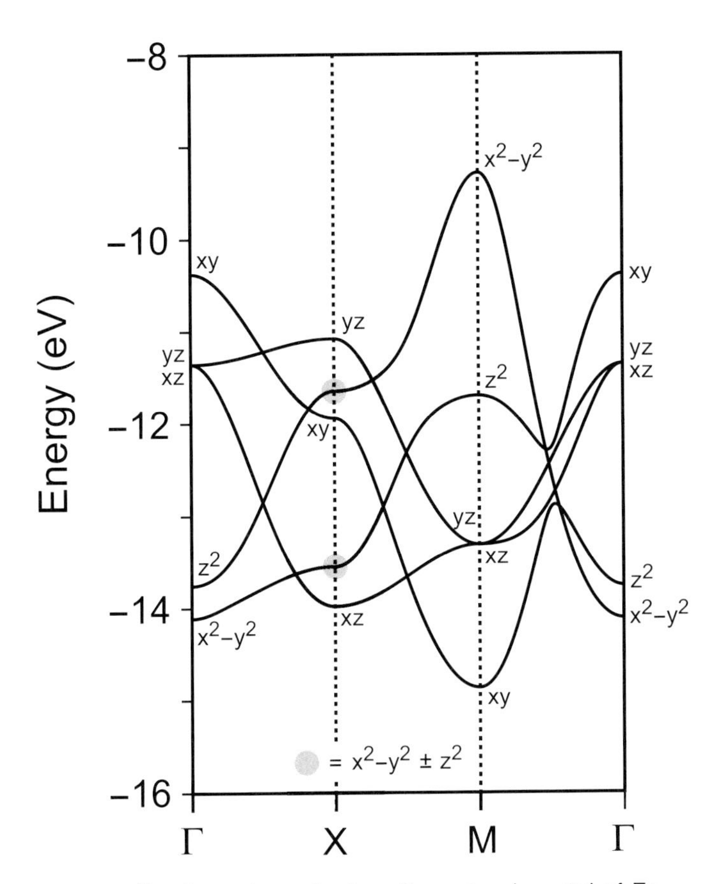

Fig. 2.15 Band structure of a two-dimensional crystal of Fe atoms spaced at 2.2 Å, each of them having five $3d$ atomic orbitals. For simplicity, the $4s$ and $4p$ functions have been suppressed.

lacking three-dimensional paper, we are forced to plot them into a single (two-dimensional) figure (Figure 2.15).

At **Γ**, the five d orbitals are no longer degenerate and the $d_{x^2-y^2}$ band is lowest in energy because it corresponds to the σ-bonding interaction (compare with Figure 2.16). The d_{xy} band, however, is the one with π^*-antibonding interaction, and it is therefore highest in energy. Both d_{xz} and d_{yz} orbitals experience δ-bonding and π^*-antibonding interactions, and thus they are degenerate at **Γ**. The d_{z^2} band (with δ-bonding interactions) lies between this degenerate set and the lowest $d_{x^2-y^2}$ band.

At **M**, the positions of $d_{x^2-y^2}$ and d_{xy} have been reversed; the $d_{x^2-y^2}$ band (see Figure 2.15) exhibits σ^*-antibonding behavior (Figure 2.16) whereas d_{xy} is of π-bonding type. The d_{xz} and d_{yz} bands are still degenerate. Their degeneracy is no longer present at the special point **X** because d_{xz} experiences

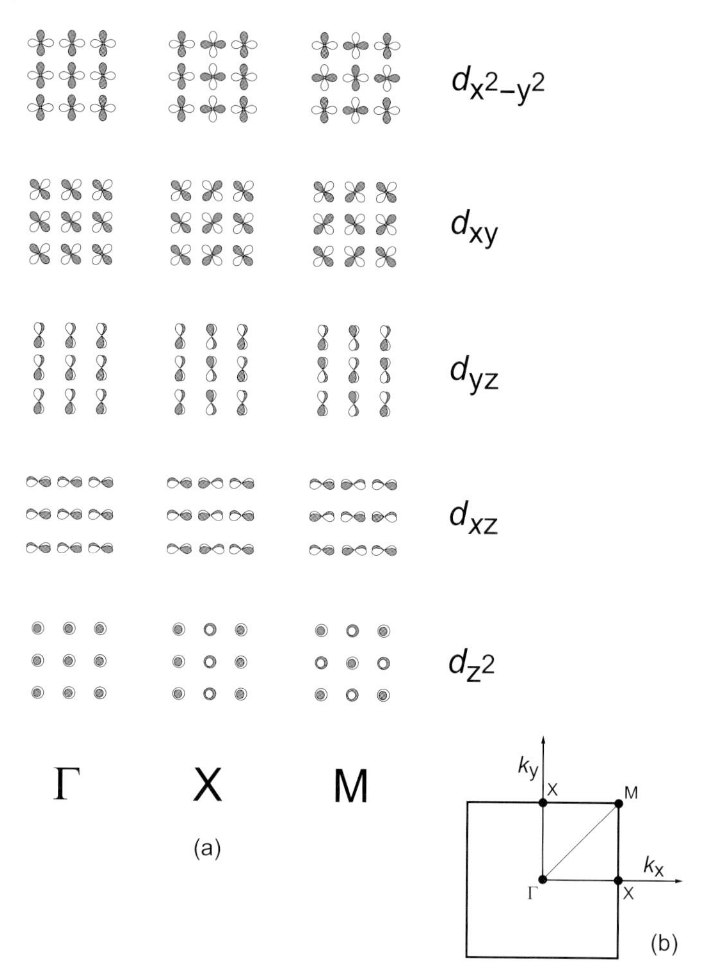

Fig. 2.16 Sketch of the five crystal-orbital phases at three different special points (a) within the Brillouin zone (b) of a two-dimensional Fe crystal; compare with Figure 2.15.

δ/π-bonding, whereas d_{yz} is of δ^*/π^*-antibonding type. Also, the formerly symmetry-independent bands $d_{x^2-y^2}$ and d_{z^2} *mix* upon going from **Γ** to **X** such that, at **X**, two linear combinations of these bands result (grey shadings at the **X** point); this mixing has not been included in the orbital icons in Figure 2.16. Back at **M**, $d_{x^2-y^2}$ and d_{z^2} are again symmetry-distinct. Between **M** and **Γ**, the degenerate sets of the bands d_{xz} and d_{yz} have slightly different energies. The $d_{x^2-y^2}$ band is the only one which keeps its orbital character, whereas the d_{xy} band turns into a d_{z^2} band, and the d_{z^2} band becomes a d_{xy} band. As a result, there is an "avoided crossing" between the latter two bands, reflecting the fact that they *strongly* interact because they have identical symmetry properties (belonging to the same irreducible representation) and they also have about the same energy.

The detailed discussion of band structures arising for three-dimensional materials is hardly ever attempted, due to their often incredibly high complexity; it *is* difficult to think in terms of reciprocal space! We will also leave the calculations of three-dimensional band structures and parts of their numerical analysis to fast (albeit dumb) computers.

2.7 Density-of-states and Basic Electron Partitioning

While the use of Bloch's theorem is the right tool to eliminate infinity through the underlying Fourier transformation, the $\boldsymbol{k}$-dependence of the crystal orbitals $\psi(\boldsymbol{k})$ makes their analysis rather difficult. First, the course of $\psi(\boldsymbol{k})$ needs to be studied for *any* $\boldsymbol{k}$; the two-dimensional band structure of Fe offered in the last section, however, stood for only *one* particular choice on how to move through reciprocal space. Despite its complexity, this band structure is only *one* of an infinite number of *more* complex band structures which one might also look at, namely by following a different route through the two-dimensional (or even three-dimensional) Brillouin zone. Second, $\psi(\boldsymbol{k})$ can (and will) adopt *complex* values, in contrast to a molecular orbital, and constructing real linear combinations from sets of degenerate $\psi(\boldsymbol{k})$s in order to sketch a band's *graphical* visualization in terms of orbital icons, is not a straightforward task. Similar technical problems are known to the computational crystallographer, simply because of the Fourier transformation and its associated reciprocal space.

What is needed, at least in the electronic-structure theory of solids, is an appropriate tool, one which averages over the entire system to make the electronic structure transparent and which brings us back into real space; there are at least two ways to accomplish this, either through averaging in real or in reciprocal space.

First, we note that the density of the electrons, $\rho(\boldsymbol{r})$, is easily accessible by experimental means (X-ray crystallography), and it is also very easy to calculate $\rho(\boldsymbol{r})$ – with real-space $\boldsymbol{r}$ coordinates – to a truly amazing precision if the crystal orbitals $\psi(\boldsymbol{k}, \boldsymbol{r})$ are accurately known; this is because the electron density is the energy-integrated square of the resulting wave function. Moreover, most parameter-free electronic-structure methods for solids are, in fact, electron-density-based schemes (see Section 2.12). On the other hand, it may come as a surprise that true understanding of chemical bonding is not immediately obvious from electron densities – contrary to common belief – because these often very closely resemble the atomic densities. Crystallographers will agree that standard crystal structure refinements are based on *atomic* densities (or scattering factors) – totally ignoring the electron-density deformation upon chemical bonding – but still result in quite acceptable data. Consequently, the analysis of experimental electron densities in terms of deformation densities arising through chemical bonding, requires a significant numerical and also intellectual effort [79]. It appears that the changes in $\rho(\boldsymbol{r})$ cannot be excessively large when chemical bonding sets in; the chemical bond is a dwarf, and this is why the LCAO approach is *so* successful in generating molecular and crystal orbitals. What is very much needed is an energy-resolved indicator of electron density and chemical bonding.

In the second approach, one deliberately sacrifices all direction-related information to yield such energy-resolved indicators, and the strategy proceeds as follows. If we were interested in the number of one-electron levels in an infinitely small energy interval dE, we would introduce the *density-of-states* $P(E)$ which, as solid-state physics teaches [53], behaves as the reciprocal gradient of the band structure. For the one-dimensional case, this is

$$P(E) \sim \left(\frac{dE}{dk}\right)^{-1} \sim v(k)^{-1}. \tag{2.42}$$

The density-of-states (or DOS), a very useful auxiliary function, is simply given by the inverse slope of the band in question. In other words, the steeper the band, the lower the DOS, or, equivalently, the flatter the band, the higher the DOS. It also transpires that the DOS is inversely proportional to the *velocity* v of an electron in a band. A very sharp DOS results from an extremely flat, atomic-like band, one which arises from an atomic orbital that does not significantly overlap with neighboring orbitals. Therefore, electrons occupied in such an orbital (band) cannot easily move through the crystal and they are slow. A wide DOS (due to a steep band), on the other hand, indicates a rapidly moving electron due to strong interatomic coupling. It goes without saying that the DOS definition is easily generalized to the two- and three-dimensional cases without any qualitative changes in its meaning.[12]

Figure 2.17 now shows the band structure of the one-dimensional hydrogen chain and its DOS side-by-side for all three atomic spacings. As mentioned before, a steep/flat band is characterized by a low/high DOS. If the band becomes more atomic-like due to a wide interatomic spacing, the DOS turns into a spike for that atomic level.

It also needs to be mentioned that there is a different style of DOS presentation in chemistry and physics. Within the latter community, it is customary

12) Whether or not density-of-levels (DOL) or density-of-states (DOS) is the more appropriate term depends on the perspective. In the first case, the one-electron picture (orbitals) prevails whereas, in the second, the calculation of a many-body wave function would be necessary; in practice, the term DOS is used throughout, even if the construction of a many-body wave function is not attempted.

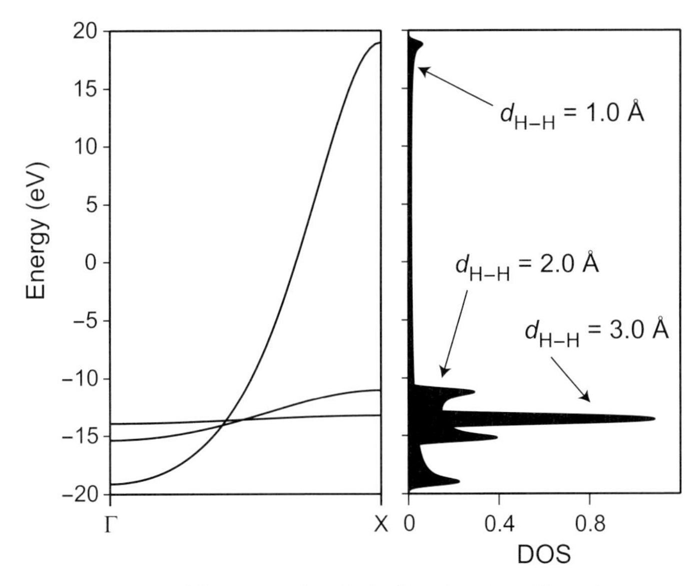

Fig. 2.17 As in Figure 2.9, but including the densities-of-states. The latter have been smoothed by a Gaussian function (see text) and they are therefore slightly wider in energy than the bands.

(and quite logical, too) to plot $P(E)$, the *dependent* quantity, as a function of E. Chemists prefer to recognize the inverse-slope relationship between $P(E)$ and $E(\psi(\mathbf{k}))$ such that $P(E)$ is plotted along the abscissa. In any case, $P(E)$ averages over the entire $\mathbf{k}$ space (unimportant for one-dimensional systems, but *very* important for higher dimensions) and in the present example with a short H–H distance of 1.0 Å (steepest band), the DOS "explodes" at lowest/highest energies; these effects go by the name of van Hoove singularities [53].

In contrast to the above sketch, which might have been performed by hand, the computer evaluates Schrödinger's $\mathbf{k}$-dependent equation for all representative $\mathbf{k}$ points in the three-dimensional Brillouin zone of real materials; the finer the mesh, the more accurate the calculation. There are different ways to accomplish this goal, and two routes have been particularly successful. The first strategy tries to find a set of special $\mathbf{k}$ points that are characteristic for the crystal symmetry [80,81], the other simply chops the whole Brillouin zone into lots of small tetrahedra [82]. Eventually, one arrives at a histogram of energy eigenvalues for all $\mathbf{k}$ points which is numerically smoothed by, for example, a Gaussian function to let the density-of-states appear as if it had been computed using an infinite number of k points, a type of artificial polishing.

When all this is done, the numerical value of the DOS at its highest *occupied* level is characteristic for the transport properties of the material. Metallic behavior will be found only for a vanishing band gap and finite DOS, whereas a large (many eV) gap between the occupied and unoccupied levels indicates insulating properties; when there is only a small gap (around 1 eV or so), the thermal activation of a few electrons from the occupied DOS into the unoccupied DOS region will induce what is called semiconducting behavior.

Before showing more complex DOS, we will first concentrate on how to characterize the electron density distribution in molecules and solids because that is where we started; this is also needed to further decompose the density-of-states and characterize the electronic structures in terms of bonding properties. Recalling the LCAO ansatz for the molecular orbital (MO) of the H_2^+ molecule as given in Equation (2.11), we may square-integrate the MO over all space and yield

$$\int \psi^* \psi d\tau = \underbrace{\int \psi^2 d\tau}_{\equiv 1} = c_1^2 \underbrace{\int \phi_1^2 d\tau}_{\equiv 1} + c_2^2 \underbrace{\int \phi_2^2 d\tau}_{\equiv 1} + 2c_1c_2 \underbrace{\int \phi_1\phi_2 d\tau}_{S_{12}}. \qquad (2.43)$$

The first three underbraced integrals are just unity because we have implicitly used normalized molecular and atomic orbitals, being able to hold just two electrons; the last integral is the overlap integral S_{12} and the last term, $2c_1c_2S_{12}$, is called the *overlap population* according to Mulliken [83]. Mulliken also heuristically partitioned the above equation into three different parts, namely

$$1 = \underbrace{c_1^2}_{\text{atom 1}} + \underbrace{c_2^2}_{\text{atom 2}} + \underbrace{2c_1c_2S_{12}}_{\text{bonding}}, \qquad (2.44)$$

such that we may work out an extremely simple interpretation for the H_2^+ molecular cation and, also, for the H_2 molecule. Given an experimental H–H distance of 0.74 Å, the overlap integral $S_{12} \equiv S_{1s1s}$ is approximately 0.64 if computed using a Slater-type orbital with an exponent of $\zeta = 1.3$; inserting the mixing coefficient $c = 1/\sqrt{2(1+S_{12})}$ from Equation (2.15) into Equation (2.44), this becomes

$$1 = \underbrace{\frac{1}{2(1+S_{12})}}_{=0.305} + \underbrace{\frac{1}{2(1+S_{12})}}_{=0.305} + \underbrace{\frac{2}{2(1+S_{12})}S_{12}}_{=0.390} \qquad (2.45)$$

for H_2^+. If we want the σ_g molecular orbital to be occupied by two electrons (as in H_2), the above equation needs to be multiplied by two, so that we have

$$2 = 0.61 + 0.61 + 0.78 \qquad (2.46)$$

for the electron distribution in the H_2 molecule, assuming totally independent electrons (see Section 2.11.3). This finding, iconized in Figure 2.18, simply indicates that the two isolated atoms with formerly one electron each have formed a chemical bond which contains 0.78 electrons; each bonded H atom still has 0.61 "leftover" electrons which are not shared in the H–H bond. The nonintegral electron numbers (remember that electrons cannot be divided) already indicates a serious interpretational problem.

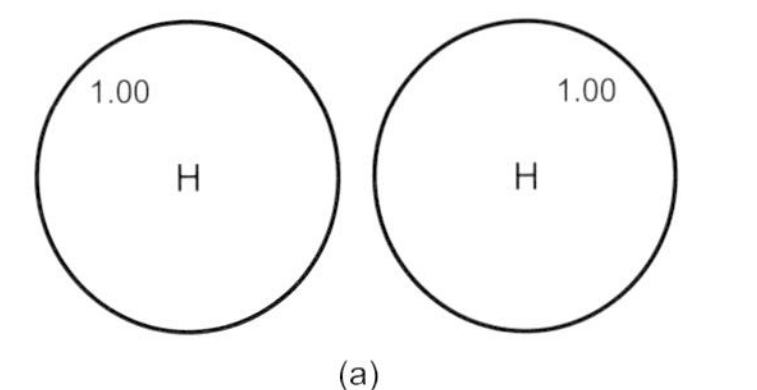

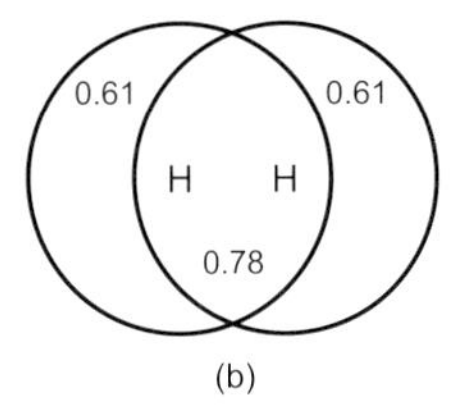

Fig. 2.18 Schematic drawing of the electron distribution in the H–H molecule before (a) and after (b) bond formation.

The above arguments form the core of Mulliken's population analysis [83], and it is customary to call the atom-centered electrons the (atomic) *net populations* (NP) which, together with the *overlap populations* (OP), add up to the total electron number; in the above example, this is $2 = NP_1 + NP_2 + OP = 0.61 + 0.61 + 0.78$. We might also cut the overlap population symmetrically and add each half to the net populations, thereby defining (atomic) *gross populations* (GP); in the H_2 example, this makes $2 = NP_1 + \frac{1}{2} \times OP + NP_2 + \frac{1}{2} \times OP = GP_1 + GP_2 = 1.00 + 1.00 = 2$, which is very simple. Within this scheme, the atomic charge q is the difference between the atomic number Z and the gross population ($q = Z - GP = 1.00 - 1.00 = 0$) such that the H atom is neutral in the H_2 molecule, as expected.

The Mulliken population analysis is easily extended to larger molecules, and it has become a standard quantum-chemical tool for this type of analysis. Its deficiencies, however, are also known. The assignments of net and overlap populations are arbitrary, and so is the symmetrical splitting of the overlap population, especially if used for highly polar covalent bonds. Multicenter bonds are hard to deal with, there are numerical problems (OPs being too large, too small, even negative), and one usually finds a strong basis-set dependency for the NPs and OPs. Alternative analyses, for example, those by Löwdin, Davidson, Jug, and Ahlrichs, have been suggested [84–87] and are in use today, and they all represent different approaches to circumnavigate the weaknesses of Mulliken's scheme, for example, by replacing the symmetrical splitting of the overlap population with one which retains the molecular dipole moment [86]. Nonetheless, the different population analyses all yield different numerical results because the fundamental problem of *any* such scheme is that, in principle, quantum-mechanical systems (such as molecules or solids) are *not exactly decomposable* by classical means, that is, *into atoms and bonds*. These approaches work nicely in the hands of a skillful theorist, but they cannot be made "exact".

What can be done, however, may still prove useful for chemical reasoning. For example, it may be shown by statistical analyses that all the atomic charges derived from various numerical population schemes are connected with a *principal component* of ionicity which accounts for more than 90% of the whole variance in the set of charges [88]. We also note that there has been a topological approach by Bader to rigorously define the sizes of atoms in molecules based on the electron densities; this recipe (see Section 2.10) works but one must accept the definitions chosen for that purpose [89].

2.8 Energy-resolved Electron and Energy Partitioning

When going over to more than one atom (or atomic orbital) per unit cell, the density-of-states must be analyzed in greater detail; the aforementioned Mulliken population analysis is still needed, but in an energy-resolved form.

If we write down the total number of electrons N in an arbitrary molecule using the LCAO expression in Equation (2.6), the generalization of Equations (2.44) and (2.45) is simply

$$N = \underbrace{\sum_{A}\sum_{\substack{\mu \\ \mu\in A}}\sum_{i}^{m} f_i c_{\mu i}^2}_{\text{net populations}} + \underbrace{2\sum_{A}\sum_{B>A}\sum_{\substack{\mu \\ \mu\in A}}\sum_{\substack{\nu \\ \nu\in B}}\sum_{i}^{m} f_i c_{\mu i} c_{\nu i} S_{\mu\nu}}_{\text{overlap populations}} \tag{2.47}$$

$$= \underbrace{\sum_{A}\sum_{\substack{\mu \\ \mu\in A}}\sum_{i}^{m} f_i \left(c_{\mu i}^2 + \sum_{B\neq A}\sum_{\substack{\nu \\ \nu\in B}} c_{\mu i} c_{\nu i} S_{\mu\nu} \right)}_{\text{gross populations}} . \tag{2.48}$$

The sum includes all mixing coefficients $c_{\mu i}$ belonging to atomic orbitals ϕ_μ centered on atoms A, and the molecular orbitals are numbered with i. We have also assumed, for simplicity, the "spin-restricted" case, meaning that there are identical molecular orbitals for "spin-up" and "spin-down" electrons (see Section 2.9), and these MOs have occupation numbers f_i with f_i = 0, 1, or 2; the "spin-unrestricted" case would simply have twice as many sums for both spin directions and spin orbitals. Although the formula looks quite powerful, nothing really new has entered if compared with the case of H_2^+; it represents the general electron partitioning of a molecule in terms of Mulliken's recipe.

The above formula does not change very much when going to the solid state. We must keep in mind, however, that the mixing coefficients become $\boldsymbol{k}$-dependent because of $c_{\mu i} \rightarrow c^*_{\mu i}(\boldsymbol{k})$ and $c_{\nu i} \rightarrow c_{\nu i}(\boldsymbol{k})$ such that an appropriate $\boldsymbol{k}$ integration must be performed. To make the notation more compact, it is customary to merge the occupation number f_i and the above mixing coefficients (eigenvectors) and convert them into the entries of a $\boldsymbol{k}$-dependent *density matrix* $P_{\mu\nu}(\boldsymbol{k})$ by writing

$$P_{\mu\nu}(\boldsymbol{k}) = \sum_i f_i c^*_{\mu i}(\boldsymbol{k}) c_{\nu i}(\boldsymbol{k}), \tag{2.49}$$

such that a subsequent integration in $\boldsymbol{k}$ space yields the desired average, namely through $\int P_{\mu\nu}(\boldsymbol{k}) d\boldsymbol{k} = P_{\mu\nu}$. Then, Equation (2.47) may be shortened to read

$$N = \underbrace{\sum_A \sum_{\substack{\mu \\ \mu \in A}} P_{\mu\mu}}_{\text{net populations}} + \underbrace{2 \sum_A \sum_{B>A} \sum_{\substack{\mu \\ \mu \in A}} \sum_{\substack{\nu \\ \nu \in B}} \mathrm{Re}[P_{\mu\nu} S_{\mu\nu}]}_{\text{overlap populations}} \tag{2.50}$$

$$= \underbrace{\sum_A \sum_{\substack{\mu \\ \mu \in A}} \left(P_{\mu\mu} + \sum_{B \neq A} \sum_{\substack{\nu \\ \nu \in B}} \mathrm{Re}[P_{\mu\nu} S_{\mu\nu}] \right)}_{\text{gross populations}}, \tag{2.51}$$

where Re characterizes the real parts of the (potentially) complex off-diagonal ($\mu \neq \nu$) entries. This is already fine, but we can make things even more transparent by partitioning the electrons in an explictly energy-dependent way, namely by a differentiation of the density matrix $P_{\mu\nu}$ to yield the *density-of-states* matrix $P_{\mu\nu}(E)$, that is

$$P_{\mu\nu} = \int^{\varepsilon_F} P_{\mu\nu}(E)\, dE \qquad \text{or} \qquad P_{\mu\nu}(E) = \sum_i f_i c^*_{\mu i} c_{\nu i} \delta(\varepsilon - \varepsilon_i). \tag{2.52}$$

In practice, the use of $P(E)$ is highly advantageous because it expands the electronic structure and also the electron distribution in terms of the energy; also, $P(E)$ is independent of the energetic position of the highest occupied band, the so-called Fermi level ε_F.[13] Thus, Equation (2.50) may reach its final formulation, namely

13) This crude simplification (Fermi level coincides with the highest occupied level) is fine for metals but not for semiconductors or insulators. For thermodynamic reasons, for the latter two the Fermi level lies exactly between the occupied region (valence band) and the unoccupied region (conduction band). Even using this improved definition, however, no change in the electron partitioning results.

$$N = \underbrace{\int^{\varepsilon_F} \sum_A \sum_{\substack{\mu \\ \mu \in A}} P_{\mu\mu}(E)\, dE}_{\text{net populations}} + \underbrace{\int^{\varepsilon_F} 2 \sum_A \sum_{B>A} \sum_{\substack{\mu \\ \mu \in A}} \sum_{\substack{\nu \\ \nu \in B}} \mathrm{Re}[P_{\mu\nu}(E) S_{\mu\nu}]\, dE}_{\text{overlap populations}} \tag{2.53}$$

$$= \underbrace{\int^{\varepsilon_F} \sum_A \sum_{\substack{\mu \\ \mu \in A}} \left(P_{\mu\mu}(E) + \sum_{B \neq A} \sum_{\substack{\nu \\ \nu \in B}} \mathrm{Re}[P_{\mu\nu}(E) S_{\mu\nu}] \right) dE,}_{\text{gross populations}} \tag{2.54}$$

$$= \int^{\varepsilon_F} \sum_A \sum_{\substack{\mu \\ \mu \in A}} \sum_B \sum_{\substack{\nu \\ \nu \in B}} P_{\mu\nu}(E) S_{\mu\nu}\, dE. \tag{2.55}$$

The last equation (2.55) indicates that the total electron number is just the energy integral of the sum of the products between the entries of the $\boldsymbol{k}$-averaged density-of-states matrix $P_{\mu\nu}(E)$ and the overlap matrix $S_{\mu\nu}$, and we might write all products in matrix form (note that $S_{ii} \equiv S_{jj} \equiv 1$):

$$\begin{pmatrix} P_{11}(E) & P_{12}(E)S_{12} & \cdots & & \cdots & P_{1\nu}(E)S_{1\nu} \\ P_{21}(E)S_{21} & P_{22}(E) & & & & \vdots \\ \vdots & & \ddots & & & \\ & & & P_{\mu\mu}(E) & & \\ \vdots & & & & \ddots & \vdots \\ P_{\nu 1}(E)S_{\nu 1} & & \cdots & & \cdots & P_{\nu\nu}(E) \end{pmatrix} \tag{2.56}$$

In order to partition the number of all electrons of an extended system into, say, atomic contributions, the sum of the elements of the above matrix (density-of-states times overlap) must be taken as in Equation (2.53), corresponding to energy-dependent gross populations. On doing so, *local* or *atom-projected* DOS plots are generated. When these are integrated, one is left with the *gross* electron populations on the specified atoms. Mathematically, the sum needed for the atom-projected DOS contains all on-site ($\mu = \nu$) entries plus half of the corresponding off-site ($\mu \neq \nu$) entries, just as for the H_2^+ or H_2 cases. The generation (or projection) of local DOS plots for extended solids is thus *identical* with the calculation of gross populations for molecules, although the former is energy-resolved. It is quite obvious that one does not necessarily have to partition the electron number N into atomic contributions alone; atomic *groups* or sets of well-defined group orbitals might also be other plausible candidates for partitioning (or projecting). Just like the overall density-of-states, local DOS are $\boldsymbol{k}$ averages over the entire Brillouin zone, having lost all directional information.

A very crude sketch of the partitioning into net and bonding electrons according to Equation (2.53) and the associated partitioning matrix, is given in Scheme 2.6. Here, the atom-centered contributions are shown in black and the between-atoms contributions are in white.

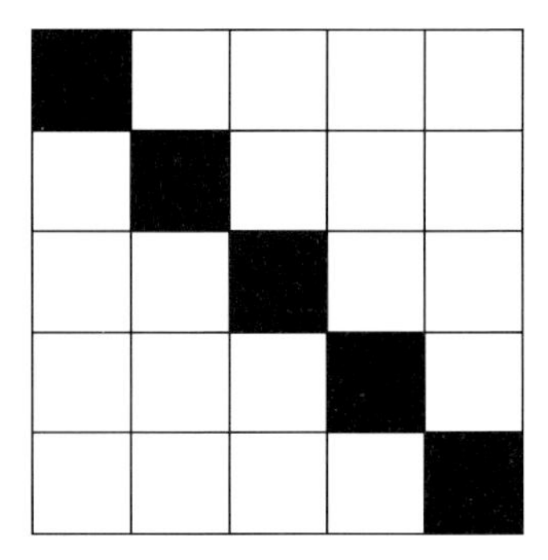

Scheme 2.6

The partitioning into net atomic and bonding electrons suggests an exclusive analysis of the chemical bonding. In fact, looking *only* at the off-diagonal entries ($\mu \neq \nu$) has turned out to be extraordinarily fruitful within solid-state quantum chemistry. Once the electronic structure is calculated, all levels must then be investigated as to their bonding proclivities (measured by $\text{Re}[P_{\mu\nu}(E)S_{\mu\nu}]$) such that an overlap population-weighted density-of-states is defined [90]. For obvious reasons, this technique has been dubbed *crystal orbital overlap population* (COOP).[14] The COOP calculation proceeds just like a DOS calculation *except* that one has to select a pair of neighboring atoms (or selected orbitals on these atoms, with $\mu \neq \nu$), and the energy integral yields the overlap population between the two atoms (orbitals), a quantum-chemical measure of the bond strength, just as in the molecular case. While positive COOPs indicate bonding interactions, negative ones stand for antibonding interactions. Nonbonding interactions will give zero COOPs.

Let us illustrate these abstract electron partitionings by a number of simple calculations, the ones we have already looked at in the preceding section. Figure 2.19 presents the band structure and DOS curves of the infinite hydrogen chain (with the atoms spaced at 2.0 Å), together with the COOP bonding analysis. At low/high energies, the DOS becomes "spiky" because of the band being flat at **Γ** and **X**, but the DOS is relatively smooth inbetween. The COOP plot shows bonding/antibonding interactions at low/high energies for the nearest H–H interactions; this is very much expected keeping the molecular orbitals of H_2^+ in mind, and it may also be derived by looking at the orbital icons given before in Figure 2.9. If we were to investigate the second-nearest H–H

14) This technique of an overlap population-weighted density-of-states would have also proceeded well by the shortcut OPW-DOS; the label OPW, however, serves as an abbreviation of a well-established (but dated) band-structure method (see Section 2.15).

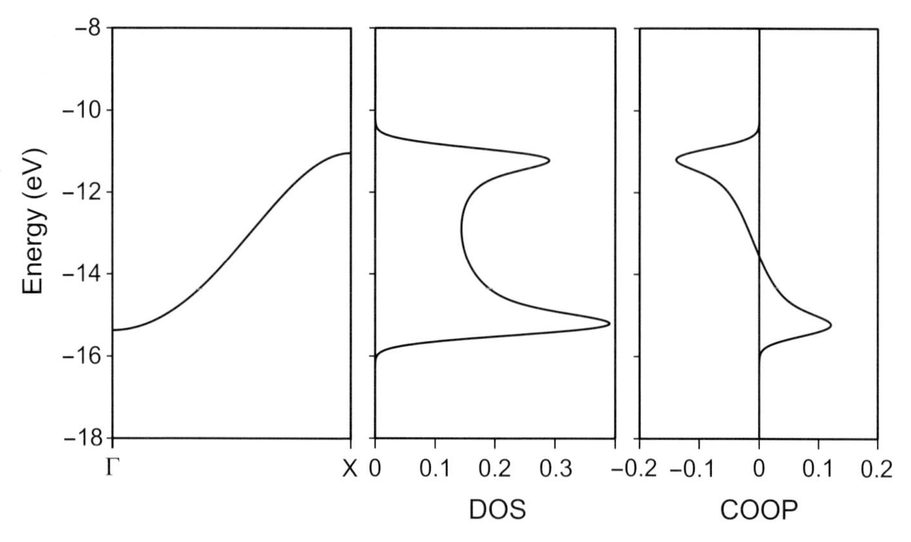

Fig. 2.19 Semiempirical (extended Hückel theory) band structure, density-of-states, and crystal orbital overlap population for a one-dimensional chain of hydrogen atoms spaced at 2.0 Å. Due to a Gaussian smoothing, DOS and COOP plots appear slightly wider in energy.

interactions (not shown), the orbital icons in Figure 2.9 then indicate bonding states *both* at low/high energies and antibonding interactions at medium energy. Because of the larger second-nearest distances, the overlap integral is *much* smaller, and so is the COOP figure.

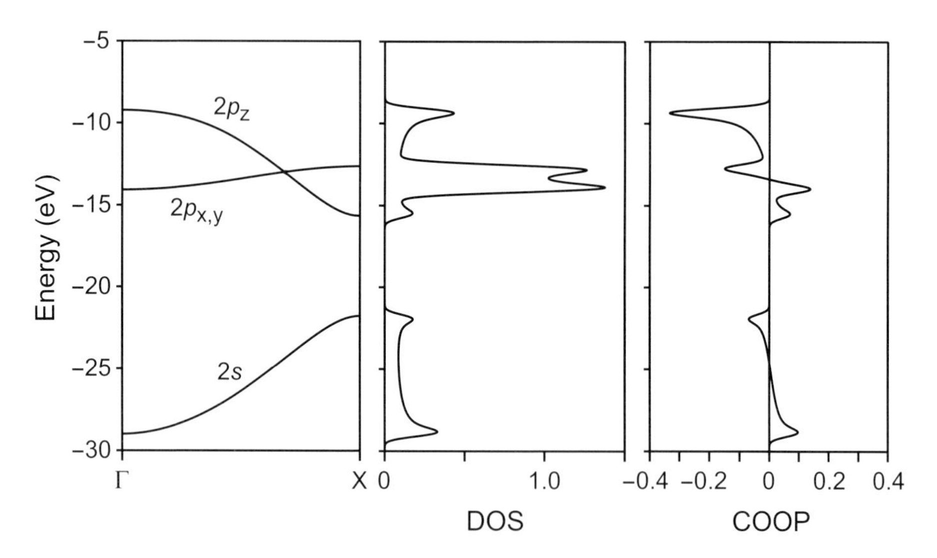

Fig. 2.20 As above, but for a chain of nitrogen atoms spaced at 2.0 Å.

Similar results are also found for the chain of nitrogen atoms given in Figure 2.20. The DOS shape results from a superposition of the 2s DOS at low energies, the $2p_z$ DOS at high energies (broad structure), and the DOS of the degenerate $2p_{x,y}$ levels also at high energies, but with a narrow shape.

In contrast to the DOS plot which just serves as a level or electron counter, a very nice feature of the COOP analysis is that it lets us *predict* the energetically advantageous electron fillings, that is, the reasonable positionings of the Fermi level. For a half-filled 2s band (ca. −25 eV), bonding results, but this will almost vanish if a second electron enters the 2s band because antibonding levels are then occupied (ca. −21 eV). A third valence electron would enter bonding 2p bands (the $2p_z$–$2p_z$ bonding interaction between −13 and −16 eV, compare with orbital icons in Figure 2.10) but the 2p states soon also become N–N antibonding when more electrons are filled in, that is, above −13 eV.

The DOS and COOP curves of the chain of vertically oriented hydrogen molecules, given in Figure 2.21, provide similarities and dissimilarities. The DOS is composed of two parts, arising from the lower σ_g and the upper σ_u^* bands, and both seem to overlap a little. On purpose, we show a computational artifact which is due to a too-strong Gaussian broadening used here; as can be seen in the band structure on the left, both bands do *not* energetically coincide. In the COOP diagram, both σ_g and σ_u^* bands indicate bonding/antibonding intermolecular interactions at lower/upper regions. The orbital icons in Figure 2.12 illustrate these features.

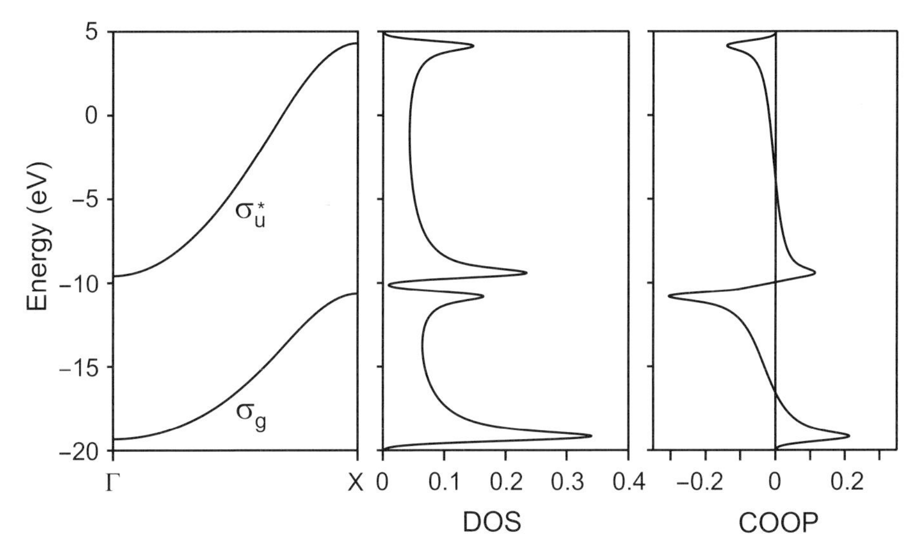

Fig. 2.21 As for Figure 2.19, but for a chain of hydrogen molecules, the molecular axis (intramolecular H–H distance = 1.0 Å) being perpendicular to the chain axis as given in Scheme 2.3; the intermolecular distance is 1.5 Å. An intermolecular COOP is shown on the right.

The COOP idea has found a plethora of applications in the framework of a popular semiempirical (see Section 2.11.1) method, and COOP is probably *the* decisive quantum-chemical tool that has convinced many solid-state (materials) chemists to eventually strive for band-structure calculations. Before the introduction of COOP, chemical bonding discussions were too often restricted to the simplistic "ionic model", no matter whether or not the chemical bonding was, in fact, ionic in nature.

With the advent of first-principles (density-functional theory, see Section 2.12) methods, another COOP derivative was developed which is no longer based on an overlap criterion. The reason is that, within parameter-free methods, the basis sets used can also be short-ranged (similar to STOs) but they may be subject to various transformations (depending on the method chosen) which will also change their shapes. In order to bypass problems due to basis-sets dependencies, an alternative partitioning is utilized. This partitioning is based on the electronic energy E, and it is the analogue of Equation (2.53), having replaced the overlap integrals by Hamiltonian matrix elements, that is,

$$E = \underbrace{\int^{\varepsilon_F} \sum_A \sum_{\substack{\mu \\ \mu\in A}} P_{\mu\mu}(E) H_{\mu\mu}(E)\, dE}_{\text{net atomic energies}} + \underbrace{\int^{\varepsilon_F} 2 \sum_A \sum_{B>A} \sum_{\substack{\mu \\ \mu\in A}} \sum_{\substack{\nu \\ \nu\in B}} \mathrm{Re}[P_{\mu\nu}(E) H_{\mu\nu}(E)]\, dE}_{\text{bonding energies}} \tag{2.57}$$

$$= \underbrace{\int^{\varepsilon_F} \sum_A \sum_{\substack{\mu \\ \mu\in A}} \left(P_{\mu\mu}(E) H_{\mu\mu}(E) + \sum_{B\neq A} \sum_{\substack{\nu \\ \nu\in B}} \mathrm{Re}[P_{\mu\nu}(E) H_{\mu\nu}(E)] \right) dE,}_{\text{gross atomic energies}} \tag{2.58}$$

$$= \int^{\varepsilon_F} \sum_A \sum_{\substack{\mu \\ \mu\in A}} \sum_B \sum_{\substack{\nu \\ \nu\in B}} P_{\mu\nu}(E) H_{\mu\nu}(E)\, dE \tag{2.59}$$

$$= \int^{\varepsilon_F} \sum_A \sum_{\substack{\mu \\ \mu\in A}} \sum_B \sum_{\substack{\nu \\ \nu\in B}} \mathrm{COHP}_{\mu\nu}(E)\, dE. \tag{2.60}$$

Similar to the COOP method, this approach has been called *crystal orbital Hamilton population* (COHP) analysis [91]. Given a short-ranged orbital basis set, the sum of all pairwise interactions rapidly converges in real space, similar to the COOP formalism. In fact, such energy-partitioning schemes have a long history in quantum chemistry [67, 92, 93], and Coulson was probably the first to set up such an expression [94].

COHP calculations have become a valuable bond-indicating tool in the parameter-free (density-functional theory-type, see Section 2.12) calculation of the electronic structures of complex materials. They look similar to COOP curves, but we note two differences between COOP and COHP. Bonding interactions (positive COOP) lead to energy lowering and thus will show up with a *negative* COHP. Likewise, an antibonding COOP (negative value) stands for an increase in energy, reflected by a *positive* COHP, such that there is an approximate plus/minus change between COOP and COHP [95]; an exact plus/minus relationship for each pair of $S_{\mu\nu}/H_{\mu\nu}$ elements does not exist, though. Nonetheless, negative COHPs are usually plotted to make them appear similar to COOPs. In addition, COOP calculations are often performed as an average over bonds, whereas COHP calculations usually include a sum of bonds.

Figure 2.22 shows the electronic structure of the square net of iron atoms, now including a DOS and COHP analysis. For this two-dimensional system, the DOS clearly exhibits a two-peak shape and also shows rather small values at lowest/highest energies. It is seemingly impossible to guess the correct electron filling of this systems *unless* the COHP is analyzed. As expected, the low-lying bands all prove to be Fe–Fe bonding whereas the high-lying bands are antibonding. If the bands were filled with electrons up to ca. −12.5 eV, i.e.,

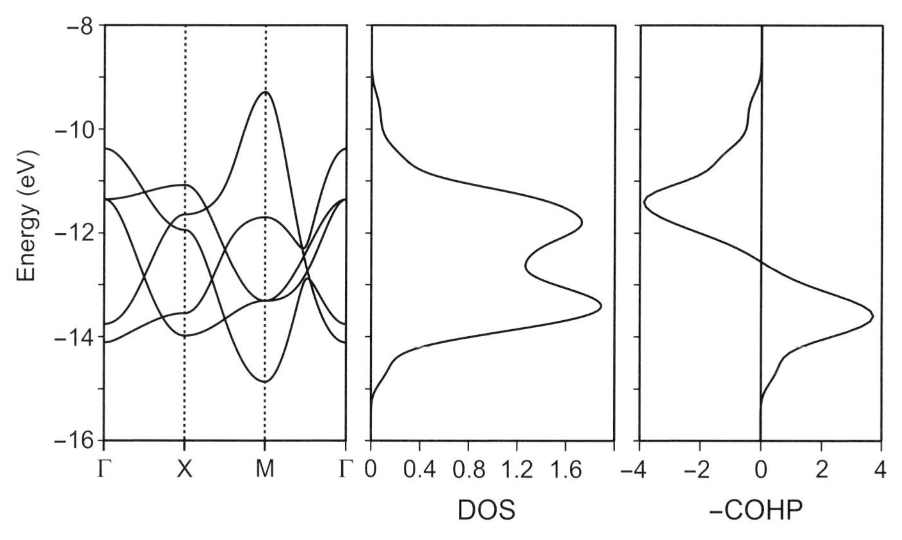

Fig. 2.22 Semiempirical band structure, density-of-states, and crystal orbital Hamilton population for a two-dimensional square lattice of iron atoms spaced at 2.2 Å.

the crossover from bonding to antibonding states, then a maximum stabilization of the two-dimensional square lattice of Fe atoms would result.

However, we need to take this result with a pinch of salt. While COHP provides a quantitative measure of bond strengths on the basis of density-functional calculations (Section 2.12), it is a partitioning of the *band-structure energy* (the sum of the energies of the Kohn–Sham orbitals), not of the total energy, in terms of orbital pair contributions. Within density-functional theory, the total energy is given by

$$E = T + E_{\mathrm{Hartree}} + E_{\mathrm{XC}} + V, \tag{2.61}$$

including the kinetic energy T of the noninteracting electrons, the Coulomb energy of the charge density (also called Hartree energy, E_{Hartree}), the electron–electron repulsion (E_{XC}) plus the electron–nuclear attraction V; this is explained in more detail in Section 2.12. The sum of all integrated COHPs (called ICOHPs) is the so-called band-structure energy or, equivalently, the sum of the Kohn–Sham (one-electron) eigenvalues,

$$\int^{\varepsilon_{\mathrm{F}}} \sum_{A} \sum_{\substack{\mu \\ \mu \in A}} \sum_{B} \sum_{\substack{\nu \\ \nu \in B}} \mathrm{COHP}_{\mu\nu}(E)\, dE = \sum_{A} \sum_{\substack{\mu \\ \mu \in A}} \sum_{B} \sum_{\substack{\nu \\ \nu \in B}} \mathrm{ICOHP}_{\mu\nu} = \sum_{i} \varepsilon_i, \tag{2.62}$$

including diagonal and off-diagonal ICOHP elements. Expanding the one-electron eigenvalues ε_i and rearranging Equation (2.62), we find

$$E = \sum_{A} \sum_{\substack{\mu \\ \mu \in A}} \sum_{B} \sum_{\substack{\nu \\ \nu \in B}} \mathrm{ICOHP}_{\mu\nu} - E_{\mathrm{Hartree}} + E_{\mathrm{XC}} - \int v_{\mathrm{XC}}(\boldsymbol{r})\rho(\boldsymbol{r})d\boldsymbol{r}, \tag{2.63}$$

including the so-called exchange–correlation potential (last entry). Thus, the ICOHP values contain contributions from all terms of the density-functional energy, and there is a direct link between the ICOHPs and the total energy of the system but, because of electron–electron interactions, not a simple one-to-one correspondence.[15]

Fortunately, this issue has so far not appeared as a limitation in practical applications of COHP, and the reason is simple. The successes of one-electron theories in quantum chemistry (see Section 2.11.1) is due in no small part to the intimate connection between the sum of one-electron eigenvalues (COHP terms) and the total energy, and this connection is one of the pillars of qualitative molecular orbital theory. Experience also reveals that the most important chemical information contained in the COHP (and also COOP) analyses results from the *shape* of these bonding indicators.

15) In addition, for solids the choice of the boundary conditions for the electrostatic potential is somewhat arbitrary and, because of this, the comparison of COHP integrals resulting from chemically strongly differing solid-state materials may be problematic. Within one particular crystal structure, however, all COHP values are well defined.

2.9 Exchange and Correlation

Let us now revisit the Hamilton operator for a many-electron system as given in Equation (2.5). It contains the kinetic energies of the electrons, their Coulomb potentials in the fields of the nuclei and, also, the electron–electron interactions which we have not further specified yet. Fortunately, the success of one-electron theories in chemistry is due to the fact that an *explicit* consideration of the electron–electron interactions may often be ignored for a qualitatively appropriate description, simply because these interactions are *implicitly* contained in some parameters; for example, one might want to recall that the spatial extent of the atomic orbitals (see Section 2.2) depends on the amount of interelectronic screening. Nonetheless, the explicit inclusion of electron–electron interactions is needed for *accurate* calculations and, just as important, also for understanding. Here is a qualitative description.

Electron–electron interactions fall into two classes, depending on the "type" of electron that is involved. We all know that each electron carries the same (elementary) charge $-e$ and is further characterized by its nonclassical angular momentum, the so-called *spin*, which can be thought of, in a classical picture, as indicating the electron's rotation around its own axis, say, clockwise or anticlockwise.[16] Thus, electrons come either as "spin-up" ($\uparrow$, or just α) or as "spin-down" ($\downarrow$, β) electrons.

The interactions between the electrons are most easily illustrated by a simple example in which we fill up degenerate atomic (or molecular) levels with more than one electron. Historically, the famous rules by Hund [97] summarize how one has to proceed in such a case. In short, whenever degenerate levels are involved, electrons occupy the degenerate orbitals not in a pairwise but in a single manner and, by doing so, they also try to maximize the total spin, that is, the sum of all individual spins.[17] For example, if we assume a set of triply degenerate $2p$ orbitals ($2p_x$, $2p_y$, $2p_z$), two electrons will occupy these $2p$ orbitals as shown in Scheme 2.7 on the next page.

16) This classical idea is *highly* questionable; spin is a function of the wave function alone, and the particle possesses only a position in space [96].

17) It is common practice to explain Hund's rules by referring to the relative energetic positioning of atomic (or molecular) *states*, not (one-electron) orbitals, and because the states represent the many-electron wave functions, they already incorporate the electron–electron interactions which they shall help to explain. Our explanation, however, starts with the much simpler one-electron picture and then adds electronic correlation in the first place and electronic exchange in the second.

p_x p_y p_z

(a)

(p_x p_y p_z)

false

(b)

Scheme 2.7

Sketch (a) shows the correct filling, and the incorrect one is given by the smaller grey sketch (b). Here we have abstained, for reasons of simplicity, from distinguishing between a spin-up (α) and a spin-down (β) electron.

Now, explicitly taking the spin into account, the above filling (a) is more correctly iconized as in Scheme 2.8:

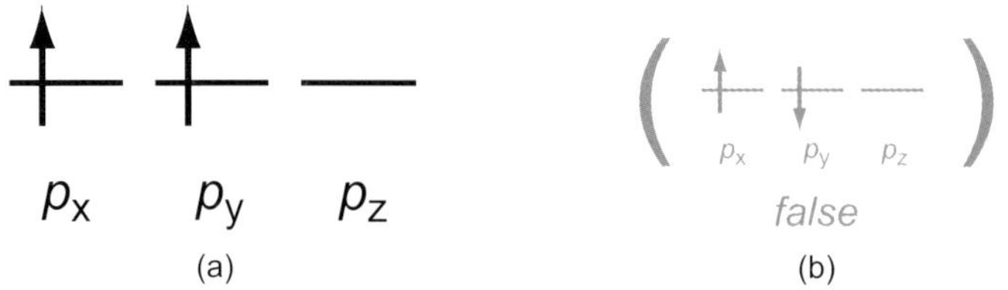

p_x p_y p_z

(a)

(p_x p_y p_z)

false

(b)

Scheme 2.8

Again, sketch (a) is correct while (b) is not. These fillings are what Hund's rules predict, but why? Simply speaking, the reason for the two electrons going into different orbitals, as sketched in Scheme 2.7(a), goes back to the interelectronic repulsion because of their *like charges*. No matter what the spins of the two electrons, their mutual Coulomb repulsion will put them as far apart from each other as possible. For the above set of three $2p$ orbitals, one electron might go into a $2p_x$ and the other one into a $2p_y$ orbital; thus, the two electrons occupy different regions in space because the two orbitals point in different directions, and the Coulomb repulsion is reduced. Note that this electrostatic repulsion, regardless of the electronic spins, and even if there were no spins at all, has been dubbed electronic *correlation* (or, simply, C) in quantum-theoretical communities. Electronic correlation is easy to understand but *very* hard to calculate for real systems. More precisely, quantum chemists distinguish between *dynamical* electron correlation – due to the dynamical character of the correlated motion of the repelling electrons – and *nondynamical* electron correlation; the latter means that different determinants making up the many-electron wave function (see Section 2.11.3) possess similar weights because they are (almost) degenerate in the highest occupied orbitals.

As can be expected, the reason for the spin maximization (Scheme 2.8(a)) goes back to the spins themselves, and it is a little more difficult to explain, although more straightforward to calculate. The laws of quantum mechanics state that a fermionic wave function, that is, a wave function for electrons or other half-spin particles, is antisymmetric with respect to the mutual exchange of its constituents (electrons); we will also touch upon this in detail in Section 2.11.3 using Hartree–Fock theory but mention that this particular electron–electron interaction goes by the name *exchange* (or, simply, X) in the literature. Alternatively, we might say that the Pauli principle [98] requires that two electrons have to differ by at least one quantum number. For illustration, Figure 2.23 shows the pair density for a combination of two spin-up (α–α) electrons and that of a spin-up/spin-down (α–β) combination. The result is derived from the noninteracting electron gas but may be extended to real systems.

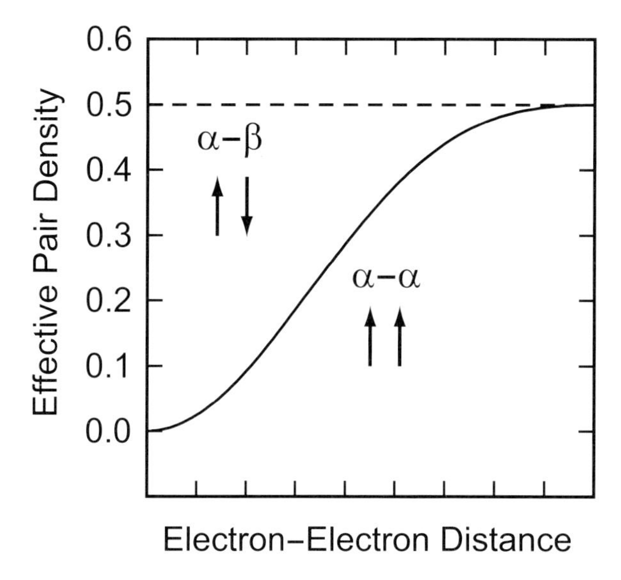

Fig. 2.23 Schematic drawing of the electronic pair density.

The figure shows that, because two electrons with the same magnetic spin quantum number are forbidden from occupying the same position due to Pauli's principle, there is a repulsion between like-spin electrons known as the exchange repulsion, generating the so-called exchange (or Fermi) hole for the α–α (and also for the β–β) pair density; this effect arises entirely from the antisymmetry of the wave function (see Section 2.11.3), and no such hole exists in the α–β curve.[18]

There are important consequences to this effect. Because of the exchange hole, like-spin electrons do not shield each other from the nucleus as well as electrons with different spins, i.e., α–α shielding is less effective than α–β

18) Here we have deliberately ignored the Coulomb hole, between all the electrons, which is due to electronic correlation (see above).

shielding. Given a system with an unequal number of α and β spins, there will also be a differential shielding of α and β electrons. If we assume, for simplicity, that the α electrons are the majority spin species (such as in the above example with two electrons going into the degenerate $2p_{x,y}$ orbitals as spin-up electrons), then they will be *less* shielded from the nucleus than any β electrons (of which there are none in the above example). Consequently the α electrons experience a higher effective nuclear charge than if they would appear as β electrons. Generally, majority spins experience a higher effective nuclear charge and will therefore move down in energy; their associated spin orbitals are more contracted.

Thus, the two electrons enter the above set of $2p$ orbitals with the same spin, thereby preferring the high-spin state, because of an *increase in electron–nuclear attraction*, as so clearly pointed out by Boyd [99]. In the triplet state (spin multiplicity $= 2S + 1$ with $S = 2 \times \frac{1}{2}$), the electron–nuclear attraction is stronger because the two electrons experience a higher effective nuclear charge. If the two electrons would appear as α *and* β electrons, building up a nonmagnetic system, the total electron–nuclear attraction would be smaller due to the better α–β shielding. Consequently (and surprisingly only at first sight), the electron–electron repulsion is *larger* for the triplet state since the electrons are more spatially contracted, closer to each other on average! A recent high-quality numerical study of the carbon atom's singlet and triplet states by means of the quantum Monte Carlo method (see Section 2.13) im-

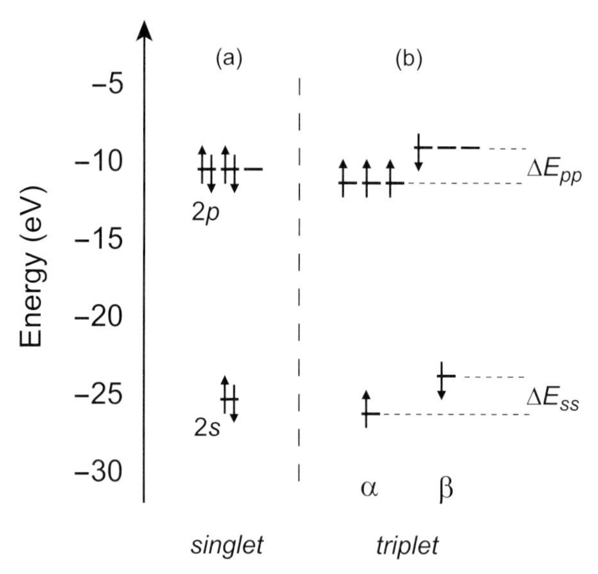

Fig. 2.24 Energy levels of atomic oxygen in a singlet excited state (a) and the triplet ground state (b).

pressively supports this interpretation [100]. Exactly the same argument also explains the relative stability of half-filled electron shells which is often mentioned in elementary textbooks on general chemistry. For half-filled shells, there is zero α–β shielding and, therefore, a maximum of electron–nuclear attraction.

For illustration, Figure 2.24 shows the changes in energy levels of atomic oxygen upon moving from a singlet to a triplet state according to density-functional (Section 2.12) calculations. On unpairing the spins – the so-called spin-polarization – the majority (α) electrons become less shielded and drop in energy; the β spins experience a higher shielding and move up in energy.

In addition, the associated *spin orbitals* (with a maximum occupation number of *one* electron) acquire different spatial extents, and they are drawn in Figure 2.25. At first sight, it is difficult to recognize that the low-lying α spin orbitals (a) are more spatially contracted than the β spin orbitals (b) but an orbital *difference* plot (c) immediately shows that the α spins dominate in the region close to the nucleus, where they experience a higher nuclear potential. The analysis may also be carried out numerically, but we will not do so here.

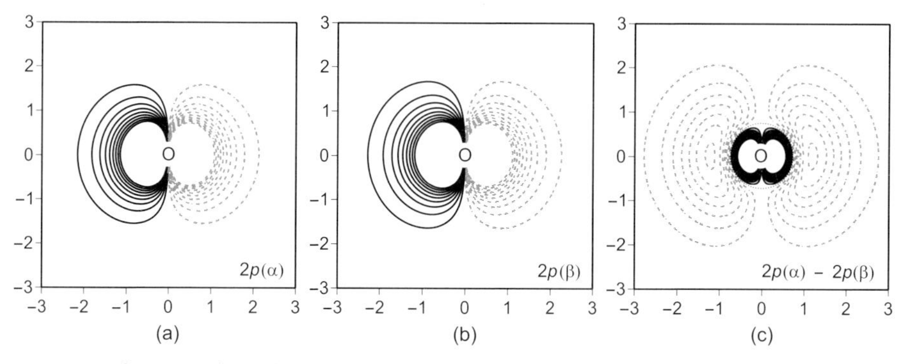

Fig. 2.25 Contour plots of the $2p$ spin orbitals ((a): α, (b): β) of triplet atomic oxygen: to distinguish these more clearly, (c) shows an AO difference plot. Positive and negative values have been plotted with solid black and dashed grey lines, respectively. All distances are in Å.

On moving down the periodic table, that is, for sulfur and selenium, there is still a preference for the triplet state, but the exchange splittings decrease by more than 50 percent; also, the changes in the spatial extents of the orbitals are now much smaller, as seen in Figure 2.26. The decrease in exchange splittings and the loss of spatial differentiation between the α/β spin orbitals is large when moving from O to Se but much smaller when moving from S to Se. The reason is found in the relatively contracted $2p$ orbitals of oxygen because there is no similarly shaped, filled orbital lying lower in energy [101]; remember that there is no $1p$ orbital! Because of this, the shielding of the $2p$ electrons is low, and they will experience a high effective nuclear charge. Therefore,

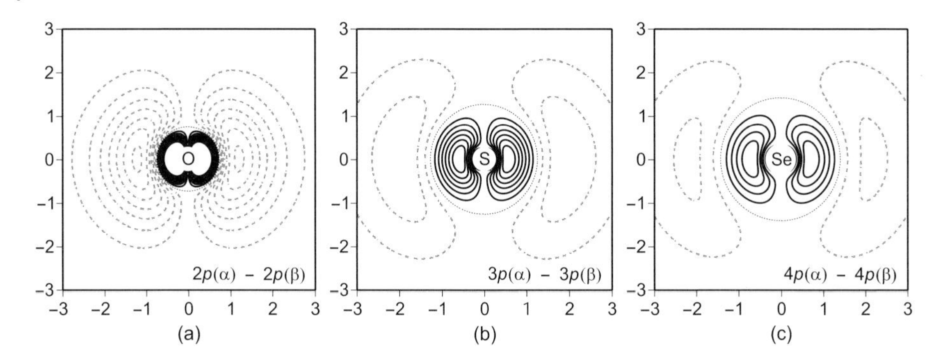

Fig. 2.26 AO difference plots for the p orbitals of triplet O (a), S (b), and Se (c).

the oxygen $2p$ electrons are closer to the nucleus such that small changes in their screening, for example, those caused by spin-polarization, have a larger impact than they do for the $3p$ (S) or $4p$ (Se) electrons. We note that the relatively contracted $2p$ orbitals are found for the entire first long period, and this effect is also responsible for the small sizes of these elements, their high electronegativities and the ease with which they can mix $2s$ and $2p$ orbitals, thereby allowing element–element multiple bonding [102].

The same arguments also apply to molecules with unpaired electrons, and we offer Figure 2.27 for illustration, comparing the π_g^* levels (π^* in short) of the diatomic molecules O_2, S_2, and Se_2 as MO difference plot. As expected, the exchange splitting decreases on moving down the group.

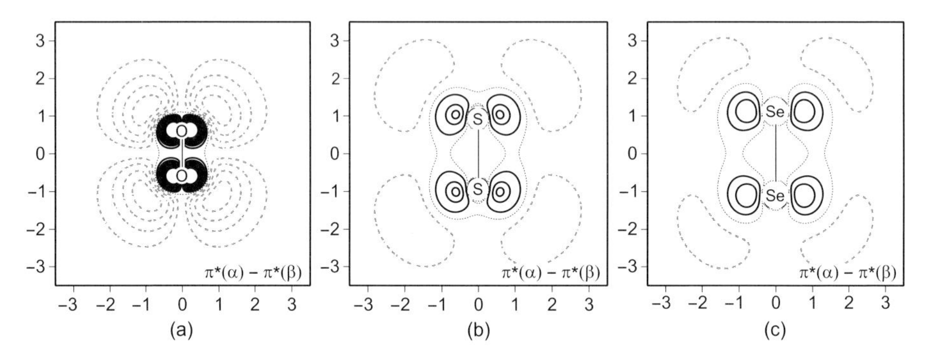

Fig. 2.27 MO difference plots for the π^* orbitals of triplet O_2 (a), S_2 (b), and Se_2 (c).

We also add that electronic exchange and the differences in electronic shielding due to spin-polarization easily explains why triplet oxygen ($^3O_2$) is more strongly bonded than singlet oxygen ($^1O_2$), an example of bond strengthening on *unpairing* spins (see Section 2.3). In the triplet state, the majority α electrons feel a stronger nuclear potential which leads to a lower total energy for the molecule such that more energy is needed to split O_2 into two O atoms. Likewise, spin-polarization increases the splitting between the highest occupied and lowest unoccupied α molecular orbitals, HOMO$^\alpha$ and LUMO$^\alpha$, and this may be interpreted in terms of the electronic structure becoming "harder" in the context of chemical hardness (see Section 2.14). According to chemical experience, stable chemical species are also chemically "harder", and $^3O_2$ is also "harder" than $^1O_2$.

Summarizing, both kinds of electron–electron interactions (exchange and correlation, X and C) have to be accounted for in order to yield a satisfying description of atoms, molecules, and solids, at least in principle. Practical experience has shown that different approaches have been followed to deal with these effects, and molecules and solids have also been treated differently. The physical nature of a solid (metal, semiconductor, insulator) *also* suggests different strategies for the proper inclusion of exchange and correlation.

2.10 Electron Localization

Before proceeding to the technical details of how we accurately (or approximately) calculate the electron–electron interactions, let us re-visit these phenomena under the aspect of chemical bonding. We have already shown how to quantify the latter, starting from quantum-chemical calculations for molecules (by use of overlap populations) and extended structures (by use of energy-resolved overlap and Hamilton populations) within Section 2.8, and the necessity to use these tools stems from the fact that both molecular-orbital and band calculations yield essentially *delocalized* representations of the electronic structures. From these, bonding characteristics in real space can be visualized – indirectly so – by first specifying a particular bond and then analyzing it using some kind of partitioning technique. That works well but, because of its indirect nature, is not favored by some people.

Driven by the desire to generate a *real-space* visualizing tool for chemical bonding, alternate real-space analyses of molecular electron densities (an observable quantity) were proposed many years ago. Unfortunately, observed (or accurately calculated) electron densities are extremely close to a superposition of atomic densities, so that the recognition of bonding electron (pair) densities is virtually impossible for the naked eye; thus, smart mathematical tricks must be introduced. The most popular approach [89] decomposes the electron density into atomic shells and electron pairs by use of the *topography* of its Laplacian $\nabla^2\rho(\boldsymbol{r})$. Despite some arbitrariness and a few remaining difficulties,[19] Bader's method works nicely for light atoms and related molecules;

19) For example, the complete set of atomic shells cannot be resolved for very heavy atoms. In addition, *any* such decomposition is somewhat arbitrary (see Section 2.7).

dependent on the particular point of view, the practice of not using molecular orbitals for understanding may either be considered fortunate or unfortunate.

Why not go another step forward and try to decorate quantum-chemically derived atomic shells and electron pairs in terms of their bonding characteristics? If that is possible, it would also have little psychological advantage, since all freshmen chemists are educated (or conditioned, as some would say) to recognize "bonding" and "nonbonding" – but certainly not antibonding! – electron pairs when drawing molecular Lewis formulae. Remember that, given the right set of magic rules for the "repulsion" of pairs of electrons, even molecular geometries are predictable for (main-group) molecules, and the simple recipe therefore goes under the name of the *valence-shell electron-pair repulsion* (VSEPR) concept contained in every introductory textbook of general chemistry.

Indeed, this route has been followed, and the tool carries the name *electron localization function* (ELF). Basically, the intention of ELF is to yield a real-space partitioning of electron densities into regions which correspond to those of the electron pairs known from the VSEPR model. Despite ELF's impressive mathematical apparatus, the underlying idea is quite simple. According to the original formulation [103] made within Hartree–Fock theory (see Section 2.11.3), the electron localization is intimately connected with the *pair probability* of like spins, $P^{\uparrow\uparrow}(\boldsymbol{r}, \boldsymbol{r}')$. As has been said in the preceding section, the spin-dependent chance of finding a pair of such like spins ($\uparrow\uparrow$ or $\downarrow\downarrow$) – nonzero at large interelectronic distances r but zero for $r = 0$ (the so-called Fermi hole) – is an exclusive function of the Pauli *exchange* repulsion. Let us therefore ask for the pair probability of finding a spin-up electron ($\uparrow$) at some position $\boldsymbol{r}'$ and with reference to a likewise spin-up electron at the initial position $\boldsymbol{r}$ provided that both spins are contained inside a sphere of radius r_{S}. This has been sketched in Figure 2.28.

The evaluation of $P^{\uparrow\uparrow}(\boldsymbol{r}, r_{\mathrm{S}})$ using Hartree–Fock theory (full inclusion of exchange interactions, see Section 2.11.3) and by spherically averaging up to r_{S} can be written as a Taylor series in r_{S}, and the first nonzero entry of the pair probability of like spins, which is quadratic in r_{S}^2, reads

$$D(\boldsymbol{r}) = \sum_i |\nabla \psi_i(\boldsymbol{r})|^2 - \frac{1}{4}\frac{(\nabla \rho(\boldsymbol{r}))^2}{\rho(\boldsymbol{r})}, \tag{2.64}$$

the sum running over all occupied molecular orbitals i. Within Equation (2.64), the first entry represents the kinetic-energy density whereas the second contains the electron density and its gradient. Its importance was recognized a little earlier than the advent of ELF [104].

The crucial *interpretative* step now consists in assuming that the *lower* the pair probability of finding a second like-spin in the proximity of the left reference spin-up electron (this is what $D(\boldsymbol{r})$ measures inside the grey region in

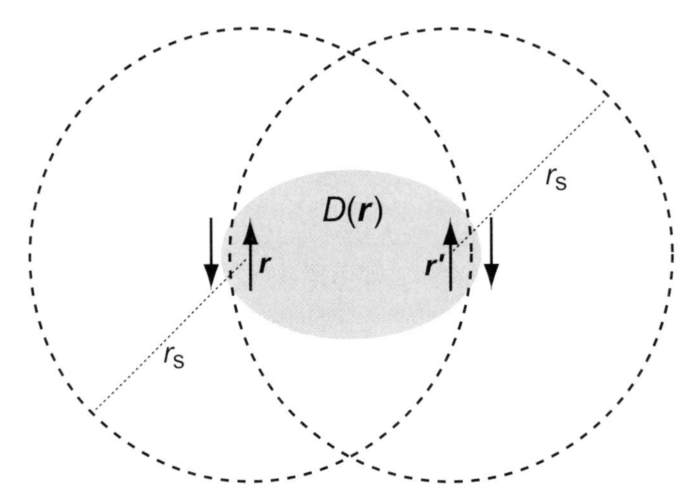

Fig. 2.28 Schematic drawing of the probability density of pairs of like spins and its relation to the localization of electrons and pairs of electrons; see text.

Figure 2.28), the *more localized* is the left (or right) spin-up electron being "spin-paired" to its neighboring spin-down electron, and vice versa. Because of this qualitatively reciprocal relationship, the electron localization function may be defined as

$$\mathrm{ELF}(\boldsymbol{r}) = \frac{1}{1 + \left(\frac{D(\boldsymbol{r})}{D_{\mathrm{h}}(\boldsymbol{r})}\right)^2}, \tag{2.65}$$

where $D_{\mathrm{h}}(\boldsymbol{r}) = \frac{3}{5}(6\pi^2)^{2/3}\rho(\boldsymbol{r})^{5/3}$ serves as the numerical reference value for the homogeneous electron gas. This definition makes ELF a *relative* measure and it also ensures that ELF is numerically bound between 0 (fully delocalized electron) and 1 (fully localized); other definitions might also do so. Within density-functional theory (see Section 2.12), reference to the pair density is avoided, and the ELF definition (Equation (2.66)) is very similar

$$\mathrm{ELF}(\boldsymbol{r}) = \frac{1}{1 + \left(\frac{t(\boldsymbol{r})}{t_{\mathrm{h}}(\boldsymbol{r})}\right)^2}, \tag{2.66}$$

with $t(\boldsymbol{r})$ as the Pauli kinetic-energy density of a closed-shell system, namely

$$t(\boldsymbol{r}) = \frac{1}{2}\sum_i |\nabla\psi_i(\boldsymbol{r})|^2 - \frac{1}{8}\frac{(\nabla\rho(\boldsymbol{r}))^2}{\rho(\boldsymbol{r})}, \tag{2.67}$$

and the reference Pauli kinetic-energy density, $t_{\mathrm{h}}(\boldsymbol{r}) = \frac{3}{10}(3\pi^2)^{2/3}\rho(\boldsymbol{r})^{5/3}$, derived from the homogeneous electron gas [105]. Further elaborations have been proposed.

A plot of a calculated (scalar) ELF for the potassium atom as a function of the distance from the nucleus is given in Figure 2.29. By its radial oscillation, the ELF nicely reflects the atomic shell structure, and the spatial regions belonging to the main quantum numbers $n = 1, 2, 3$ and 4 are easily recognizable; adopting the typical ELF jargon, ELF maxima ("basins") are separated by ELF minima ("separatrices"). Within the maxima, considering the electrons to become more "localized" (that is, spatially restricted) makes immediate sense, and numerical integration of the number of electrons contained in those shells produces reasonable values [106]. Nonetheless, the interpretation of the electrons being mostly "localized" at the origin and *outside* the potassium atom is rather tricky.

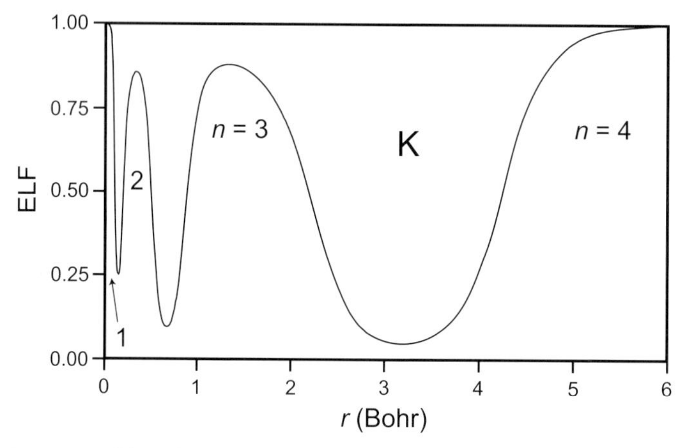

Fig. 2.29 The electron localization function of the potassium atom, adapted from Ref. [106].

When related calculations are carried out for molecules (and also solids) [107, 108], one either plots color-coded two-dimensional ELF($\mathbf{r}$) slices to represent the local ELF values, or one draws three-dimensional isosurfaces referring to a chosen ELF value, usually around 0.8 or close to it. With the advent of powerful computer graphics, such representations produce aesthetically compelling color figures from which ELF "attractors" (bonded and nonbonded electron pairs as well as atomic shells) show up and reflect the underlying molecular symmetry. All-electron calculations are generally preferable for this purpose [106].

And yet there are interpretational difficulties. It is easily tolerable that ELF is a nonobservable quantity (no problem for chemists)[20] but, despite a diametrically different treatment of electronic exchange, extended Hückel-derived

20) It is certainly possible to calculate an ELF from an experimental charge density but the definition for the pair probability as given in Equation (2.64) would require the further theoretical generation of numerical orbitals ψ_i [106].

(see Section 2.11.1) molecular ELFs are qualitatively *identical* with Hartree–Fock-based ELFs [109] which is difficult to understand, at first sight. The clue to this bizarre finding, together with the answer to the question whether [109] or not [108, 110] ELF *really* measures the Pauli repulsion, has already been given by the conclusive work of Burdett and McCormick [111]. Their perceptive numerical analysis shows that the ELF behavior is mostly dominated by the kinetic-energy density (first entry in Equation (2.64)) and, therefore, by the *nodal properties* of the system's occupied orbitals. High ELF values occur for points in space where there is significant electron density but no (or few) nodes passing through. Low ELF values are found when there is little electron density or when there are enough nodes from the occupied orbitals passing through, such that the nodal contributions (first entry in Equation (2.64)) overcome the density-based (second entry) ones to the ELF.

Another ELF difficulty does not go back to the function itself but to an over-interpretation sometimes found when ELF is utilized for a chemical bonding analysis. In particular, there is an oversimplifying tendency to rephrase the universe of chemical bonding *entirely* into effects of electron pairing, and this is clearly wrong. To give two examples, this notion already appears problematic for, say, hydrogen-bridging bonds or the bonding in metals with (mostly) delocalized electrons although both kinds of chemical bonds lower the total energies of these systems; otherwise, we would not call them bonds. But even for covalent bonds, chemical bonding does not necessarily go back to spin-paired electrons. While it is true that a covalent single bond (e.g., H_3C–CH_3) contains two such spin-paired localized electrons, the triplet oxygen molecule covered in Section 2.9 stabilizes because two π electrons *unpair* and gain potential energy by an enlarged effective nuclear charge, and the latter effect is due to the less effective shielding between these spin-up electrons; thinking in terms of electron pairing to rationalize this and other systems does not capture the essence of chemical bond formation, because the presence/absence of electron pairs does not necessarily (see also Section 2.3) infer strengthened/weakened chemical bonding; an energy criterion, however, will give the correct answer. The latter does not appear accidentally, because the strength of chemical bonding may be *defined* in terms of energy.

2.11
How to Deal with Exchange and Correlation

Not too surprisingly, the quality of the Hamiltonian chosen for the analysis of a quantum-chemical problem depends on the effort that is used for dealing with the electron–electron interactions. There are also other important criteria (for example, the shape of the electronic potential, later covered in

Section 2.15) but exchange and correlation typically serve as to characterize the worth of the computational approach; the (sometimes questionable) distinction between (semi-)empirical and *ab initio* methods also goes back – to some extent – to this question. Let us group the different approaches, a bit simplified, and see how they relate with each other.

2.11.1 Ignoring it or Pretending to do so

This strategy, being rather radical in spirit, simply sets the last entry of Equation (2.5) to *zero* and therefore has the ideal consequence that each of the N electrons moves completely independently of each other; thus, the Hamiltonian nicely simplifies to a sum of N one-electron Hamiltonians $\hat{h}_i$ as in

$$\mathcal{H} = \sum_{i=1}^{N} \hat{h}_i = -\sum_{i=1}^{N} \left(\frac{1}{2}\nabla_i^2 - \sum_{A=1}^{M} \frac{Z_A}{r_{iA}} \right) . \tag{2.68}$$

This will lead to a set of N orbitals ψ_i, and the total wave function Ψ is then constructed as a simple product of these one-electron wave functions.

Although the simplification is rather drastic, it may still yield a qualitatively instructive picture; for the solid state, the famous Sommerfeld model of the free electrons [53] with its many useful conclusions belongs to this class of theories. Quantitatively, however, a Hamilton operator as given in Equation (2.68) does not lead anywhere because it is far too primitive for chemical questions, which can be easily illustrated for, say, the caesium atom. The above $\hat{h}_i$ obviously means that the 6*s* valence electron moves in a nuclear potential generated by 55 protons but this very electron does *not* sense the remaining 54 core-like electrons which (should) lie deeper in energy, which must be quantitatively incorrect.

In order to achieve a more reasonable $\hat{h}_i$, one might want to smuggle electron–electron interactions into the problem, such that the N electrons are effectively – but not mathematically! – *dependent* on each other, to be accomplished by adding an external *one-electron* potential v_i to $\hat{h}_i$ or just by modifying Z_A to become an *effective* nuclear charge.[21] An alternative strategy has been followed in semiempirical quantum chemistry. We first agree in that we are only interested in the exclusive description of the important valence orbitals. Second, we reconsider the secular determinant (Equation (2.10)) which suggests that we derive meaningful entries for the interaction ($H_{\mu\nu}$) and overlap ($S_{\mu\nu}$) integrals, not necessarily by the explicit solution of $\hat{h}_i$ from Equation (2.68), but through another parametrization.

The most influential parametrization is probably given by extended Hückel (EH) theory [54], a very useful semiempirical method which has found wide-

21) A general approach using the same idea is covered in Section 2.15.2.

spread use in theoretical organic, organometallic, inorganic and also solid-state chemistry. Here, the above overlap matrix elements $S_{\mu\nu}$ are calculated for the given geometry using Slater-type orbitals which, as we know, are excellent approximations for the true atomic valence functions. The computational effort when using STOs is negligible since the total number of integrals which must be calculated is not too large; only the valence orbitals are dealt with. Physicists should note that extended Hückel theory is *not* similar to the simplistic Hückel theory (see Section 2.1) because the EH Hamiltonian is set up as follows:

To start with, the diagonal (on-site) matrix elements $H_{\mu\mu}$ are taken as the experimental ionization potentials derived from spectroscopic data of the valence levels. The nondiagonal (off-site) interaction elements $H_{\mu\nu}$, however, are approximated as some arithmetic average of the $H_{\mu\mu}$ values of the interacting atomic orbitals according to the Wolfsberg–Helmholz formula [112]:

$$H_{\mu\nu} = k\frac{H_{\mu\mu} + H_{\nu\nu}}{2}S_{\mu\nu} \tag{2.69}$$

Thus, the strength of the interaction depends on the overlap S, and the other proportionality constant k – which has nothing to do with reciprocal space! – is chosen as 1.75; it may also be an adjustable parameter. There is also a weighted, improved form of the Wolfsberg–Helmholz equation that helps to suppress so-called counterintuitive orbital mixing [113], a weakness of EH theory and also other minimum basis-set methods.

Extended Hückel theory is well-known for its flexibility and the ease with which its results lend themselves to chemical interpretation, no matter whether the objects of study are molecules or solids, and a plethora of chemical (bonding) information has been deduced from it, at least qualitatively, and in many cases semiquantitatively. EH theory, however, is not at all made for total-energy calculations or structure optimizations because the electronic potential is too primitive to yield acceptable molecular geometries; the latter weakness may be cured, though [114].

When applied to solid-state materials, EH theory is conceptionally equivalent to the *empirical tight-binding method* (including overlap) of the solid-state physics community (see also Section 2.15.1). The method can be further simplified by setting all overlap integrals to zero (except $S_{\mu\mu}$ which is kept at unity) and also by dropping the $H_{\mu\nu}$ integrals of all second-nearest (σ-bonded) neighbors; *then*, it turns into the Hückel method.

Coming back to the question of electron–electron interactions, exchange and correlation should not be contained in the EH approach. *Formally*, they are not, and the technical and interpretational simplicity of the method is due to the electrons being mathematically independent. Therefore, the EH total energy is simply the sum of all one-electron eigenvalues ε_i for n orbitals,

$$E = \sum_{i=1}^{n} f_i \varepsilon_i, \qquad (2.70)$$

with occupation numbers f_i (= 0, 1, or 2); note that this formula only holds for the case of independent electrons. In reality, however, the use of parametrized integrals, the values of which originate from *experimental* data, *implicitly* introduces exchange and correlation into the calculation, thereby increasing the applicability of EH theory to the real world; still, Equation (2.70) is valid because the mathematical structure is that of an independent-electron theory.

Instead of relying on experimental data for the ionization potentials, the essential EH energy ($H_{\mu\mu}$) and orbital contraction (ζ) parameters can also be deduced from theoretical calculations [115,116]. Recently, a complete set of EH parameters has been derived from atomic Hartree–Fock–Slater calculations (an early form of density-functional theory, see Section 2.12) which were also adjusted to fit some experimental data. The parameter set thus derived [117] includes exchange, some correlation, and also the influences of relativity;[22] for convenience, we include these data in Table 2.1. These parameters may be used to study the trends in the periodic table and, also, to perform simple calculations. Other sets of EH parameters, from very different sources, are also available. These then typically include better basis sets (such as double-ζ parameters for d orbitals) although they are less self-consistent for the whole periodic table.

22) Effects of (special) relativity occur for heavy elements due to the very high velocity of the electrons in close proximity to the nuclei. Because of the resulting *mass increase* of these electrons, there are direct relativistic effects, such as the radial stabilization/contraction of the s levels, and indirect ones, such as the radial destabilization/expansion of the p and d levels; also, spin-orbit coupling is due to relativity. Relativistic effects are known to be strongest for the element gold, and they are also responsible for the increased stability of the low oxidation states of the heavier main-group elements [118].

Tab. 2.1 Extended Hückel energy parameters $H_{\mu\mu}$ (eV) and orbitals exponents ζ for all elements of the periodic table [117] based on a statistically optimized k proportionality constant of 1.3861; note that ζ(H) is unity instead of the often-used 1.30 value.

		s levels			*p* levels			*d* levels			*f* levels		
Atom	Z	n	$H_{\mu\mu}$	ζ	n	$H_{\mu\mu}$	ζ	n	$H_{\mu\mu}$	ζ	n	$H_{\mu\mu}$	ζ
H	1	1	−13.60	1.000									
He	2	1	−24.59	1.647									
Li	3	2	−5.39	0.653	2	−3.54	0.531						
Be	4	2	−9.32	1.037	2	−5.60	0.899						
B	5	2	−14.69	1.399	2	−8.30	1.269						
C	6	2	−20.34	1.721	2	−11.26	1.611						
N	7	2	−26.69	2.035	2	−14.53	1.940						
O	8	2	−28.54	2.240	2	−13.62	2.048						
F	9	2	−36.37	2.564	2	−17.42	2.402						
Ne	10	2	−44.92	2.881	2	−21.57	2.742						
Na	11	3	−5.14	0.868	3	−3.09	0.615						
Mg	12	3	−7.65	1.194	3	−4.28	0.881						
Al	13	3	−11.84	1.514	3	−5.99	1.166						
Si	14	3	−17.74	1.758	3	−8.15	1.434						
P	15	3	−19.80	1.986	3	−10.49	1.676						
S	16	3	−20.94	2.136	3	−10.36	1.772						
Cl	17	3	−25.45	2.362	3	−12.97	2.018						
Ar	18	3	−30.17	2.580	3	−15.76	2.250						
K	19	4	−4.34	0.936	4	−2.79	0.691						
Ca	20	4	−6.11	1.211	4	−3.74	0.933						
Sc	21	4	−6.56	1.287	4	−3.91	0.983	3	−9.45	2.434			
Ti	22	4	−6.83	1.342	4	−3.97	1.010	3	−10.50	2.644			
V	23	4	−6.75	1.357	4	−3.84	0.995	3	−10.37	2.781			
Cr	24	4	−6.77	1.380	4	−3.74	0.978	3	−10.64	2.978			
Mn	25	4	−7.43	1.476	4	−4.06	1.064	3	−13.02	3.221			
Fe	26	4	−7.90	1.547	4	−4.23	1.111	3	−14.81	3.454			
Co	27	4	−7.88	1.565	4	−4.14	1.100	3	−14.82	3.602			
Ni	28	4	−7.64	1.553	4	−3.96	1.059	3	−13.82	3.702			
Cu	29	4	−7.73	1.579	4	−3.91	1.053	3	−14.10	3.896			
Zn	30	4	−9.39	1.778	4	−4.66	1.245						
Ga	31	4	−13.23	2.068	4	−6.00	1.507						
Ge	32	4	−16.58	2.270	4	−7.90	1.768						
As	33	4	−19.83	2.455	4	−9.79	1.982						
Se	34	4	−20.66	2.568	4	−9.75	2.055						
Br	35	4	−24.17	2.752	4	−11.81	2.265						
Kr	36	4	−27.78	2.930	4	−14.00	2.462						
Rb	37	5	−4.18	1.096	5	−2.66	0.799						
Sr	38	5	−5.70	1.366	5	−3.49	1.042						
Y	39	5	−6.22	1.461	5	−3.70	1.110	4	−7.17	2.158			
Zr	40	5	−6.63	1.539	5	−3.87	1.165	4	−8.74	2.383			
Nb	41	5	−6.76	1.593	5	−3.82	1.174	4	−10.26	2.626			
Mo	42	5	−7.09	1.658	5	−3.95	1.219	4	−11.92	2.824			
Tc	43	5	−7.28	1.693	5	−4.04	1.249	4	−12.59	2.934			
Ru	44	5	−7.36	1.735	5	−3.98	1.251	4	−14.21	3.152			
Rh	45	5	−7.46	1.767	5	−3.99	1.262	4	−15.33	3.311			
Pd	46	5	−6.44	1.626	5	−3.52	1.122	4	−8.34	3.086			
Ag	47	5	−7.58	1.818	5	−3.97	1.272	4	−17.40	3.617			
Cd	48	5	−8.99	1.990	5	−4.69	1.460						
In	49	5	−11.95	2.465	5	−5.79	1.685						
Sn	50	5	−14.52	2.404	5	−7.34	1.913						
Sb	51	5	−16.64	2.549	5	−8.61	2.078						
Te	52	5	−17.70	2.658	5	−9.01	2.172						
I	53	5	−20.08	2.808	5	−10.45	2.339						
Xe	54	5	−22.74	2.960	5	−12.13	2.507						
Cs	55	6	−3.89	1.199	6	−2.54	0.892						

continued on next page

Tab. 2.1 (continued)

Atom	Z	*s* levels n	$H_{\mu\mu}$	ζ	*p* levels n	$H_{\mu\mu}$	ζ	*d* levels n	$H_{\mu\mu}$	ζ	*f* levels n	$H_{\mu\mu}$	ζ
Ba	56	6	−5.21	1.452	6	−3.33	1.140						
La	57	6	−5.58	1.533	6	−3.48	1.198	5	−7.83	2.274	4	−10.70	4.416
Ce	58	6	−5.54	1.538	6	−3.44	1.193	5	−7.59	2.291	4	−20.78	4.948
Pr	59	6	−5.47	1.516	6	−3.45	1.183	5	−4.94	2.056	4	−16.41	4.898
Nd	60	6	−5.53	1.532	6	−3.47	1.192	5	−4.88	2.072	4	−17.23	5.074
Pm	61	6	−5.58	1.549	6	−3.49	1.202	5	−4.81	2.086	4	−18.01	5.247
Sm	62	6	−5.64	1.565	6	−3.51	1.212	5	−4.75	2.100	4	−18.76	5.415
Eu	63	6	−5.67	1.576	6	−3.51	1.215	5	−4.62	2.098	4	−19.16	5.568
Gd	64	6	−6.15	1.670	6	−3.72	1.287	5	−7.92	2.486	4	−27.59	5.989
Tb	65	6	−5.86	1.619	6	−3.60	1.246	5	−4.62	2.138	4	−20.90	5.904
Dy	66	6	−5.94	1.636	6	−3.63	1.257	5	−4.57	2.147	4	−21.52	6.060
Ho	67	6	−6.02	1.654	6	−3.67	1.269	5	−4.52	2.157	4	−22.13	6.216
Er	68	6	−6.11	1.672	6	−3.70	1.281	5	−4.49	2.167	4	−22.76	6.370
Tm	69	6	−6.18	1.690	6	−3.74	1.293	5	−4.44	2.173	4	−23.31	6.521
Yb	70	6	−6.25	1.706	6	−3.77	1.303	5	−4.38	2.175	4	−23.76	6.669
Lu	71	6	−5.43	1.665	6	−3.37	1.217	5	−5.43	2.380			
Hf	72	6	−6.83	1.841	6	−3.91	1.382	5	−8.42	2.770			
Ta	73	6	−7.55	1.955	6	−4.26	1.486	5	−10.79	3.019			
W	74	6	−7.86	2.019	6	−4.38	1.530	5	−12.29	3.194			
Re	75	6	−7.83	2.045	6	−4.29	1.528	5	−13.05	3.324			
Os	76	6	−8.44	2.136	6	−4.58	1.610	5	−15.45	3.524			
Ir	77	6	−8.97	2.217	6	−4.83	1.681	5	−17.78	3.708			
Pt	78	6	−8.96	2.265	6	−4.69	1.676	5	−19.70	3.900			
Au	79	6	−9.23	2.319	6	−4.79	1.713	5	−21.57	4.053			
Hg	80	6	−10.44	2.431	6	−5.52	1.867						
Tl	81	6	−12.40	2.578	6	−6.11	1.990						
Pb	82	6	−14.48	2.724	6	−7.42	2.184						
Bi	83	6	−14.66	2.787	6	−7.29	2.215						
Po	84	6	−16.57	2.931	6	−8.42	2.383						
At	85	6	−19.90	3.116	6	−10.70	2.620						
Rn	86	6	−20.45	3.205	6	−10.75	2.687						
Fr	87	7	−4.07	1.416	7	−2.64	1.060						
Ra	88	7	−5.28	1.634	7	−3.41	1.301						
Ac	89	7	−5.17	1.654	7	−3.25	1.289	6	−6.97	2.374	5	−7.93	3.796
Th	90	7	−6.31	1.838	7	−3.92	1.473	6	−9.95	2.658	5	−16.69	4.361
Pa	91	7	−5.89	1.777	7	−3.70	1.412	6	−8.09	2.571	5	−20.37	4.554
U	92	7	−6.19	1.825	7	−3.88	1.459	6	−9.12	2.650	5	−23.37	4.770
Np	93	7	−6.27	1.845	7	−3.92	1.474	6	−9.24	2.694	5	−25.26	4.941
Pu	94	7	−6.03	1.798	7	−3.84	1.437	6	−6.32	2.512	5	−22.86	4.988
Am	95	7	−5.97	1.801	7	−3.78	1.432	6	−6.10	2.517	5	−23.81	5.130
Cm	96	7	−5.99	1.841	7	−3.69	1.442	6	−8.43	2.735	5	−27.87	5.348
Bk	97	7	−6.20	1.846	7	−3.91	1.470	6	−6.29	2.592	5	−27.53	5.460
Cf	98	7	−6.28	1.865	7	−3.96	1.484	6	−6.32	2.621	5	−29.22	5.614
Es	99	7	−6.42	1.889	7	−4.04	1.505	6	−6.44	2.659	5	−31.20	5.774
Fm	00	7	−6.50	1.907	7	−4.08	1.519	6	−6.46	2.685	5	−32.87	5.922
Md	101	7	−6.58	1.924	7	−4.13	1.532	6	−6.47	2.709	5	−34.53	6.069
No	102	7	−6.65	1.940	7	−4.16	1.544	6	−6.46	2.730	5	−36.13	6.213
Lr	103	7	−8.50	2.129	7	−4.90	1.720	6	−9.40	2.990			
Rf	104	7	−6.00	1.920	7	−3.50	1.450	6	−8.10	2.970			

2.11.2 The Hartree Approximation

If electron–electron interactions must be calculated explicitly, the last entry of Equation (2.5) impedes the separation of the Hamiltonian because each electron is influenced by the positions of all the others. In order to derive a set of one-electron equations, one idea that came to the mind of Hartree [119] is to think of each electron as moving in the *field* built up by all other electrons. For such a scenario, an *approximate* electron–electron interaction for a given electron only depends on the position of this particular electron *alone* because the other electrons merely constitute the electronic "sea" in which this electron propagates. That is equivalent to writing, in atomic units,

$$\sum_{i=1}^{N}\sum_{j>i}^{N}\frac{1}{r_{ij}} \approx \sum_{i=1}^{N} v_i(\boldsymbol{r}_i) = \sum_{i=1}^{N}\int \frac{\rho(\boldsymbol{r}_j) - |\psi_i(\boldsymbol{r}_j)|^2}{|\boldsymbol{r}_i - \boldsymbol{r}_j|} d\boldsymbol{r}_j, \tag{2.71}$$

yielding a set of "effective" one-electron potentials v_i and orbitals ψ_i for each electron i. Note that, first, each electron only sees the electron density ρ generated by the other $N-1$ electrons (because its own contribution has been subtracted) and, second, the potential is *not exactly known at the beginning* because it is *constructed* from the set of those orbitals ψ_i which one wishes to determine. Nonetheless, if one is lucky enough to find these wave functions, it is still allowed to express the total wave function Ψ as the Hartree product of the one-electron wave functions ψ_i, i.e.,

$$\Psi = \psi_1(\boldsymbol{r}_1)\psi_2(\boldsymbol{r}_2)\cdots\psi_N(\boldsymbol{r}_N). \tag{2.72}$$

The important conceptional advantage of the Hartree approximation is that one may think of statistically independent electrons as occupying different orbitals with integer occupations numbers f_i between 0 and 2; the electrons are *not* independent, though! Because the calculation of any ψ_i depends on the electron–electron interaction which *itself* is defined in terms of ψ_i (see right-hand side of Equation (2.71)), the solution of the Hartree method must be sought for *iteratively*. Starting with a, hopefully, good guess for the orbitals (for example, those of independent electrons), a new potential (including electron–electron interaction) *develops* upon calculation, thereby generating a set of new orbitals; the process is repeated until input and output orbitals ψ_i and ψ_i' no longer differ. This is the *self-consistency cycle* which has to be paid for by including electron–electron interactions but, still, mathematically treating the electrons almost independently.

For completeness, we will just mention that the extended Hückel method *also* allows one to iterate towards self-consistency, namely if the Coulomb integrals $H_{\mu\mu}$ are allowed to better reflect differing atomic potentials (chemists should think of different oxidation states). Then, these integrals may vary according to

$$H_{\mu\mu} = Aq^2 + Bq + C, \tag{2.73}$$

using empirically determined parameters (A, B, C) upon minimization of all atomic charges q.

The Hartree method is a conceptionally important step for things to come, especially concerning density-functional theory (see Section 2.12), simply because of the idea of self-consistency. The systematic inclusion of the electronic spin and, therefore, electronic exchange, leads to yet another method, which is the most important method of molecular quantum chemistry.

2.11.3
The Hartree–Fock Approximation

Within the Hartree method, the electronic spin does not appear explicitly except for the fact that no more than two electrons may go into a single orbital. The existence of the Pauli exclusion principle, however, needs to be accounted for in order to go beyond the Hartree method, and that is what the Hartree–Fock method [120] is all about. We first formulate an arbitrary three-dimensional orbital ϕ for electron i by writing it as the product of a purely space-dependent part ξ and a spin function (spinor), α or β, characterizing spin-up or spin-down electron, for example $\phi_i(\boldsymbol{x}_i) = \xi_i(\boldsymbol{r}_i)\alpha_i$; here we use $\boldsymbol{x}$ to indicate a variable which includes both space ($\boldsymbol{r}$) and spin (α). A Hartree-like product wave function between two one-electron wave functions ϕ_1 and ϕ_2 could then be written as

$$\Psi' = \xi_1(\boldsymbol{r}_1)\alpha_1\xi_2(\boldsymbol{r}_2)\beta_2 \qquad \text{but also as} \qquad \Psi'' = \xi_1(\boldsymbol{r}_2)\alpha_2\xi_2(\boldsymbol{r}_1)\beta_1, \tag{2.74}$$

since both are equally suitable. Now, because any fermionic wave function must be antisymmetric (= change sign) with respect to an exchange of electrons – which is an even stricter version of Pauli's principle – the wave function can only read

$$\Psi = \Psi' - \Psi'' = \xi_1(\boldsymbol{r}_1)\alpha_1\xi_2(\boldsymbol{r}_2)\beta_2 - \xi_1(\boldsymbol{r}_2)\alpha_2\xi_2(\boldsymbol{r}_1)\beta_1 = \begin{vmatrix} \xi_1(\boldsymbol{r}_1)\alpha_1 & \xi_2(\boldsymbol{r}_1)\beta_1 \\ \xi_1(\boldsymbol{r}_2)\alpha_2 & \xi_2(\boldsymbol{r}_2)\beta_2 \end{vmatrix} \tag{2.75}$$

which is usually called a Slater determinant [121] and is an antisymmetrized product wave function [51]. We now switch back to the ϕ notation and write it, generally, as

$$\Psi(\boldsymbol{x}_1 \cdots \boldsymbol{x}_N) = \frac{1}{\sqrt{N!}} \begin{vmatrix} \phi_1(\boldsymbol{x}_1) & \phi_2(\boldsymbol{x}_1) & \phi_3(\boldsymbol{x}_1) & \cdots \\ \phi_1(\boldsymbol{x}_2) & \phi_2(\boldsymbol{x}_2) & \phi_3(\boldsymbol{x}_2) & \cdots \\ \phi_1(\boldsymbol{x}_3) & \phi_2(\boldsymbol{x}_3) & \phi_3(\boldsymbol{x}_3) & \cdots \\ \vdots & \vdots & \vdots & \end{vmatrix} \tag{2.76}$$

for a total of N electrons – the pre-factor takes care of the normalization – because this automatically guarantees that two electrons with the same spin *cannot* go into the same orbital. In this case, two columns of the Slater determinant would be identical, giving a zero determinant and a collapse of Ψ. This nice mathematical feature of the determinant allows for the compact formulation of the Hartree–Fock equations. The iterative, self-consistent solution of the Hartree–Fock equations proceeds in quite a similar way to the Hartree case, except that they are mathematically *much* more demanding. This is because the antisymmetrization leads to a *nonlocal* (dependent on two variables) exchange potential. A further distinction has to be made between closed-shell (pairwise assignment of electrons into orbitals) and open-shell (nonpairwise) systems (see also below), but this is more a technical issue.

Historically, the Hartree–Fock (HF) or self-consistent field (SCF) method [51] proved to be extremely powerful in the case of molecular quantum chemistry after it had been shown by Roothaan [122] how the HF integro-differential equations could be turned into algebraic equations – namely by the introduction of a set of (spatial) basis functions – and then solved by numerical matrix routines. In fact, HF theory has become the undisputed standard *ab initio* method for molecules [61] upon which further improvements can be built. Nonetheless, the absolute number of integrals to be evaluated, soon becomes rather large even for medium-sized molecular systems (the HF method formally "scales" with the fourth order of the number of electrons or orbitals, N^4) and, therefore, a couple of semiempirical methods have historically originated from the HF method by parametrizing (or neglecting) specific integrals. Most of these methods carry the "NDO" (neglect of differential overlap) label in their names (such as NDDO, INDO, CNDO, MNDO, SINDO, etc.), and this collection of simplified HF-based methods [58,123] represents a more systematic approach (if compared to methods such as the extended Hückel) which can speed up the computational performance but keep the accuracy close to the *ab initio* reference. The success of these techniques has been particularly good within *organic* molecular chemistry, so that we do not have to go into the details here.

Even if no integral parametrizations are introduced and the HF equations are all correctly solved, the method eventually turns out to be theoretically incomplete. Despite the correct treatment of electronic exchange (X) within Hartree–Fock theory, electronic correlation (C) is totally missing. This is easily shown for the case of the H_2 molecule in which we use the bonding solution of the H_2^+ molecular ion ($\psi_+ \equiv \sigma_g$ from Equation (2.15)) to build up an antisymmetrized molecular wave function. This means that we put *both* electrons (r_1 and r_2) of the H_2 molecule into the same ψ_+ orbital, and Pauli's principle is obeyed by means of the α/β spinors. Neglecting orbital overlap and any pre-factors, for simplicity, the so-called Hund–Mulliken [124] (another name for the molecular-orbital, MO, or Hartree–Fock, HF, approach) wave function becomes

$$\Psi^{MO} \sim \psi_+(\boldsymbol{r}_1)\psi_+(\boldsymbol{r}_2)(\alpha_1\beta_2 - \beta_1\alpha_2) \tag{2.77}$$

$$\begin{aligned} &\sim (\phi_1(\boldsymbol{r}_1) + \phi_2(\boldsymbol{r}_1))(\phi_1(\boldsymbol{r}_2) + \phi_2(\boldsymbol{r}_2))(\alpha_1\beta_2 - \beta_1\alpha_2) \\ &\sim [\phi_1(\boldsymbol{r}_1)\phi_1(\boldsymbol{r}_2) + \phi_2(\boldsymbol{r}_1)\phi_2(\boldsymbol{r}_2) + \phi_1(\boldsymbol{r}_1)\phi_2(\boldsymbol{r}_2) + \phi_2(\boldsymbol{r}_1)\phi_1(\boldsymbol{r}_2)] \\ &\quad \times (\alpha_1\beta_2 - \beta_1\alpha_2). \end{aligned} \tag{2.78}$$

This wave function can be iconized, from left to right, as a mix of the four electronic configurations

$$H^-H^+ \quad \text{plus} \quad H^+H^- \quad \text{plus} \quad \text{H–H} \quad \text{plus} \quad \text{H–H}.$$

For example, the $\phi_1(\boldsymbol{r}_1)\phi_1(\boldsymbol{r}_2)$ summand indicates that both electrons of H_2 are found in the 1s atomic orbital of the left H atom, whereas the other 1s orbital of the right atom is empty, thus H^-H^+. The other icons are self-explanatory. Let us contrast this solution with the valence-bond (VB) solution for H_2 which itself rests on the Heitler–London ansatz for the H_2^+ ion. Here one starts from loosely interacting H atoms, which are weakly perturbing each other, such that the atomic orbitals themselves are good approximations for the wave function.[23] In the valence-bond case, the solution for the hydrogen molecule is

$$\Psi^{VB} \sim [\phi_1(\boldsymbol{r}_1)\phi_2(\boldsymbol{r}_2) + \phi_2(\boldsymbol{r}_1)\phi_1(\boldsymbol{r}_2)](\alpha_1\beta_2 - \beta_1\alpha_2), \tag{2.79}$$

which consists of only two electronic configurations, namely H–H plus another H–H.

A comparison of Equations (2.78) and (2.79) yields that in both approaches, MO ($\equiv$ Hartree–Fock) and VB, the Pauli exchange has been correctly included. The difference between the two is solely given by their amount of electronic *correlation*. In the MO approach, the electrons are completely uncorrelated (independent), and they may even go into the same atomic orbital, albeit with different spins, thus producing ionic states (H^-H^+). The MO approach therefore does *not* take care of the energy penalty due to the Coulomb repulsion between the two electrons (see Section 2.9). Because the electronic Coulomb correlation has been completely ignored, the correlation energy E^{corr} may be *defined* as the difference between the correct energy and that of the Hartree–Fock solution, that is

$$E^{corr} \equiv E - E^{HF} \equiv E - E^{MO}. \tag{2.80}$$

In contrast, the ionic states do not exist in the VB solution, which is a fully correlated wave function.

23) In fact, the first solution of the H_2^+ problem did *not* follow the route we have sketched in Section 2.1 by starting from an LCAO procedure, but was focused on this very idea of two very weakly coupled hydrogen atoms [125].

In reality, an optimum wave function will be either closer to the molecular orbital (Hund–Mulliken or Hartree–Fock) or to the valence bond (Heitler–London) result, depending on the amount of electronic correlation. It transpires that, extremely fortunately for molecular quantum chemistry, which relies heavily on Hartree–Fock theory [61], the correlation is not too strong in most light-atom molecules of interest (say, hydrocarbons), and even if considerable correlation is involved, the Hartree–Fock wave function is still good enough for structure optimization. Thus, theoretical geometry data almost coincide with those obtained from experimental (mostly X-ray diffractional) sources, and the errors are of the order of 0.01 Å for interatomic distances and 1–2° for angles [126]. Also, molecular orbital (Hartree–Fock) theory is straightforward for computer programming, and this, too, explains its enormous success for molecular quantum chemistry and, also, the smaller impact of valence-bond theory. To some, MO theory is also easier to understand than VB theory, but this depends on one's point of view.[24]

Nonetheless, MO theory often gives the wrong products of molecular dissociation, simply because the MO wave function is pretty good for equilibrium distances but not for torn-apart molecules which would better be described by a more correlated VB function. To be fair, this is not the case for *unrestricted* Hartree–Fock (UHF) theory [128] where α and β electrons are allowed to enter different one-electron orbitals (because they experience different effective nuclear charges, see Section 2.9), but then other problems manifest. Although the UHF wave function of H_2, for example, correctly dissociates into two H atoms, it suffers from spin contamination by falsely mixing a high-lying triplet state into the singlet ground state [51,58].

If correlation needs to be included into Hartree–Fock theory, people either simply apply perturbation theory (according to Møller–Plesset) [51,129] up to a certain order, or they allow the electrons to enter higher-lying orbitals, thereby mixing other electronic configurations (that is, other distributions of the electrons over the set of orbitals) into the final wave function. For the H_2 molecule, one might mix a little antibonding σ_u^* into the final wave function which would correspond to a "double excitation" of the two electrons from σ_g into σ_u^*; although the latter MO is higher in energy, the *interelectronic* repulsion is reduced in this orbital. We will not go into the details of these two approaches but merely mention that the latter one, dubbed *configuration interaction* (CI), is a systematic way to *fully* include all missing correlation [51,58,61]. For the simplest case of H_2, a pedagogical look at configuration interaction shows that just five configurations are needed to account for more than 90% of the total correlation energy, and the contributions of "left-right", "in-out"

24) There is a recent beautiful discussion on the pluses and minuses of both valence-bond and molecular-orbital theory for the calculation and understanding of chemistry, and it is really worth reading [127].

and "angular" correlation between the two electrons may be nicely visualized [130].

Nonetheless, CI is computationally too demanding for most molecules because there are simply too many electrons, such that, in practice, restricted (that is, simplified) versions of CI come into play which, however, suffer from the problem of *size-consistency* [51]. That is to say that the amount of electronic correlation included also depends on the number of electrons in the system. Remedies for this serious problem exist, for example, by making use of electron-pair theories or the coupled cluster approximation [51]. This is explained in more detail in Section 2.13.

The problem of size-consistency is particularly frightful for restricted CI calculations on solids where the correlation energy falsely arrives at *zero*, which is the worst possible result [126]. The difficulties of Hartree–Fock theory for the solid state or, more correctly, for solid *metals*, however, show up much earlier. When Hartree–Fock theory is applied to a gas of noninteracting electrons (free electrons) which comes close to the electronic situation in simple metals, such as the alkali elements, the band energy takes the following analytical form [131]:

$$E(k) = \frac{\hbar^2 k^2}{2m} - \frac{e^2 k_\mathrm{F}}{2\pi}\left(2 + \frac{k_\mathrm{F}^2 - k^2}{k_\mathrm{F} k}\ln\left|\frac{k_\mathrm{F}+k}{k_\mathrm{F}-k}\right|\right) \tag{2.81}$$

At the Fermi energy or, equivalently, at the corresponding quantum number $k = k_\mathrm{F}$, the derivate of $E(k)$ with respect to k goes to *infinity*, such that the inverse slope E vs. k goes to zero. Thus, on recalling Equation (2.42), the density-of-states (DOS) *vanishes* at the Fermi level as schematically depicted in Figure 2.30.

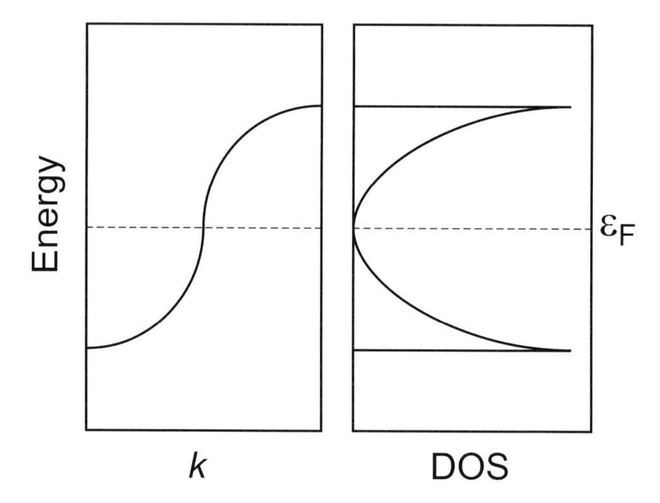

Fig. 2.30 Schematic drawing of a one-dimensional band structure and density-of-states for free electrons (simple metals) according to Hartree–Fock theory.

Because of the vanishing DOS at the Fermi level, Hartree–Fock theory predicts simple metals (more correctly: electron-gas-like materials) to be insulating. The origin of this flaw is the consequence of having completely neglected electronic correlation, and the latter is decisively important in such metallic solids (but not in molecules). To numerically correct the problem, the electron *gas* must be replaced by a Fermi *liquid* in the theoretical description. Admittedly, we have shown an extreme case because, indeed, Hartree–Fock theory can be elegantly formulated and carried out for three-dimensional periodic systems [132], and the calculation of physical properties of crystalline materials is straightforward [133]. As may be guessed from the above correlation problem showing up in metallic systems, periodic Hartree–Fock theory may perform perfectly well for real materials which behave as semiconductors or insulators [134–136], and for these HF theory may be an excellent choice. The progress of solid-state quantum theory, however, has been historically linked with the electronic structure of *metals*, so the treatment of exchange and correlation has followed another route. But that is how one may trace back, in a very simplified way, the origin of the theoretical schism between the strategy taken in molecular quantum chemistry on the one hand (Hartree–Fock theory and beyond) and solid-state theory on the other, the latter heavily relying on density-functional theory. This approach is covered in the next section.

2.12 Density-functional Theory

The aforementioned wave function-related difficulties in incorporating exchange and correlation, especially for the treatment of metallic solids, can be elegantly solved – or circumvented, as some would say, to an astonishingly accurate degree – by moving away from the N-electron (Hartree–Fock style) wave function Ψ and its corresponding Schrödinger equation and replacing them by a computational scheme which rests, instead, on the electron density $\rho(\boldsymbol{r})$.

Such ρ-based theories date back, in fact, to the early days of quantum mechanics, and the pioneering work of Thomas [137] and Fermi [138] provides a method of "statistically" describing the distribution of the electrons in an atom. Without going into detail, the Thomas–Fermi (TF) total energy of an atom with a nucleus charged Z may be directly given as

$$E^{\mathrm{TF}}\{\rho(\boldsymbol{r})\} = 2.871 \int \rho^{5/3}(\boldsymbol{r})d\boldsymbol{r} - Z\int \frac{\rho(\boldsymbol{r})}{r}d\boldsymbol{r} + \frac{1}{2}\iint \frac{\rho(\boldsymbol{r})\rho(\boldsymbol{r}')}{|\boldsymbol{r}-\boldsymbol{r}'|}d\boldsymbol{r}d\boldsymbol{r}'. \quad (2.82)$$

The curly brackets around the electron density $\rho(\boldsymbol{r})$ indicate that the total energy is a *functional* of the electron density, and this anticipates an important concept of density-functional theory: A functional is a function, of which the variable is itself a function; that is, E is a functional of ρ because ρ is a function of $\boldsymbol{r}$, thus $E\{\rho(\boldsymbol{r})\}$.

The first entry of Equation (2.82), which is just the kinetic energy, makes it clear that Thomas–Fermi theory allows the *explicit* – albeit approximate! – calculation of an atom's total energy which numerically is $-0.7687Z^{7/3}$, based upon the knowledge of the electron density $\rho(\boldsymbol{r})$ only; unfortunately, the accuracy is rather low. Other weaknesses of TF theory are that there is no atomic shell structure and also that the decay of the electron density, upon moving away from the nucleus, is just too slow. In addition, the errors for the atomic *total* energies are relatively large, in particular for light atoms, and the situation worsens on going to molecules which are all *unstable* because chemical bonding is impossible within TF theory [139]; this is called the nonbinding theorem.

Another use of the electron density was introduced by Slater [140] quite early on, as a method of approximating the nonlocal exchange potential which is so tedious to calculate within the framework of Hartree–Fock theory (see Section 2.11.3). With reference to an electron-gas model of uniform density, Slater replaced the nonlocal exchange potential by the so-called Xα *local* potential

$$v_{X\alpha}(\boldsymbol{r}) = -\frac{3}{2}\alpha\left(\frac{3}{\pi}\rho(\boldsymbol{r})\right)^{1/3}, \tag{2.83}$$

thereby introducing the Hartree–Fock–Slater (HFS) method which is much easier to handle than HF theory. However, HFS theory does not include electronic correlation, just as HF theory. The α parameter has its own history, ranging from 1 to 2/3, and it slightly differs from atom to atom [141] and may be looked upon as just another variational parameter. Recall that the set of extended Hückel parameters in Table 2.1 are based on HFS optimizations that were fitted to experimental quantities.

In contrast, density-functional theory (DFT) is an exact theory, at least in principle, and it includes both electronic exchange and correlation [142]. It is essentially based on two theorems of Hohenberg and Kohn [143] stating that the wave function of the ground state Ψ is a unique functional of the electron density $\rho(\boldsymbol{r})$ because the potential energy appearing in the Schrödinger equation is also totally determined by $\rho(\boldsymbol{r})$. In addition, there is a variational principle such that the search for the lowest-energy wave function Ψ_0 can be replaced by the search for the lowest-energy electron density $\rho_0(\boldsymbol{r})$. Thus, because $\Psi \equiv \Psi\{\rho(\boldsymbol{r})\}$ (the wave function is a function of the electron density which itself is a function of the spatial coordinate), the ground-state energy E_0 is given as the minimum of the functional $E\{\rho(\boldsymbol{r})\}$. It really is that simple.

A little consideration is needed at this point. Recall that within Thomas–Fermi theory, the kinetic energy (and, therefore, the total energy) can be directly calculated, but only *approximately*, if the electron density is known. It would be preferable, however, to *correctly* calculate the kinetic energy because it makes up a very large part (a leading part, in fact) of the total energy. For a system of noninteracting electrons, this can be done by introducing *orbitals*, and the kinetic energy T_0 (the 0 subscript now indicates the noninteracting system) then becomes

$$T_0 = \sum_{i=1}^{N} f_i \left\langle \psi_i \left| -\frac{1}{2}\nabla^2 \right| \psi_i \right\rangle , \tag{2.84}$$

while the electron density is

$$\rho(\boldsymbol{r}) = \sum_{i=1}^{N} f_i |\psi_i(\boldsymbol{x})|^2 , \tag{2.85}$$

with $\boldsymbol{x}$ indicating space and spin coordinates; the f_i are the occupation numbers of the orbitals. Note that this is an *indirect* – albeit accurate! – calculation of T for *noninteracting* electrons, therefore designated T_0. The orbitals merely serve as *tools* to get to the kinetic energy.

The ingenious trick used by Kohn and Sham [144] is the introduction of a *reference system of noninteracting electrons with the same (spin) density*. For illustration, one might think of gradually reducing all electron–electron interactions to zero and, in order to keep the same (spin) density, introduce a little (spin-dependent) extra potential to achieve that goal; by this means, the interacting system has become a *quasi-noninteracting* system. Formally, the functional of the total energy is expressed as

$$E\{\rho(\boldsymbol{r})\} = T_0\{\rho(\boldsymbol{r})\} + \frac{1}{2}\iint \frac{\rho(\boldsymbol{r})\rho(\boldsymbol{r}')}{|\boldsymbol{r}-\boldsymbol{r}'|} d\boldsymbol{r}d\boldsymbol{r}' + E_{\mathrm{XC}}\{\rho(\boldsymbol{r})\} + \int \rho(\boldsymbol{r})v(\boldsymbol{r})d\boldsymbol{r}, \tag{2.86}$$

including the kinetic energy of the system of noninteracting electrons with the same density, a classic Coulomb energy of the charge density, an energy functional for exchange and correlation, and the attractive Coulomb potential provided by the fixed nuclei. The numerical success of DFT is a logical consequence of its very smart strategy in treating the large energetic parts (kinetic energy) as accurately as possible – by using orbitals – and approximating the small ones (exchange–correlation energies); all residual errors have been effectively dumped into these. The price to be paid for this *accurate but indirect* calculation is similar to that for Hartree theory. Within DFT, the solution of the many-particle problem has been replaced by the *self-consistent, iterative* solution of N Kohn–Sham equations

$$\left[-\frac{1}{2}\nabla^2 + v_{\mathrm{eff}}(\boldsymbol{r})\right] \psi_i(\boldsymbol{r}) = \varepsilon_i \psi_i(\boldsymbol{r}), \tag{2.87}$$

with an *effective* one-particle potential of the form

$$v_{\text{eff}}(\boldsymbol{r}) = v(\boldsymbol{r}) + \int \frac{\rho(\boldsymbol{r}')}{|\boldsymbol{r} - \boldsymbol{r}'|} d\boldsymbol{r}' + v_{\text{XC}}(\boldsymbol{r}), \tag{2.88}$$

which consists of the potential from the nuclei, a Hartree-style potential and the potential for exchange and correlation, v_{XC}. The latter is defined as

$$v_{\text{XC}}(\boldsymbol{r}) = \frac{\partial E_{\text{XC}}\{\rho(\boldsymbol{r})\}}{\partial \rho(\boldsymbol{r})}, \tag{2.89}$$

and the electron density is given by Equation (2.85). By solving a set of Hartree-like equations (but with a more general potential including both exchange and correlation) in an iterative, self-consistent manner, DFT has a couple of computational and also interpretational advantages. It is fast, truly size-consistent, and it remains an orbital theory, not a complicated many-body theory, although it is rather difficult to understand what these Kohn–Sham orbitals *really* mean [145, 146]. Just as in Hartree theory, the "one-electron" eigenvalues ε_i do *not* sum up to the total energy. On the contrary, these eigenvalues merely serve as Lagrange multipliers in combination with the orbitals ψ_i such that the ε_i should *not* be interpreted as one-particle energies (in computational practice, however, this is *very* often done). The same argument also holds for semiempirical electronic-structure theories.

Consequently, DFT is restricted to ground-state properties. For example, band gaps of semiconductors are notoriously underestimated [142] because they are related to the properties of excited states. Nonetheless, DFT-inspired techniques which also deal with excited states have been developed. These either go by the name of time-dependent density-functional theory (TD-DFT), often for molecular properties [147], or are performed in the context of many-body perturbation theory for solids such as Hedin's GW approximation [148].

Technically, a regular DFT calculation may start from a guess for the first (Hartree-like) potential used for solving Schrödinger's equation; after the first iteration, there results a new electron density from which the Hartree and the exchange–correlation potentials are generated to yield a new potential, if one has some idea of the size of the exchange–correlation energy. The process is repeated until self-consistency is achieved, just as in Hartree theory.

As has been alluded to by mentioning the electronic spin, an extension of density-functional theory to spin-density-functional theory is possible by simply decomposing the total electron density $\rho(\boldsymbol{r})$ into the spin densities $\rho^{\alpha}(\boldsymbol{r})$ and $\rho^{\beta}(\boldsymbol{r})$ although there are some recent doubts as to the validity of the entire idea [149,150]. In practice, spin-DFT works very well. Spin-DFT is needed whenever the electrons feel an external magnetic field or are subject to spontaneous spin-polarization, such as in magnetic atoms, molecules, and solids (for example, itinerant magnets, see Section 3.5). Spin-DFT also allows different spatial orbitals for a better description of the majority/minority spins because of different effective nuclear potentials, already covered in Section 2.9.

2.12.1 Exchange–Correlation Functionals

For scientific theories, being exact *in principle* seems to be a nice euphemism for being approximate *in practice*. Density-functional theory suffers from the same fate, and any DFT calculation can only be as reliable as the incorporated parametrization scheme for exchange and correlation. Indeed, the search for reliable exchange–correlation functionals is the greatest challenge to DFT.

The first and, possibly, most influential and also very successful approximation [144] is given by the *local-density approximation* (LDA),

$$E_{\mathrm{XC}}^{\mathrm{LDA}}\{\rho(\boldsymbol{r})\} \equiv \int \rho(\boldsymbol{r})\varepsilon_{\mathrm{XC}}^{\mathrm{hom}}\{\rho(\boldsymbol{r})\}d\boldsymbol{r}, \tag{2.90}$$

in which $\varepsilon_{\mathrm{XC}}^{\mathrm{hom}}\{\rho(\boldsymbol{r})\}$ is the exchange–correlation energy per particle of a homogeneous gas of electrons of density $\rho(\boldsymbol{r})$. While the exchange part can be expressed analytically, the numerical data for the correlation part may be derived from perturbation theory or from an exact many-particle calculation (Monte Carlo-type) of the free electron gas. Eventually, these data are tabulated for numerical use.

Clearly, this approximation is *so* crude that it seemed to be only applicable to systems with *slowly* varying electron densities. Consequently, its use for atoms and, particularly for molecules, appeared to be out of the question, and an accurate description of chemical bonding was not expected at all at the very beginning [144]. Computational practice, however, revealed surprisingly good results (atomic shell structure, molecular bond lengths, and bonding energies), probably due to large error cancellations in the exchange part. A careful analysis yields that it is only the angular average which enters the evaluation of the exchange energy [151]. In fact, even more approximate versions of DFT have proved to be quite useful [152, 153].

At the same time, the LDA gave an *a posteriori* justification of the old Xα method by Slater, because the latter is a special LDA variant without correlation. The corresponding spin-dependent version of the LDA is called a local spin-density approximation (LSDA or LSD or just spin-polarized LDA), and even now when people talk of LDA functionals, they always refer to its generalized form for systems with (potentially) unpaired spins. Among the most influential LDA parametrizations, the one of von Barth and Hedin (BH) [154] and the one of Vosko, Wilk and Nusair (VWN) [155] are certainly worth mentioning. The latter is based on the very accurate Monte Carlo-type calculations of Ceperley and Alder [156] for the uniform electron gas, as indicated above.

At the present time, the (probably) most accurate representation of the correlation energy is due to Perdew and Wang [157].

The LDA success story has been particularly evident in the case of solids, and most experimental lattice parameters, for example, are reproduced (slightly underestimated, in fact) to within only a few percent. Likewise, the magnetic moments of the itinerant metals are often perfectly reproduced. Any remaining LDA errors are mostly due to an underestimation/overestimation of E_X/E_C. Although the correlation energy of the electron gas can be calculated quite accurately (to within 1%) [156, 157], an LDA calculation typically results in too-small exchange energies (by ca. 10%) but also too-large correlation energies (by a factor of two), partially cancelling each other [158]. It can also be shown that, within a local approximation, an electron falsely interacts with itself, and the corresponding self-interaction-corrected (SIC) version of the LDA has been invented to remove that weakness [142, 159]. Hartree–Fock theory does not suffer from this unphysical interaction.

Another, more recent, route has been followed by taking the *gradient* of the electron density into account for the development of the so-called gradient-corrected functionals. The idea is also to include $\partial\rho(\boldsymbol{r})/\partial\boldsymbol{r}$, as well as $\rho(\boldsymbol{r})$, to better describe the exchange hole which would improve the LDA especially in regions of low density. The most popular method goes by the name of *generalized gradient approximation* (GGA), and the exchange–correlation energy is schematically written as

$$E_{\mathrm{XC}}^{\mathrm{GGA}}\{\rho(\boldsymbol{r})\} \equiv \int \rho(\boldsymbol{r})F\{\rho(\boldsymbol{r}), \nabla\rho(\boldsymbol{r})\}d\boldsymbol{r}. \tag{2.91}$$

Functionals such as the GGA have enormously popularized the DFT-based methods in the (Hartree–Fock-oriented) quantum-chemical community because the GGA stands for significantly improved energetic results – as well as better geometries – especially when it comes to bond breaking and bond formation. Here, the expectations of "chemical accuracy" [160] have been rather high and, indeed, some functionals that are part of the GGA route seem to achieve that goal. Historically, the following steps have been taken:

With respect to the *exchange* part of $E_{\mathrm{XC}}^{\mathrm{GGA}}$, a gradient correction to the LDA exchange functional was proposed in 1986 by Perdew and Wang (PW86) [161], and another correction was developed in 1988 by Becke (B88 or simply B); the latter [162] is (in part) fitted to exact exchange data of the noble gases and has proved very popular in computational practice. With regard to the *correlation* part, one must mention Perdew's 1986 gradient correction (P86) to the LDA [163] and the later (1991) gradient correction by Perdew and Wang (PW91) [157, 164]; both have found extensive use. A quite different route has been taken by Lee, Yang and Parr who introduced a true gradient-based correlation *functional* (not an LDA correction) through a numerical fit to the data of the helium atom; it goes by the abbreviation LYP [165]. Upon combining such gradient corrections (or gradient-based functionals) for exchange and correlation, the first part of the abbreviation refers to exchange, the second one to correlation. Thus, merging Becke's exchange correction to the LDA with the correlation functional of Lee, Yang and Parr will result in a "BLYP" calculation.

In addition, the desire to further improve the numerical accuracy of DFT-based methods has made people deliberately *mix* contributions of exchange and correlation which arise from very different sources. One of the most widespread *hybrid* functionals, dubbed B3LYP, is a three-parameter ansatz [166] including contributions from the LDA, Hartree–Fock theory (!), Becke's gradient-based exchange correction and the LYP correlation functional, a most remarkable blend indeed. The term "hybrid" alludes to the peaceful merging of DFT- and HF-based ingredients. Because of space limitations, we will not go into any more detail but instead refer to the specialized chemical literature [58, 167] which, however, is a bit more focused at the molecular regime in contrast to solid state and materials science [168]. Contrary to the original intention, it is probably fair to say that *empirical* knowledge about the particular system under study is helpful in choosing the "best" density-functional parametrization. On the other hand, the various GGA approaches are not *that* different with regard to the final results they deliver; a numerical comparison of these GGA functionals for the test case of the H_2 molecule evidences similar trends, namely a slight overestimation of the bonding energy and also an overestimation of the H–H equilibrium distance, both due to the self-interaction error. Its consideration by a SIC functional, however, results in an underestimation of the equilibrium distance which is even worse [169].

Viewed from the point of view of theoretical solid-state (metal) physics, this community (as opposed to the chemists) has probably been more conservative with respect to GGA-based approaches, and many theoretical calculations are still performed using the simple LDA. It does have deficiencies, but the method is quite transparent. An exception is made with respect to the GGA scheme by Perdew, Burke and Ernzerhof (PBE) which has seemingly become another standard [170] and is not far from PW91. Such gradient-based functionals have cured a serious LDA flaw when it comes to total-energy differences for magnetic materials [171]. For example, an LDA calculation will falsely predict the ground state of metallic iron to be the nonmagnetic face-centered cubic (fcc) structure, and not the ferromagnetic body-centered cubic (bcc) one that is known from experiment. The correct order of ferromagnetic bcc-Fe lying lower in energy than nonmagnetic fcc-Fe, however, is nicely given by a GGA calculation, depicted in Figure 2.31.

Nonetheless, both LDA and GGA fail, although to a different extent, if electronic correlation becomes strongly dominant, as in antiferromagnetic insula-

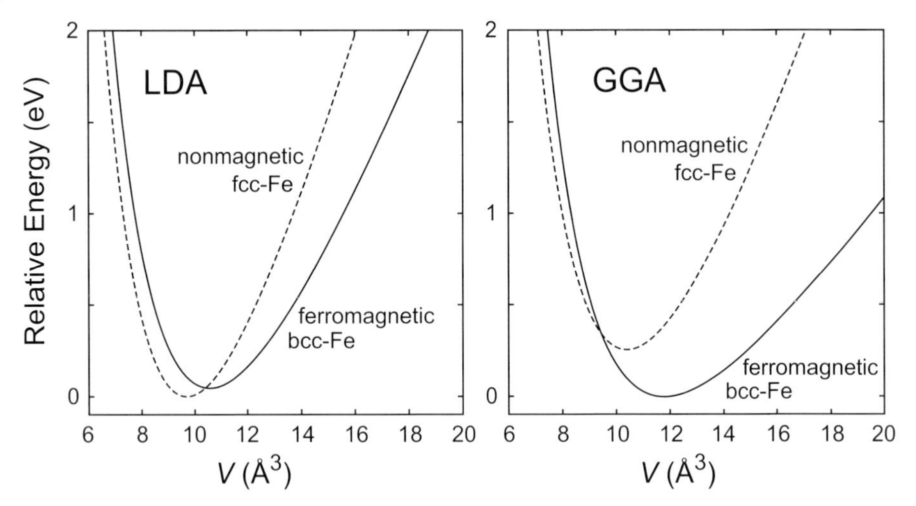

Fig. 2.31 Energies of two allotropes of metallic iron as given by the LDA and the GGA, using a pseudopotential method and plane waves.

tors such as CoO which falsely results as a *metal* by any bare DFT calculation. In reality, electron correlation leads to a *localization* of the electrons [172], and the metallic nature is lost. Fortunately, the DFT error of yielding the metallic state in the calculation is immediately obvious and can therefore be taken care of. It has been known for a long time that such an underestimation of electronic correlation can be easily corrected by introducing an additional electron repulsion parameter U, and this is the origin of a DFT patch called LDA+U method (see also Section 2.13). On the other hand, these and other corrections have not necessarily improved the acceptance of DFT within the *ab initio* quantum-chemical community because these patches have given density-functional theory a somewhat semiempirical touch, probably unwittingly so. Also, it should be noted that an *ab initio* approach to density-functional theory, on the basis of electron-pair sum rules, reveals remarkable findings with respect to widely used gradient-corrected functionals, especially when it comes to spin-related phenomena [173]. Summarizing, a couple of fairly reliable DFT parametrizations are available but any systematic improvement of the functionals for exchange and correlation remains difficult.

2.13 Beyond Density-functional Theory

Without being too harsh on density-functional theory, there are clear weaknesses, especially if effects of strong electronic correlation come into play. When the Coulomb interactions between the electrons are becoming large, and this behavior may become dominant for some transition/rare-earth metals and their compounds (such as CoO), the idea of a weakly correlated electron gas is no longer justified. Therefore, a couple of techniques have been developed in order to go beyond the aforementioned DFT limits. Unfortunately, it is rather difficult to see the wood for all the theoretical trees and, also, a large number of contradictory claims have been made as to individual computational superiority. Because I do not intend to fight with the various schools, the following account cannot at all be considered exhaustive (an impossible task), and will not allude to any kind of ranking of "accuracy". I merely say that these methods are available and/or under development.

As has been alluded to already (see Section 2.12.1), some insulating transition-metal oxides (also called Mott insulators) are falsely predicted as being metals when their electronic structures are calculated on the basis of the LDA because the full amount of the local Coulomb repulsion experienced by the electrons within the d orbitals is underestimated, and the inclusion of some extra repulsion U is needed to theoretically change them into insulators. This is how the LDA+U method is typically justified [174]. The method works well for transition-metal oxides and may also be used for "correlated" metals,[25] in particular, the f elements. One may well ask whether a method that goes by the noble name of *first principles* or *ab initio* should be augmented by the introduction of a somewhat arbitrary U energy parameter, but that is matter of taste. The calculation of U requires additional approximations (the so-called constrained LDA method may deal with the problem [175]) and the spatial extent of the basis orbitals effectively determines the size of U, as expected by any quantum chemist.

Another method, dubbed *dynamical mean-field theory* (DMFT), has been proposed as the many-body generalization of the LDA and also the LDA+U [176]. In the language of its proponents, DMFT recaptures the correct quasiparticle physics and the corresponding energetics [177]. What makes DMFT dynamical is that, in the limit of a weakly correlated system, it will converge to the LDA solution whereas, for heavily correlated materials, it will eventually turn into the LDA+U method. The DMFT equations are based on the spectral function of the solid and are difficult to solve since they refer to a local impurity model where the amount of correlation essentially depends on the atomic environment. Currently, both approximate as well as exact methods have been developed to handle the DMFT equations but their results may differ quantitatively. When the DMFT band structure is found, it is characterized by *changes* in the occupations of the bands due to electronic correlation, and this also changes the electron density. Upon feeding the newly found electron density into the LDA scheme, one arrives at a self-consistent variant of

25) Here we use the jargon of the theoretical physics community which often classifies LDA-challenging materials as "correlated".

DMFT. A remarkable DMFT test case is given by the correlated $5f$ element plutonium where the LDA fails both for reproducing the equilibrium volume and the magnetic properties [178], in contrast to DMFT. An important *psychological* advantage of DMFT is that it has increased the mutual understanding of two very different theoretical communities.[26]

Yet another prominent approach in the consideration of electronic correlation is given by the *quantum Monte Carlo* (QMC) method [179] which, depending on the amount of correlation, comes in two "flavors". Given a not too correlated system, a trial many-body wavefunction is set up such that its main part consists of Slater determinants of spin-up and spin-down orbitals, which originate from either Hartree–Fock or DFT (LDA) theory, thereby ensuring Pauli's principle. These are then augmented by additional functions, taking care of the correlated motion of pairs of electrons. Because the latter functions are used for energy minimization, the method goes under the name *variational* QMC [180]. If the starting HF or DFT approximations are not accurate enough, the problem is rewritten on the basis of the imaginary-time Schrödinger equation which is solved by a numerical simulation (i.e., a random diffusion in multi-dimensional space), the central idea of the *diffusion* QMC method. The antisymmetry of the final wave function must be ensured by the so-called fixed-node approximation. Just as for DFT, the computational cost scales with the third power of the system, but it explodes with the nuclear charge. This is why diffusion QMC calculations are then performed on the basis of pseudopotentials [181] (see also Section 2.15.2).

The question for a more *systematic* inclusion of electronic correlation brings us back to the realm of molecular quantum chemistry [51, 182]. Recall that (see Section 2.11.3) the *exact* solution (configuration interaction, CI) is found on the basis of the self-consistent Hartree–Fock wave function, namely by the excitation of the electrons into the virtual, unoccupied molecular orbitals. Unfortunately, the ultimate goal of *full* CI is obtainable for very small systems only, and *restricted* CI is size-inconsistent: the amount of electron correlation depends on the size of the system (Section 2.11.3). Thus, size-consistent but perturbative approaches (Møller–Plesset theory) are often used, and the simplest practical procedure (of second order, thus dubbed MP2 [129]) already scales with the fifth order of the system's size N, in contrast to Hartree–Fock theory ($\sim N^4$). The accuracy of these methods may be systematically improved by going up to higher orders but this makes the calculations even more expensive and slow (MP3 $\sim N^6$, MP4 $\sim N^7$). Fortunately, restricted CI can be mathematically rephrased in the form of the so-called *coupled clus-*

26) These are the true many-body community, studying highly correlated materials by means of analytical but approximate tools, and the band-structure community, mostly focusing on weakly correlated materials by means of accurate numerical first-principles (i.e., density-functional) methods.

ter (CC) method [183] which, although nonvariational,[27] retains the important property of size consistency. When including single (S) and double (D) substitutions of the electrons, the thus-derived CCSD method scales with N^6; a slightly less accurate but both size-consistent *and* variational method, QCISD ($\equiv$ quadratic configuration interaction including single and double substitutions), has become another quantum-chemical standard [184].

It has been realized that the bad scaling behavior of these accurate approaches does *not* go back to the underlying physics of the very short-ranged electron correlation which decays with the sixth order of the interelectronic distance, at least for nonmetallic systems. The bad scaling behavior is, in fact, a consequence of the *delocalized* ansatz (LCAO) of the molecular orbital approach, not of the "nearsightedness" [158] of the phenomena themselves; this is explained in greater detail in Section 2.15.5 dealing with the so-called *order N* methods. Because it is possible to re-transform the set of orthogonal delocalized Hartree–Fock orbitals into nonorthogonal localized molecular orbitals, *ab initio* quantum-chemical methods making use of fast MP2 [185] and CCSD [186] schemes which scale *linearly* with the number of electrons have been developed, a major step in electronic-structure theory.

The nonorthogonal orbitals mentioned above come close to the objects of chemical intuition, namely covalently bonded and nonbonded electron pairs, at least for many molecules. A chemically justified way to define a system's wave function is indeed possible by an antisymmetrized product of electron group functions or generalized group functions [187]; probably the most important case of group functions are the two-electron functions (geminals, opposed to orbitals) [188], and they can also be given a linear scaling [189]. The geminal approach belongs to the family of pair theories [190] and its success is due to the particular importance of pair-electron correlations [191]. The essential computational problem, however, is *how* to optimize a wave function in the form of a group product, for example, as an antisymmetrized product of strongly orthogonal group functions [192].

The final challenge is how to bring the highly accurate methods of molecular quantum chemistry into the solid state. It is certainly possible to run an accurate calculation for a cluster *embedded* in a potential that is characteristic for the infinite environment, but the use of Bloch's theorem and periodic boundary conditions requires, for example, either Gaussians adapted to the translational symmetry [193] or local (Wannier) functions [194] although the latter are obtained *after* having solved the $\boldsymbol{k}$-dependent Hartree–Fock equations. Alternative approaches targeted at achieving local orbitals directly in

27) A nonvariational method does not behave according to the variational principle, such that a lowered energy is not necessarily connected with a better wave function, and vice versa. For example, semiempirical extended Hückel theory is nonvariational due to its internal parametrization.

real space have been proposed [195]. There are also advanced *cyclic* cluster models which combine the virtues of the local (cluster) approach and the delocalized approach with periodic boundary conditions [196].

The above-mentioned group functions can also be useful for calculating the electronic structure of solids. Low-dimensional systems have been investigated on the basis of geminals [197] and three-dimensional periodic systems by making use of nonorthogonal basis sets [198]. For one-dimensional periodic systems comprised of nonequidistant and equidistant hydrogen (and other) atoms, a general variational formalism for electron groups [199] quantitatively recovers that the correlation energy is larger in the nonequidistant system, in accordance with intuition. There is a lot of exciting research going on in this rapidly evolving field.

2.14
Absolute Electronegativity and Hardness

The probably most general chemical reaction is given by

$$\text{Lewis acid} + \text{Lewis base} \rightleftharpoons \text{Lewis acid–base complex},$$

in which the Lewis acid/base represents the electron acceptor/donor. Adopting this language, nonelemental molecules and solid-state materials may be looked upon as finite and infinite examples of acid–base complexes. This classification can be greatly augmented by the qualitative description of *hard* and *soft* acids and bases (HSAB) as done by Pearson [200], nowadays an essential topic of general chemistry.

Accordingly, a soft acid is a big species (cation), not too highly charged, whose valence shell is easily polarizable. In contrast, a hard acid is well represented by a small and highly charged species (cation) which is difficult to polarize. Likewise, a soft base is an easily polarizable, easily oxidizable species (anion) with a small electronegativity and low-lying orbitals. Consequently, a hard base is generated by a species (anion) which is difficult to polarize or oxidize, with a high electronegativity and high-lying empty orbitals. Pearson's hard-prefers-hard and soft-prefers-soft rule for compound formation has become textbook material [201]. The rule immediately explains – in contrast to Pauling's electronegativity argument – why the solid-state reaction

$$\text{LiI} + \text{CsF} \rightleftharpoons \text{LiF} + \text{CsI}$$

is highly exothermic ($\Delta H_R \approx -130$ kJ/mol) on going to the right side, the "hard-hard" and "soft-soft" one, although the maximum electronegativity difference (CsF) is found for the educt side.

It turns out that the above statements can be expressed in terms of quantum-mechanical quantities using density-functional theory, and that is the reason why we include them here. Both *absolute electronegativity* χ and also *absolute hardness* η were defined [202] in terms of the ionization potential I and the electron affinity A of a system having total energy E and total electron number N according to

$$\begin{aligned} \chi &= -\mu && (2.92) \\ &= -\left(\frac{\partial E}{\partial N}\right)_{N^0} && (2.93) \\ &\approx \frac{1}{2}(I+A), && (2.94) \end{aligned}$$

where μ is the chemical potential and N^0 is the number of electrons within the neutral molecular species, and

$$\begin{aligned} \eta &= \frac{1}{2}\left(\frac{\partial^2 E}{\partial N^2}\right)_{N^0} && (2.95) \\ &= -\frac{1}{2}\left(\frac{\partial \chi}{\partial N}\right)_{N^0} && (2.96) \\ &\approx \frac{1}{2}(I-A) \geq 0. && (2.97) \end{aligned}$$

The last line is only valid if the first ionization potential is larger than the corresponding electron affinity but, in fact, there are *no* atomic or molecular counterexamples. If a negative absolute hardness would exist for an arbitrary system X, however, then 2X were unstable with respect to a decay into X^+ and X^- (spontaneous disproportionation), reflecting an attractive (!) electron–electron correlation which is difficult to imagine. The size and energetic positioning of χ and η may be visualized by referring to the positions of HOMO and LUMO of some molecule, as shown in Figure 2.32. The negative absolute electronegativity $(-\chi)$ lies in the middle between HOMO and LUMO, and the HOMO-LUMO gap is twice the absolute hardness η. Molecules with a wide HOMO-LUMO gap are therefore "harder" than those with a narrow gap.

To investigate how the total energy E of the ground state changes with the electron number N, one may now express E as a power series of N while keeping everything else constant to find

$$\begin{aligned} E &= E^0 + \left(\frac{\partial E}{\partial N}\right)_{N^0} dN + \frac{1}{2}\left(\frac{\partial^2 E}{\partial N^2}\right)_{N^0} dN^2 + \ldots && (2.98) \\ &= E^0 + \mu dN + \eta dN^2 + \ldots && (2.99) \end{aligned}$$

Thus, the change in the ground-state energy is reflected by the chemical potential μ which equals the slope of the function E vs. N at N^0 and measures the escaping tendency of an electron. The electronegativity therefore plays a

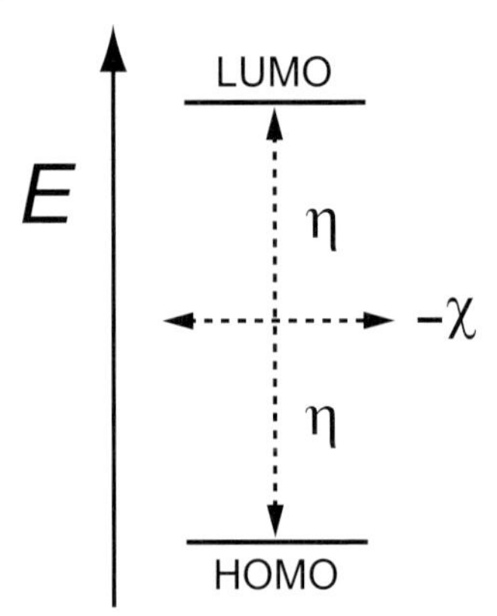

Fig. 2.32 Schematic orbital energy diagram of an arbitrary molecule in the proximity of the highest occupied molecular orbital (HOMO) and the lowest unoccupied molecular orbital (LUMO); also see text.

similar role in the DFT variational principle which the energy plays in wave-function theory [142], a most interesting discovery!

Figure 2.33(a) and (b) depict experimental E vs. N functions for two atoms. Although the two atoms could not be more different in terms of chemistry and physics (a light nonmetal, P, and a heavy metal, Pb), the shapes of the two functions are indeed similar. In Figure 2.33(c), their shapes are generalized

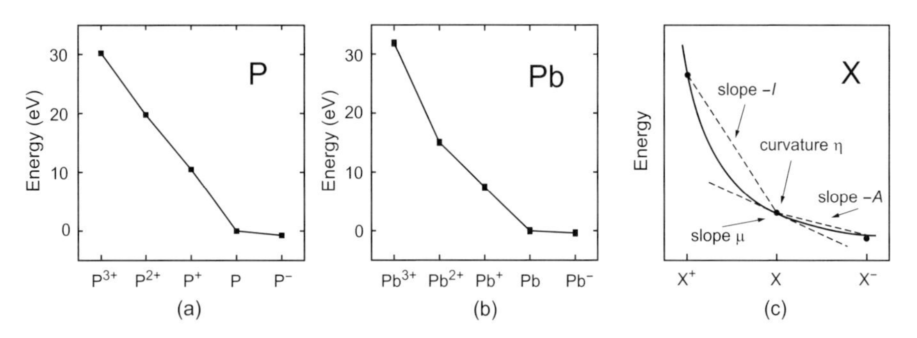

Fig. 2.33 Total energy E of an atomic/molecular system ((a): atomic P; (b): atomic Pb; (c): arbitrary system X) as a function of the electron number N.

into a more schematic E vs. N function, and its slope (μ) for the neutral system X is approximated by taking the arithmetic average of the slope on the left side ($= -I$) and the slope on the right side ($= -A$) as done before; Mulliken's expression for the electronegativity is immediately regained.[28]

28) One problem, however, becomes obvious when investigating fractional electron numbers as a time average. Because the E vs. N curve is a series of straight line segments with slope discontinuities at integral N (see Figure 2.33(a) and (b)), the chemical potential *jumps* because of irregularities in the exchange–correlation potential. This is also known as the band-gap problem in density-functional theory [203].

The second derivative η, measuring the curvature, is of comparable interest as it stands for the electronic tendency of a system to disproportionate (see discussion above) and the sensitivity of the electronegativity to change in the number of electrons. If the second derivative is greater than zero, the ground state E is *concave upwards* with changing particle number, and the system will not decompose into charged pieces (i.e., will not disproportionate). The finite-difference approximation for the curvature η is

$$\eta \approx \frac{1}{2}(E^{+} + E^{-}) - E^{0}, \tag{2.100}$$

and given a positive η, the system X stays at E^0 instead of breaking up into two pieces, X^+ and X^-, having energies of E^+ and E^-. It seems obvious to think of the curvature value η as some kind of "resistance indicator" of the system against an electronic (chemical) attack.

Corresponding tabulations of atomic and ionic absolute electronegativities and absolute hardnesses have been generated [204] using numerical estimates of electron affinities and ionization potentials from gas-phase measurements. In general, these data are in fascinating agreement with chemical knowledge; for convenience, we include them in Tables 2.2 and 2.3.

The χ and η values also allow a huge amount of chemical information to be rationalized based on the principles of electronic structure [201]. In addition, the approximate formulae

$$\Delta N = \frac{\chi_{\mathrm{A}} - \chi_{\mathrm{B}}}{2(\eta_{\mathrm{A}} + \eta_{\mathrm{B}})} \tag{2.101}$$

and

$$\Delta E = -\frac{1}{4}\frac{(\chi_{\mathrm{A}} - \chi_{\mathrm{B}})^2}{\eta_{\mathrm{A}} + \eta_{\mathrm{B}}}, \tag{2.102}$$

introduced by Parr and Pearson [202] yield semiquantitative measures of electron transfer ΔN and energy transfer ΔE during a chemical reaction between atoms A and B, due to the equalization of their chemical potentials.

For periodic systems, an *ad hoc* partitioning of their absolute hardnesses into atomic contributions results in the definition of reactivity indices for such extended compounds [205] which proves useful in order to understand the chemical behavior of solid-state materials [206]. For example, the differing reactivities of various polymorphs of lithium aluminate may be rationalized in terms of their acid–base behavior [207], and the quantum-chemical explanation is in accordance with the principle of Le Chatelier.

Tab. 2.2 Absolute electronegativities and absolute hardnesses for atoms [204].

Atom	χ (eV)	η (eV)	Atom	χ (eV)	η (eV)
Ag	4.44	3.14	Mo	3.9	3.1
Al	3.23	2.77	N	7.30	7.23
As	5.3	4.5	Na	2.85	2.30
Au	5.77	3.46	Nb	4.0	3.0
B	4.29	4.01	Ni	4.40	3.25
Ba	2.4	2.9	O	7.54	6.08
Be	4.9	4.5	Os	4.9	3.8
Bi	4.69	3.74	P	5.62	4.88
Br	7.59	4.22	Pb	3.90	3.53
C	6.27	5.00	Pd	4.45	3.89
Ca	2.2	4.0	Pt	5.6	3.5
Cd	4.33	4.66	Rb	2.34	1.85
Cl	8.30	4.68	Re	4.02	3.87
Co	4.3	3.6	Rh	4.30	3.16
Cr	3.72	3.06	Ru	4.5	3.0
Cs	2.18	1.71	S	6.22	4.14
Cu	4.48	3.25	Sb	4.85	3.80
F	10.41	7.01	Sc	3.34	3.20
Fe	4.06	3.81	Se	5.89	3.87
Ga	3.2	2.9	Si	4.77	3.38
Ge	4.6	3.4	Sn	4.30	3.05
H	7.18	6.43	Sr	2.0	3.7
Hf	3.8	3.0	Ta	4.11	3.79
Hg	4.91	5.54	Te	5.49	3.52
I	6.76	3.69	Ti	3.45	3.37
In	3.1	2.8	Tl	3.2	2.9
Ir	5.4	3.8	V	3.6	3.1
K	2.42	1.92	W	4.40	3.58
La	3.1	2.6	Y	3.19	3.19
Li	3.01	2.39	Zn	4.45	4.94
Mg	3.75	3.90	Zr	3.64	3.21
Mn	3.72	3.72			

Tab. 2.3 Absolute electronegativities and absolute hardnesses for monoatomic cations [204].

Ion	χ (eV)	η (eV)	Ion	χ (eV)	η (eV)
Ag^{+}	14.53	6.96	Mn^{3+}	42.4	8.8
Ag^{2+}	28.2	6.7	Mo^{2+}	21.60	5.51
Al^{3+}	74.22	45.77	Mo^{3+}	36.8	9.6
Au^{+}	14.9	5.6	Na^{+}	26.21	21.08
Au^{3+}	45.8	8.4	Nb^{2+}	19.68	5.36
B^{3+}	148.65	110.72	Nb^{3+}	31.7	6.6
Be^{2+}	86.05	67.84	Ni^{2+}	26.67	8.50
Br^{+}	16.8	5.0	Ni^{3+}	45.0	9.9
Ca^{2+}	31.39	19.52	Os^{2+}	22.3	5.7
Cd^{2+}	27.20	10.29	Os^{3+}	35.2	7.5
Ce^{3+}	28.48	8.28	Pb^{2+}	23.49	8.46
Cl^{+}	18.39	5.42	Pd^{2+}	26.18	6.75
Co^{+}	12.46	4.60	Pt^{2+}	27.2	8.0
Co^{2+}	25.28	8.22	Rb^{+}	15.77	11.55
Co^{3+}	42.4	8.9	Re^{3+}	33.7	7.8
Cr^{2+}	23.73	7.23	Rh^{+}	12.77	5.31
Cr^{3+}	40.0	9.1	Rh^{2+}	24.57	6.49
Cs^{+}	14.5	10.6	Rh^{3+}	42.2	11.2
Cu^{+}	14.01	6.28	Ru^{2+}	22.62	5.86
Cu^{2+}	28.56	8.27	Ru^{3+}	39.2	10.7
Fe^{2+}	23.42	7.24	Sc^{2+}	18.78	5.98
Fe^{3+}	42.73	12.08	Sc^{3+}	49.11	24.36
Ga^{3+}	47	17	Sn^{2+}	22.57	7.49
Ge^{2+}	25.08	9.15	Sr^{2+}	27.3	16.3
Hf^{2+}	19.1	4.2	Ti^{2+}	20.54	6.96
Hf^{3+}	28.3	5.0	Ti^{3+}	35.38	7.89
Hg^{2+}	26.5	7.7	Tl^{+}	13.27	7.16
I^{+}	14.79	4.34	Tl^{3+}	40.3	10.4
In^{3+}	41	13	V^{2+}	21.98	7.33
Ir^{+}	13.0	3.9	V^{3+}	38.01	8.70
Ir^{3+}	37.4	7.9	W^{2+}	20.9	4.5
K^{+}	17.99	13.64	W^{3+}	32.4	7.0
La^{3+}	34.57	15.39	Y^{2+}	16.38	4.14
Li^{+}	40.52	35.12	Y^{3+}	41.2	20.6
Lu^{3+}	33.08	12.12	Zn^{2+}	28.84	10.88
Mg^{2+}	47.59	32.55	Zr^{2+}	18.06	4.93
Mn^{2+}	24.66	9.02	Zr^{3+}	28.65	5.68

Let us close this section with some qualitative thoughts. Because the electronegativity plays such an important role in chemical understanding, it may also be utilized to order the plethora of chemical substances in terms of their bonding characters. The original idea is quite old and goes back to van Arkel [208] and Ketelaar [209], and we offer such a van Arkel–Ketelaar triangle in Figure 2.34. Quite obviously, the electronegativity difference $\Delta\chi$ is an excellent measure of ionic character, whereas the mean electronegativity $\langle\chi\rangle$ allows one to separate metallic and covalent systems (bottom). Here, a metalloid entry of a semiconductor (Si) stands for the crossover. Zintl phases (such as NaTl, SiGe, and many others [2]) represent the border between metallic and ionic materials, and the "polymeric" systems, that is, those which are nonmolecular but not really ionic either (such as silicates, borates, phosphates, etc.), symbolize polar covalent species.

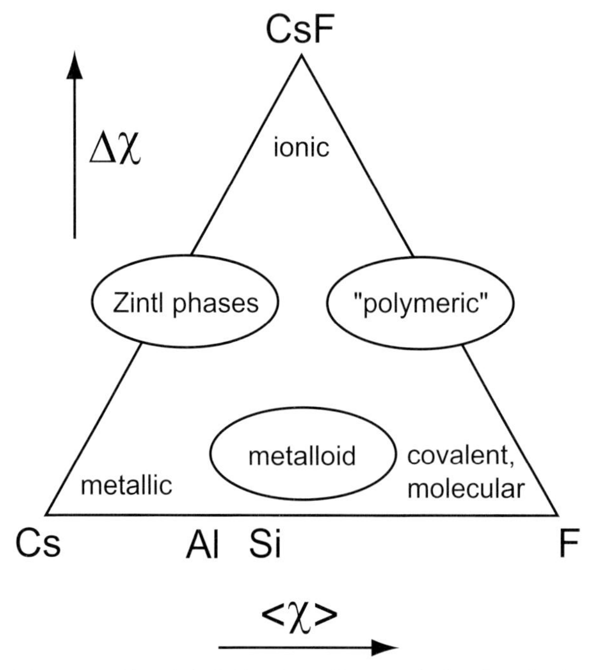

Fig. 2.34 Van Arkel–Ketelaar diagram indicating the bonding character as a function of electronegativity difference $\Delta\chi$ and mean electronegativity $\langle\chi\rangle$; also see text.

2.15 Potentials and Basis Sets in Solids

If we know the translational symmetry of an extended solid (Bloch's theorem) and also have a trustworthy strategy on how to deal with the nuclear potential and the electron–electron interactions (by, say, a semiempirical method or by DFT), we are ready to explicitly *calculate* the band structure of any *real* material. Although we have done this before for idealized systems (see sketches in Section 2.6), let us now attack the problem once again, but in more general terms. For real materials, one needs to solve Schrödinger's equation using the *true* potential $v(\boldsymbol{r})$, namely,

$$\left[-\frac{1}{2}\nabla^2 + v(\boldsymbol{r})\right] \psi(\boldsymbol{k},\boldsymbol{r}) = E(\boldsymbol{k})\psi(\boldsymbol{k},\boldsymbol{r}), \tag{2.103}$$

and it will soon become obvious how the potential (or its choice) effectively determines both the *band-structure method* and its associated *basis functions*, in a somewhat similar way to the realm of molecular quantum chemistry. The *type* of basis functions, however, very often strongly differs between the molecular world and the world of infinite molecules (that is, extended solids) simply because of the underlying translational symmetry which needs to be taken care of.

Because translationally invariant systems are extended, *local* basis functions such as Slater-type or Gaussian-type orbitals appear as ill-fitting, at first sight, to the computational problem. Although this preconception is not correct, the translational symmetry indeed suggests another straightforward set of basis functions. Bloch's theorem itself *contains* the natural basis function for a translationally invariant solid, namely a *plane wave* ($e^{i\boldsymbol{k}\boldsymbol{r}}$) which is simply the symmetry-adapted wave function to a crystal's boundary conditions. This also means that *any* kind of crystal wave function $\psi_n(\boldsymbol{k},\boldsymbol{r})$ can be expressed as a so-called *plane-wave expansion* which is nothing but a linear combination of various exponential functions, the delocalized LCAO analogue to Equation (2.6). Within solid-state physics, this is usually written as

$$\psi_n(\boldsymbol{k},\boldsymbol{r}) = \sum_{\boldsymbol{K}} c_n(\boldsymbol{k},\boldsymbol{K}) e^{i(\boldsymbol{k}+\boldsymbol{K})\boldsymbol{r}} = e^{i\boldsymbol{k}\boldsymbol{r}} \sum_{\boldsymbol{K}} c_n(\boldsymbol{k},\boldsymbol{K}) e^{i\boldsymbol{K}\boldsymbol{r}}. \tag{2.104}$$

Here, $\boldsymbol{K}$ is a reciprocal lattice vector (see Section 2.5), and the mixing coefficients c_n must be sought, either analytically or numerically.

We reiterate that, just as in LCAO-MO theory, the originally unknown function $\psi_n(\boldsymbol{k},\boldsymbol{r})$ is expressed by a set of known functions, but the latter are entirely delocalized and not localized as atomic orbitals. While this ansatz ultimately leads us away from the notion of a solid-state material being composed of atoms with their associated atomic functions (this ansatz is a *very* nonchemical idea), it admittedly has certain advantages in terms of mathematics. Because of their simplicity, plane waves are easy to deal with computationally, especially if they are applied to likewise simple (nonchemical) potentials. For example, the so-called "jellium solid" of solid-state physics, with uniform charge everywhere, is characterized by analytical plane-wave-like Hartree solutions. For realistic potentials, however, plane waves may become troublesome, which we will illustrate shortly.

Figure 2.35 shows the 3*s* radial function (radial part of the atomic orbital) of an arbitrary atom (Na in this case) according to a self-consistent atomic calculation. On each side, there are two radial nodes[29] close to the origin characteristic for the 3*s* electron's high kinetic energy which itself reflects the large nuclear potential.

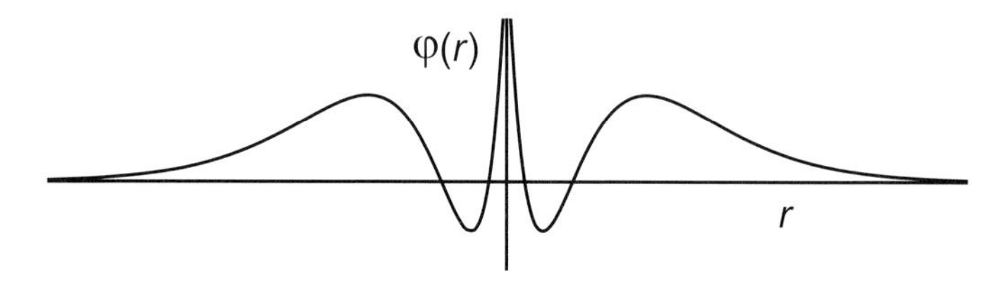

Fig. 2.35 Radial part of the 3*s* atomic orbital of the Na atom.

Outside the atom, there is a smooth exponential decay of the valence function which might be approximated by a nodeless STO. If this atom is made part of a solid, such as in crystalline $Na_{(s)}$, the new situation will also reflect, just as for the isolated atom, very different regions with either steep potential valleys because of a high Coulomb potential (close to the atomic core, with wiggled core functions) or a rather smooth potential in the bonding region, dominated by the overlapping valence functions. This has been sketched in Figure 2.36. It goes without saying that the course of the symmetry-adapted

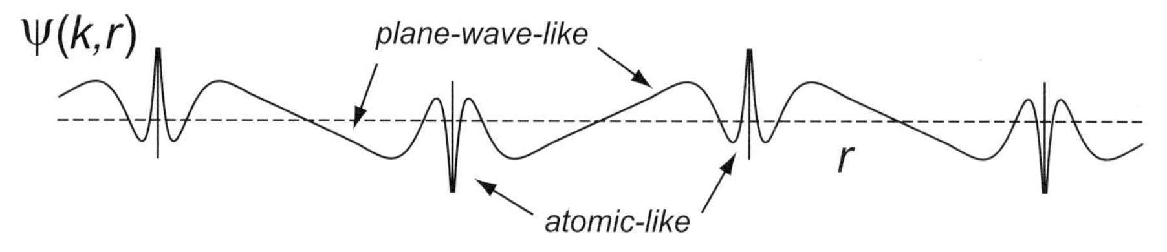

Fig. 2.36 Schematic drawing of a 3*s*-derived Bloch function of a one-dimensional crystal of Na atoms at the **X** point. Close to the Na nuclei, $\psi(k,r)$ oscillates rapidly; in the bonding region, $\psi(k,r)$ is a smooth function. If one neglects the atomic-like region, the Bloch function resembles a simple cosine function.

29) For a given atomic orbital, the number of nodes is $n - l - 1$ where n and l stand for the main and angular-momentum quantum numbers. Thus, the 3*s* atomic orbital must have $3 - 0 - 1 = 2$ nodes.

wave function (Bloch function) of a one-dimensional Na crystal at the **X** point reflects both types of potential. The formerly local 3*s* atomic function has mutated into a fully delocalized one which, and this is important, must possess enough variational freedom to *simultaneously* exhibit core-like behavior close to the atom, but also typical plane-wave behavior between the atoms. This is easy to understand but not easy to accomplish numerically using plane waves because the wiggled core-like functions require substantial effort. Why is this? The strongly wiggled core functions are high-energy functions (recall that they stand for the electron's high kinetic energy close to the nucleus) and a plane-wave expansion (Equation (2.104)) of such a core function must contain high-energy (or, high-frequency) contributions with a large $\boldsymbol{k} + \boldsymbol{K}$ argument in the exponential part; recall that $E = k^2$ in the free-electron Sommerfeld model, see Section 2.5. Thus, the size (or energy) of the plane-wave basis set which is proportional to Z^3 will be tolerable for a band-structure calculation on, say, diamond containing the light carbon atom ($Z = 6$). For heavier atoms (such as lead with $Z = 82$), however, it will computationally explode and make realistic calculations almost impossible unless some trick is introduced.

Three different approaches have been followed to solve this clue, and they form the backbones of all existing band-structure methods in terms of their nuclear potentials. Somewhat simplified, one may either ignore the core functions (empirical tight-binding approaches), one may modify the potential, thereby also ignoring the core functions (pseudopotential approach), or one may modify the basis sets and split the functions into core and beyond-core functions (cellular approaches and successors) [210].

2.15.1 Empirical Tight-binding and Nonempirical Relatives

In short, the **tight-binding (TB)** method is based on the assumption that the atoms within the solid-state material under consideration engage in almost *no* chemical interaction. The semantically probably misleading term "tight-binding" therefore does *not* allude to bonding between the atoms! On the contrary, one thinks of the *electrons* as being very tightly bound to the atoms they "belong to" such that the atomic wave functions in the solid must be extremely close to the original, energy-independent atomic wave functions ϕ_μ. If that is so, let us then expand an arbitrary crystal orbital $\psi_j(\boldsymbol{k}, \boldsymbol{r})$ as a Bloch sum (with a translation vector $\boldsymbol{T}$) over a set of local atomic orbitals (such as STOs), that is

$$\psi_j(\boldsymbol{k}, \boldsymbol{r}) = \sum_{\boldsymbol{T}} e^{i\boldsymbol{k}\boldsymbol{T}} \sum_{A} \sum_{\substack{\mu=1 \\ \mu \in A}}^{n} c_{\mu j}(\boldsymbol{k}) \phi_\mu(\boldsymbol{r} - \boldsymbol{T}). \qquad (2.105)$$

Having thus utilized the valence eigenfunctions of the free atoms as basis functions for the Bloch expansion, the reader will have no difficulty in recognizing that this is just the solid-state, $\boldsymbol{k}$-dependent analogue of the molecular LCAO method which we saw in Equation (2.6)! Ironically, upon neglecting atom–atom interactions (or, equivalently, chemical bonding) at the very beginning, this method eventually turns out to be *closest* to chemical understanding because it retains the idea of atoms making up the solid. Practical experience also shows that, in contrast to the initial concept, the method can also be used for almost free-electron-like materials, giving it an enormous potential and also appreciation by the solid-state physics community [53,211].

The performance of any TB implementation crucially depends on the basis set and the accuracies of Hamiltonian and overlap matrix elements; the latter need to be explicitly calculated because neighboring atomic functions are not necessarily orthogonal to each other. In practice, $H_{\mu\nu}$ and $S_{\mu\nu}$ are either approximated or fitted to experimental data, thereby establishing the **empirical tight-binding** method, as it is usually called, within solid-state physics. Nonempirical TB-related approaches are covered at the end of this section. It goes without saying that the TB method is not only similar to the quantum-chemical extended Hückel theory (based on Slater-type orbitals and a recipe to approximate $H_{\mu\nu}$ by calculating $S_{\mu\nu}$, see Section 2.11.1), but is virtually identical to it. We note that the derivation of the band structures of one- and two-dimensional systems covered in Section 2.6 was based on such a tight-binding approach. In particular, the above Bloch expansion in terms of atomic orbitals simplifies to

$$\psi(k) = \sum_T e^{ikT} \phi(T) \tag{2.106}$$

when applied to a one-dimensional system and one atom/orbital per unit cell, and this has been used in Section 2.6 to draw the band structure of the one-dimensional chain of hydrogen atoms. Thus, we have already used the method without informing the reader of its name.

The greatest advantage of the empirical tight-binding method in terms of numerical computation is given by the extremely small (often minimum, in fact) basis sets. A calculation of silicon in the diamond structure would need only four orbitals per Si atom ($3s$, $3p_x$, $3p_y$, $3p_z$) such that an 8×8 matrix (because there are two Si atoms per primitive unit cell) would have to be diagonalized. Likewise, body-centered cubic iron, with one Fe atom per primitive unit cell, would be very well described by six basis functions ($4s$, $3d_{x^2-y^2}$, $3d_{z^2}$, $3d_{xy}$, $3d_{yz}$, $3d_{xz}$), giving only a 6×6 matrix;[30] this makes the method also quite transparent and chemically intuitive, especially when it comes to electron (or

30) Note that this primitive unit cell with one Fe atom is only half as large as the crystallographic *I*-centered unit cell of bcc-Fe, depicted in Figure 1.2(c). Also, the primitive unit cell of Si is only half as large as the crystallographic *F*-centered unit cell of the diamond structure.

energy) partitionings which are simple to derive because of the *local* nature of the orbitals (see Section 2.8). Without any disrespect to competing methods, probably no other approach has shed more light on quantum-*chemical* problems than the tight-binding method in the form of the extended Hückel approach. The algebraic eigenvalue problem corresponds to a matrix diagonalization because, as was mentioned at the beginning, the atomic orbitals are energy-independent. This is different for other methods (see Section 2.15.3).

In turn, the advantages may emerge as disadvantages – unfortunately, they really do – because the energy-independence of the small basis sets makes the latter somewhat inflexible to changing potentials. If carried out using semiempirical parametrizations, the latter can always be questioned, and it seems difficult to describe quantitatively, for example, electron–electron interactions (let alone spin-dependent phenomena) in a convincing manner. Without any disrespect, these details are almost impossible to cover using an empirical method.

This is the motivation for the development of **nonempirical tight-binding-**related schemes which, just as before, start with a set of local atomic orbitals but now explicitly evaluate all necessary integrals in an *ab initio* manner. For example, **Hartree–Fock** theory may be brought to the solid state and is then typically based on local Gaussian functions [132]; no other simplifications go into the Hamilton operator. Nonetheless, whenever very large systems must be investigated, the speed of Hartree–Fock for periodic systems may be significantly improved using a proper self-consistent parametrization of the various integrals, and this goal has been achieved for many elements [212]. Clearly, this is somewhere between empirical and nonempirical tight-binding.

Within density-functional theory, a linear combination of overlapping nonorthogonal orbitals from first principles may be utilized to arrive at at **full-potential local orbital (FPLO)** method [213], and this $\boldsymbol{k}$-dependent LCAO approach comes close to full-potential APW-based methods (see Sections 2.15.3 and 2.15.4) in terms of numerical accuracy, although FPLO is much faster simply because of the locality of the basis set. Even faster, due to a strongly simplified potential, is a parameter-free (density-functional) tight-binding method called **TB-LMTO-ASA**, derived through localization of a delocalized basis set (see Section 2.15.4).

2.15.2 Pseudopotentials

As has been said before, plane waves might serve as general and straightforward basis functions if there were not the rapid oscillations of the atomic wave functions close to the nuclei. If these oscillations are artificially suppressed, such as in the free-electron model, plane waves are the optimum choice. Since these oscillations, however, reflect the atomic nature, that is, the *chemistry* of the individual atoms, the plane-wave solutions merely reflect the translational symmetry of the material, and all the chemistry is lost. This is totally unacceptable, at least for chemical questions.

Given that the low-lying core orbitals ϕ_{core} of an atom are known from an atomic calculation, however, it is possible to construct plane waves representing the valence levels by forcing these plane waves to be *orthogonal* to the core levels for a specified $\boldsymbol{k}$, namely by writing

$$\phi^{\mathrm{OPW}}(\boldsymbol{k},\boldsymbol{r}) = e^{i\boldsymbol{k}\boldsymbol{r}} + \sum_{\mathrm{core}} c_{\mathrm{core}}(\boldsymbol{k})\phi_{\mathrm{core}}(\boldsymbol{k},\boldsymbol{r}), \tag{2.107}$$

and this is the original approach of the **orthogonalized plane wave (OPW)** method which uses a $\boldsymbol{k}$-weighted combination of OPWs for expanding the Bloch function [214]. Thus, an OPW oscillates at the core and behaves as a plane wave in the outer regions of the atom. OPW calculations were used in the early days for the calculation of band structures of semiconductors, despite some inherent mathematical problems.

A more systematical way to deal with the oscillations of the core functions is given by the **pseudopotential** approach. Here, the idea is to get rid of these oscillations completely by replacing the strong ion–electron potential by a much weaker (and physically slightly dubious) pseudopotential. The atom may then be regarded as a small perturbation of the electron gas. The mathematical motivation is impressive but it must be based on a convincing physical or chemical argument; here is one. On the one side, the valence electrons experience a high nuclear charge which is somewhat screened by the core electrons. On the other side, the valence electrons *cannot enter the core region* mostly because of Pauli's principle. Remember that all core levels are filled already such that there is a strong repulsion between valence and core electrons, and that is also why valence and core orbitals must be orthogonal to each other. The two opposite effects (Coulomb attraction of the valence electrons by the nucleus *but* Pauli repulsion of the valence electrons by the core electrons) cancel each other, almost completely, and only a weak pseudopotential is left. This beautiful argument was first formulated by Hellmann [3,215], and pseudopotential theory – in the mid 1930s called the **combined approximation method** – was born. This milestone of blending the concept of the periodic table with quantum mechanics is depicted in Figure 2.37. Similar but independent ideas were derived, at about the same time, by Gombás [216]. Pseudopotentials nowadays also go by the somewhat more pompous name of **effective core potentials**.

In the late 1950s, pseudopotentials were re-invented for the solid state by replacing the OPW orthogonality recipe with an effective potential, the above pseudopotential [217,218], and the method has blossomed ever since. Indeed, pseudopotential theory is *not* a specialty of the solid state because a large part

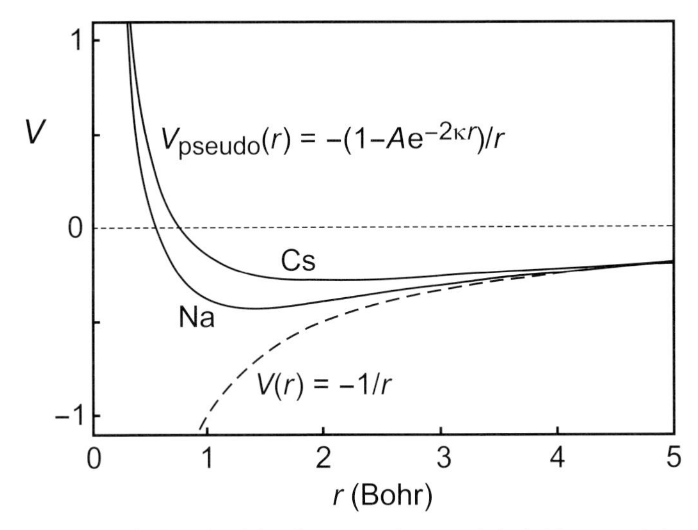

Fig. 2.37 A sketch of the first pseudopotentials (with potential parameters A and κ) for two alkali metals as envisaged by Hellmann.

of today's molecular quantum chemistry is (and has been) performed on the basis of pseudopotentials [219]. In contrast to Hellmann's simple, *local* pseudopotential (Figure 2.37) for the alkali metals

$$v_{\text{pseudo}}(r) = -\frac{1}{r}\left(1 - Ae^{-2\kappa r}\right), \tag{2.108}$$

with $A = 1.826/1.672$ and $\kappa = 0.533/0.333$ for Na/Cs, modern pseudopotentials are either *semilocal* or *nonlocal* in nature. This means that one allows the pseudopotential to be angular-momentum dependent such that the effective potential seen by s, p, d and f levels, differs.[31] Mathematically, a projection operator $\mathcal{P}_l$ for each l channel ensures, in the semilocal case, that each orbital feels its own correct pseudopotential,

$$v_{\text{pseudo}}(r) = \sum_{l=0}^{\infty} v_l(r)\mathcal{P}_l, \tag{2.109}$$

with a particular $v_l(r)$ for a given s, p, d and f level. In the nonlocal case, the treatment of the individual levels may be entirely different.

Within the molecular quantum-chemical regime, various strategies on how to generate sets of pseudopotentials have been followed, such as the *shape-consistent* pseudopotentials for which the pseudo-valence orbital is identical

31) In Section 2.9, we have already illustrated the different effective potentials acting upon $2p$ (strong), $3p$ (weaker) and $4p$ orbitals (even weaker), and this is because a valence p orbital is most strongly "pushed out" by a filled core p function. Since there is no $1p$ function, only the $2p$ pseudopotential is close to the real potential.

with the actual valence function beyond some critical (core) radius r_c, but is nodeless inside. Alternate powerful approaches – energy-consistent pseudopotentials, generalized relativistic effective core potentials, and several others – plus additional refinements have brought pseudopotential theory to maturity [220]. For the solid state, the motivation to use pseudopotential theory [221] is driven by the necessity to improve the convergence of the plane-wave expansion, that is, to make the plane-wave basis set extremely compact (see discussion above). To do so, nonlocal but *norm-conserving* pseudopotentials (because they retain the same charge density in the atom's inner part) may be constructed from first principles [222]. Adopting the pseudopotential jargon, the smaller the plane-wave basis set that can be used with that pseudopotential, the *softer* the pseudopotential. For some elements (alkali metals, transition metals) the plane-wave basis sets are still pretty large, but this problem can be circumvented by introducing *ultra-soft* pseudopotentials [223]; for these, the norm-conservation is abandoned which, in turn, needs less plane waves in the final calculation but significantly more effort to generate the pseudopotential in the first place.

There are many advantages of pseudopotential theory, in particular for the solid state. For obvious reasons, the pseudopotential carries basic *chemical* experience into the physical equations. Atoms having the same number of valence electrons exhibit similar pseudopotentials, the quantum-chemical corroboration of the chemists' ancient decision to define groups of elements in the periodic table, decades before quantum mechanics was established. The extreme simplicity of plane waves translates into mathematical advantages (speed), and deficiencies of local atomic functions (so-called basis-set superposition errors) are unknown for the position-independent plane waves, unlike traditional molecular quantum-chemistry schemes based on STOs or GTOs. The size of the basis set can be nicely controlled by setting an energy cut-off E for the plane-wave expansion (Equation (2.104)) including only those coefficients with $E \leq (\boldsymbol{k} + \boldsymbol{K})^2$.

The disadvantage of the pseudopotential (plane-wave) approach can be summarized as follows. First, all information for the core-like wavefunctions and their associated electron density is lost, trivially so. Second, an essentially delocalized plane-wave basis set poses difficulties when it comes to questions of *chemical interpretation* in terms of atoms and bonds. In addition, there has always been some concern about the arbitrariness of pseudopotentials because there is certainly more than *one* pseudopotential (in fact, an infinite number) for a given atom. Recall that the partitioning of a quantum-mechanical system into subsystems is invalid because the electrons *cannot* be distinguished; in principle, there are no "core" or "valence" electrons. Nonetheless, pseudopotential plane-wave calculations have established themselves as accurate and powerful tools for electronic-structure theory, both for molecules and, in particular, for solids [224]. In addition, they have paved the way for molecular dynamics from first principles, briefly covered in Section 2.17.

2.15.3
Cellular (Augmentation) Methods

A straightforward way to calculate the electronic band structure of a solid would be to solve Schrödinger's equation for an atom in a one-atom (primitive) unit cell and then try to match the energy-dependent atomic functions with neighboring functions (located on atoms in the neighboring unit cells) at the boundary between the unit cells. This early approach by Wigner and Seitz [225], dubbed the **cellular method**, sounds elegant but is computationally quite demanding. It also contains two of the essential ingredients of all other methods yet to come. First, the wave functions sought for are characterized by an explicit energy-*dependence* and, second, the real atomic potential is *no longer spherically symmetric*, schematically sketched for a two-dimensional example in Figure 2.38(a).

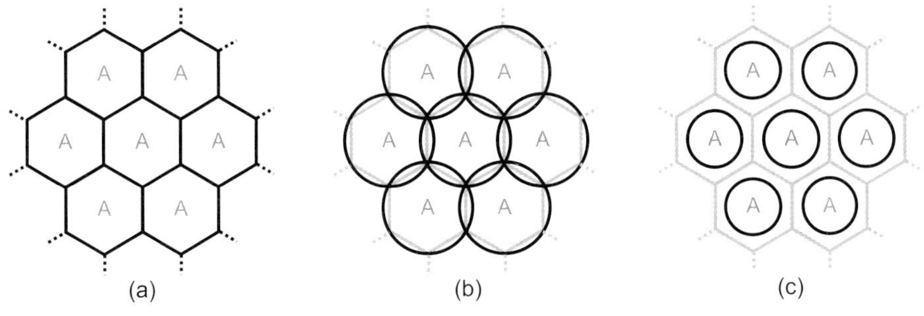

Fig. 2.38 A two-dimensional lattice of A atoms (a) with hexagonal potentials; the primitive unit cells are replaced by overlapping circles (b); a muffin-tin potential (c) for the hexagonal lattice, composed of circular potentials (atomic regions) with flat potentials (interstitial regions) inbetween.

Upon replacing the polyhedral cells (a) by overlapping spheres (Figure 2.38(b)), the problem has already become slightly easier to solve, but it still suffers from a discontinuous potential derivative at the sphere boundaries; thus, a more comfortable situation is enforced by setting up a *muffin-tin* potential (c), with an atomic-like shape close to the nucleus and a zero potential inbetween.

This is the simplified, two-region potential, on which the **augmented-plane wave (APW)** method by Slater [226, 227] is essentially based. It stands for the rigorous approach which correctly describes *both* the spherical, atomic-like region close to the nucleus *and also* the flat region between the nuclei, by treating the two regions differently, in the spirit of the above muffin-tin idea.

To do so, a hybrid wave function (the augmented plane wave) is constructed, composed of atomic functions (called partial waves, with spherical harmonics Y_{lm}) inside the atomic, muffin-tin sphere r_{MT} and a single plane wave outside the atomic sphere, that is

$$\begin{aligned}\phi(E,\boldsymbol{k},\boldsymbol{r}) &= \textstyle\sum_{lm} c_{lm} R_l(E,r) Y_{lm}(\vartheta,\phi) = \chi_{\mathrm{inside}} && \text{if } r < r_{\mathrm{MT}} \\ &= e^{i\boldsymbol{kr}} = \chi_{\mathrm{outside}} && \text{if } r \geq r_{\mathrm{MT}}.\end{aligned} \tag{2.110}$$

The strongly oscillating partial waves form the inner wave function χ_{inside} and the plane wave represents the outer smooth wave function χ_{outside}. Note that there is no direct $\boldsymbol{k}$-dependence for the atomic-like functions inside but only an *indirect* one. At the muffin-tin border r_{MT}, χ_{inside} and χ_{outside} are glued together – a process called *augmentation* – and this is equivalent to finding the correct mixing coefficients c_{lm} for the different orbitals. Eventually, all augmented plane waves are used to construct a corresponding Bloch sum up to a certain $\boldsymbol{K}$.

Because the augmentation trick is so fundamental to the APW method and its successors, we plot it schematically in Figure 2.39. While $\chi_{\mathrm{inside}} = \chi_{\mathrm{outside}}$

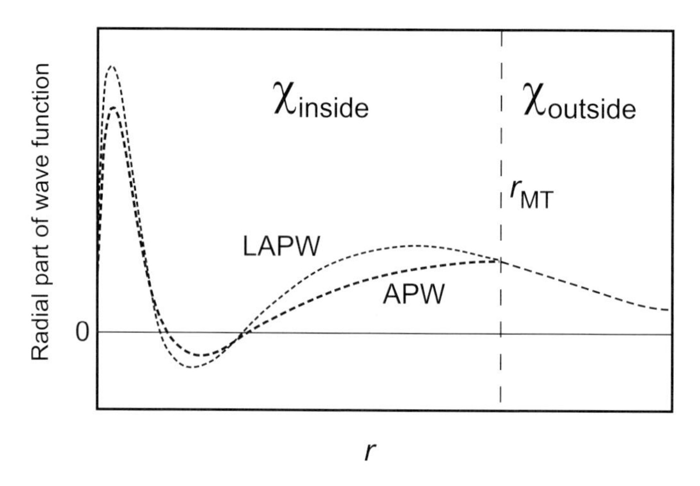

Fig. 2.39 Schematic drawing of the augmentation process at the muffin-tin border utilized in the APW and LAPW methods.

can be easily accomplished at the muffin-tin border (in fact, this is only correct for an infinite sum over all angular momenta, namely, $s, p, d, f, g, h \ldots \infty$), there is, in principle, a discontinuity in the slope at r_{MT}. In addition, an augmented plane wave will always be an energy-*dependent* basis function since it is based on the energy-*dependent* partial waves from the inside of the muffin-tin potential. This is an advantage, physically and chemically, because the basis set automatically *adjusts* to the respective potentials on construction of the APW such that there is more variational freedom when compared to those fixed basis sets mentioned before (STOs, GTOs). Mathematically, it is a serious drawback because it renders the direct diagonalization of the secular matrix (Equation (2.10)) impossible: their elements, $H_{\mu\nu}(E,\boldsymbol{k}) - ES(E,\boldsymbol{k})$, are also energy-dependent! Therefore, a nonlinear eigenvalue problem needs to be solved by numerically searching – one by one! – the roots of $|H - ES|$ for each $\boldsymbol{k}$, so that the APW method is very slow.

For completeness, we add that there exists an APW-related method based on the integral form of the Schrödinger equation which goes back to Korringa, Kohn and Rostoker, shortened to the **KKR** method [228, 229]; it also goes by the name the **Green's function** method. Here, the electronic-structure problem is cast into scattering theory by imagining an electron beam, given by a plane wave, travelling through a crystal. The wave is scattered by the atoms, and a bound atomic state is detected whenever the interference is destructive. This approach is equivalent to a basis-set method including energy- and potential-dependent basis-set functions. Just as in the APW method, a muffin-tin potential leads to a straightforward solution of the KKR equations.

2.15.4 Linear Methods

As has been said before, a partial-wave method such as APW, in spite of being physically (and chemically) advantageous, suffers from the energy-dependence of its basis functions $\phi(E)$. Albeit highly accurate, this makes it computationally demanding and quite slow. In order to combine the virtues of the energy-independent approaches (speed) and those of the energy-dependent ones (accuracy), **linear methods** were introduced by Andersen [230].

These accelerated and simplified approaches also start from the notion of a two-region potential in which the inner function χ_{inside} within the atomic sphere is given by the partial waves $R_l(E,r)Y_{lm}(\vartheta,\varphi)$ but then the radial part of these energy-dependent partial waves is written, for each l and each atomic sphere, as a Taylor series of the energy. For brevity, we shorten $R_l(E,r)$ to $R_l(E)$ and write

$$\begin{aligned} R_l(E) &= R_l(E_\nu) + \left.\frac{\partial R_l}{\partial E}\right|_{E_\nu}(E - E_\nu) + O(E - E_\nu)^2 \\ &= R_l + \dot{R}_l(E - E_\nu) + O(E - E_\nu)^2. \end{aligned} \tag{2.111}$$

R_l and its energy derivative $\dot{R}_l$ fulfill the differential equations

$$\left(-\frac{1}{2}\nabla^2 + v\right) R_l = E_\nu R_l \qquad \text{and} \qquad \left(-\frac{1}{2}\nabla^2 + v\right) \dot{R}_l = E_\nu \dot{R}_l + R_l \tag{2.112}$$

and are orthogonal to each other. The central idea of the linear methods is to *truncate* the above Taylor expansion after the second term which is linear in energy, hence its name. In the vicinity (about one Rydberg) of the expansion energy E_ν, which can be chosen to lie close to the Fermi level, the residual (linear) terms of the Taylor expansion – which have now become energy-*independent* but only within a small energy frame – are sufficient to yield accurate band structures. However, a direct diagonalization of the Hamiltonian is straightforward because $R_l \neq f(E)$.

The linear methods come in different varieties and qualities. In the **linearized augmented plane wave (LAPW)** method, the linearized partial waves are augmented by plane waves at the atomic sphere. This is also schematically included in Figure 2.39. As a "hole in one", the linearized augmented plane wave is continuous *and* differentiable at the crossover from partial wave to plane wave because they match in both value *and* slope. Despite the use of an atomic sphere resembling the APW muffin-tin idea, the LAPW method is *not* restricted to muffin-tin potentials. The atomic spheres are merely used to *define* the LAPW basis set (energy-independent but made from energy-dependent partial waves), and the so-called **full-potential linearized augmented plane wave** (FP-LAPW or, shorter, **FLAPW**) approach, without any shape approximations for the potential and the charge density, is now considered to be one of the most accurate methods available for performing state-of-the-art band-structure calculations. It also needs to be said that the convergence of an LAPW calculation can profit from "local orbitals" [231–233], similar to inner polarization functions confined to atomic spheres, and they increase the size of the basis set just slightly. Further technical details touch upon a partitioning of the levels which can be grouped into valence, semicore and core levels, and these require different computational treatment.

In the case of the **linearized muffin-tin orbital (LMTO)** method, which turns out to be the linearized form of the KKR approach, the outer envelope functions are replaced not by plane waves but by analytical *Hankel* functions, irregular solutions of the Laplace equation corresponding to a vanishing kinetic energy in the interstitial region. Despite its name, LMTO is *not* restricted to muffin-tin potentials; on the contrary, the LMTO method can be brought to the same accuracy as the LAPW method (including **full-potential** offsets such as **FLMTO**). Nonetheless, the majority of all LMTO calculations has been performed using the atomic-spheres-approximation (ASA) in which muffin-tin spheres are blown up until overlapping and volume-filling spheres are utilized, similar to the sketch in Figure 2.38(b); also, the potential within the atomic spheres is spheridized. In loosely packed structures, so-called empty spheres (partial waves without nuclei) must be introduced to ensure volume filling and to increase variational freedom, nowadays an automatic procedure.

Within the ASA, the LMTO method not only acquires incredible speed – comparable to a semiempirical method – but both the Hamiltonian and the overlap matrix automatically decompose into solely structure-dependent structure constants (which only have to be calculated once) and potential parameters. The calculation of structural deformations, however, is then sacrificed in favor of speed. In addition, the LMTO basis set can be further transformed into an extremely localized, short-ranged basis set, mathematically analogous to the electrostatic screening of the long-ranged potential of a charge multipole [234]. This local basis set is sometimes called second-generation LMTOs. Because of this exact transformation between the conventional LMTO basis set and the short-ranged one, tight-binding LMTO theory using the ASA – **TB-LMTO-ASA** – has made tight-binding band-structure calculations (see Section 2.15.1) possible from first principles, including the succeeding interpretational techniques based on electron and/or energy partitioning (see Section 2.8). This immediately explains the popularity of the TB-LMTO-ASA method amongst solid-state chemists, as no other method, with the exception of semiempirical extended Hückel theory, has brought more chemical insight into the quantum chemistry of solid-state materials.

Within third-generation LMTO theory, also called **NMTO** theory, the Taylor series is not truncated after the second term but extended to go up to Nth order, thereby increasing its accuracy [235]. The **augmented spherical wave (ASW)** method [236] closely resembles LMTO theory despite a few technical differences (chosen energy for linearization and exact shape of the analytical envelope function).

2.15.5 Modern Developments

The above approaches, especially the various kinds of pseudopotential and linearized augmented-wave methods, have become well-established tools within theoretical solid-state science. During the last decade, new approaches have been developed, which promise both accuracy and even faster performance. Here are two developments which show high potential because they enable one to describe the chemistry and physics of complex and, in particular, *large* systems.

At first sight, the pseudopotential approach and the different partial-wave methods do not seem to have very much in common. In the first approach, the inner, atom-like wave functions are discarded altogether and replaced by a much weaker potential. In the second group, the outer wave function augments exactly these atom-like partial waves. The **projector-augmented wave (PAW)** method by Blöchl [237], however, combines the two ideas into a unified electronic-structure method. Without going into detail, the PAW method can be looked upon as a pseudopotential method in which the pseudopotential *instantaneously* adapts to the electronic environment. This is because the PAW method is, in fact, a complete all-electron method, and its internal pseudopotential only pops up as an auxiliary quantity in the calculation without special physical meaning. Thus, the pseudopotential is generated, on the fly, from an all-electron augmented-wave scheme, thereby avoiding the annoying transferability problem of pseudopotential theory. Numerically, PAW utilizes a transformation between the true wave functions (with their complete nodal structure) onto auxiliary wave functions (which are numerically convenient) being subject to a rapidly convergent plane-wave expansion. The avoidance of the transferability problem is particularly important whenever calculations of high-spin atoms are to be made. It also opens the way towards *ab initio* molecular dynamics performed using an all-electron method (see Section 2.17).

On the other hand, there is an obvious demand for electronic-structure methods from first principles which are able to treat *huge* chemical systems composed of, say, several thousands of atoms. Let us think of nano- and biomaterials, just to name two important research directions. Unfortunately, standard methods will not do here simply because their *scaling behavior* with respect to the number of atoms is insufficient. We recall that the time needed to diagonalize a DFT-based Hamiltonian goes with the third power of the size of the system. In the real world, mathematical tricks may fortunately be used to bring this cubic scaling down to something of the order of N^{2-3}. Nonetheless, this is still too demanding for systems larger than, say, 100–1000 atoms, and no real help can be expected from better computer architectures. Faster computers will *not* do it.

As any bright chemist or physicist might have guessed already, the bad scaling behavior does *not* stem from the chemistry (or physics) itself, it is a consequence of the mathematical apparatus. Consider a large one-dimensional system composed of many different atoms, schematically depicted in the following Scheme 2.9:

···C–**D**–E–P–E–R–E–N–N–E–N–I–L–N–I–S–I–S–O–L–I–D–U–M–V–**W**–X...

Scheme 2.9

No matter what kind of chemical bonding interactions exist in this very system, it is clear that, for example, atom D on the far left will certainly not interact very strongly with atom W on the far right because they are too far apart from each other. In fact, chemical intuition lets us assume, safely, that each of the atoms only "sees" its immediate environment. Thus, the idea of *locality* or "nearsightedness" [158] pops up. The latter, however, has been mathematically destroyed, unnecessarily so it seems, by the totally delocalized construction of a molecular or crystal orbital from atomic-like ones, the LCAO scheme. The same holds true for an extended plane wave which is mathematically simple but a total disaster with respect to the idea of locality for which we are striving. It is of course possible to eventually derive orthogonal *localized* orbitals from totally delocalized ones (physicists might think of so-called

Wannier functions whilst chemists should remember VSEPR-like functions containing electron pairs), but a more direct way of bypassing the standard (delocalized) route is needed.

At the present time, several different approaches (Fermi operator, divide-and-conquer, orbital minimization, and others) have been proposed and reviewed, and they all try to utilize the locality principle, for example, by modifying the density matrix in order to make it a sparse matrix [238]. All methods either go by the name of *linear-scaling* approaches (not to be confused with the linear methods of band theory) or simply by the **order-*N*** term [239]. We will give it some meaning by means of a simple example. Imagine we had a set of *super*-localized atomic basis functions, and these basis functions had the property of *vanishing* totally at some critical distance r_c (say, a few Å) apart from the atomic center. Since these basis functions would no longer extend infinitely into space, their super-contracted properties would ensure that, besides finite overlap integrals of a given atom O in Scheme 2.9 with its immediate neighbors S and L, all other overlaps with second-nearest neighbors I and I and beyond would be *exactly* zero. Then, the entire secular determinant will indeed turn into a sparse determinant because the vast majority of all the off-diagonal entries would vanish completely (Scheme 2.10). By shortening $H_{AB} - S_{AB}E$ as H_{AB}, we may write:

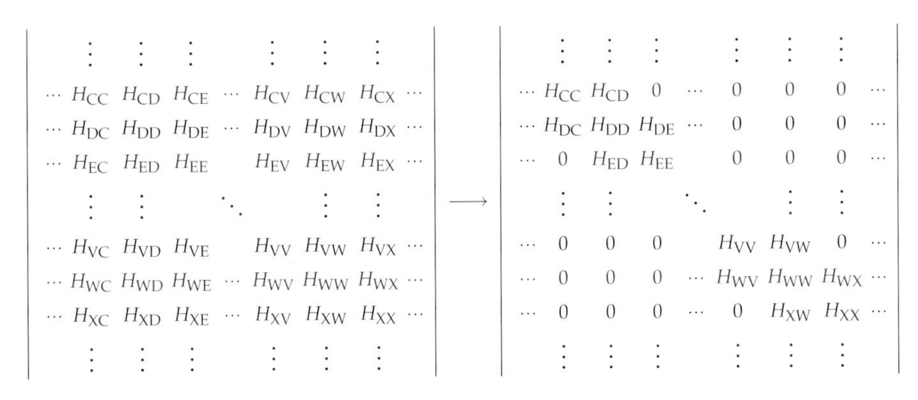

Scheme 2.10

As a consequence, the computer only needs to store the nonzero ($\approx$ diagonal) elements and may eliminate the operations on the zero ($\approx$ off-diagonal) elements, thereby greatly speeding up the calculation. Such a strategy using super-localized orbitals [240] eventually leads to a kind of "Hückelization" of the electronic-structure calculation, not in terms of accuracy but in terms of efficiency. In contrast to traditional semiempirical methods, however, it is not based on the parametrization of integrals but may be carried out in the framework of regular Hartree–Fock or density-functional theory, given that the system allows a local description. Note, however, that far-reaching

Coulomb interactions such as those occurring in ionic crystals, are in conflict with the above recipe, not because of mathematical but for intrinsically physical reasons.

As expected, the performance of all competing order-N methods depends on the system under investigation, the accuracy needed, the amount of experimental information available, the questions that need to be answered, and also computer-related parameters (processors, parallel architectures, etc.). All approaches have their pluses and minuses. It is also clear that the increased mathematical effort will "pay off" (*if* at all) only beyond a critical number of atoms (around 100–1000 or so); below that, the normal route with cubic scaling is faster. Nonetheless, the same locality arguments may be used to derive *linear-scaling* methods for the extremely efficient calculation of electronic correlation (see Section 2.13).

2.16 Structure Optimization

Contrary to popular belief, a significant proportion of experimentally reported crystal structures refined from X-ray and neutron diffraction data contain slightly false atomic positions. If electronic-structure calculations are based on these, problems are inevitable, and the theory will be blamed for inaccuracies that, in fact, are due to sloppy experiments. Keeping the atoms fixed in their positions will always be a risky thing if the atomic positions have not been determined by a *perfectly reliable* experimental method.[32] In other cases, no experimental positional parameters are available at all, and one is left to make a rough guess at the atomic structure.

In such a case, *structure optimization* is mandatory and may be introduced by calculating the interatomic forces using electronic-structure theory; the alert theorist always checks the starting geometry. If the electronic structure of the molecule or solid and its wave function Ψ are known, the famous Hellmann–Feynman theorem [3,241] gives access to the force $\boldsymbol{F}$ between two nuclei separated by a distance $\boldsymbol{R}$

$$\boldsymbol{F} = -\frac{dE}{d\boldsymbol{R}} = -\left\langle \Psi \left| \frac{\partial \mathcal{H}}{\partial \boldsymbol{R}} \right| \Psi \right\rangle , \tag{2.113}$$

where $\mathcal{H}$ is the Hamilton operator derived from the original Born–Oppenheimer approximation. Remember that $\mathcal{H}$ (Equation (2.5)) did not include the

32) This comment does *not* intend to criticize crystal-structure determination as a whole. On the contrary, an *accurately* determined crystal structure may possess an advantage over a theoretical structure because the former yields atomic coordinates which represent time averages over all atomic vibrations.

kinetic energies of the nuclei and is an *electronic* operator. Provided a high-quality wave function (or, alternatively, an accurate electron density by DFT), the force is equivalent to the Coulomb force of a given nucleus on the other nucleus and the electronic charge cloud $\rho(\boldsymbol{r})$. Unfortunately, small errors in the calculation of the latter translate into large errors for the forces. Having successfully calculated all interatomic forces, the atoms are slightly shifted along the direction of the forces, and then the electronic-structure calculations, together with the force calculations, are repeated until all interatomic forces vanish. In practice, the calculations are stopped when a lower force limit has been reached. Hopefully, the structure is truly optimized and not stuck in some false minimum!

This computational scheme of structure optimization [242, 243] by moving the atomic positions based on interatomic forces, can be made an automatic procedure, and it is part of many electronic-structure packages for molecules and solids. The numerical strategies used, however, may differ considerably from method to method. It must also be emphasized that each single force calculation does not infer any dynamics because all atoms are standing still for the given $\mathcal{H}$. With respect to solids, the use of *full-potential* schemes is mandatory because of the high sensitivity of the force calculation to residual errors. Although simplified (e.g., atomic-spheres approximation, ASA) schemes are quite acceptable in determining total energies, they fail for the calculation of reliable forces. Whenever plane waves are used, so-called Pulay forces [244], which are force corrections due to the derivative of the basis set with respect to the atomic position, vanish completely [245], a nice advantage of this delocalized basis set.

2.17 Molecular Dynamics

Up to this point, all electronic-structure methods have been concerned with the movement of the electrons only. This approach was justified by the idea of light electrons moving in the field of the *fixed* nuclei with large masses, and this simplification is a very reasonable one for most solids, especially if the properties to be predicted do not depend on the atomic movement. For example, to theoretically characterize insulators, semiconductors, and metals, atomic dynamics is unimportant; likewise, magnetic phenomena (linked to the electronic spin) are largely independent of atomic movement.

Whenever *dynamical* properties are of interest, the motions of the nuclei must be taken care of from the very beginning, but the sheer computational complexity of describing *both* electrons and nuclei renders a full quantum-mechanical approach extremely difficult, even today. Historically, two different approaches linked with statistical mechanics have been pursued, and both branches may be considered specific arts of computer simulations, being computational blends derived from atomic physics, quantum chemistry, electronic structure theory, classical as well as statistical mechanics/thermodynamics, and also computer science [246,247].

In the first approach, dubbed the **Monte Carlo** technique [248], a chain of configurations of the particles (atomic or molecular nuclei) is generated by a random (hence the name) procedure. We will not discuss this stochastic approach any further but merely mention that the method represents a random walk through phase space, and the new particle positions are either accepted or rejected by an energy criterion. Given a *very* large number of configurations, thermodynamical equilibrium is eventually achieved such that structural as well as thermodynamical properties may be calculated from all the configurations.

The other approach, called **molecular dynamics** [249], is essentially based on Newton's mechanics utilized to describe the nuclear motions. Thus, quantum mechanics is discarded (at least for the moment), and classical mechanics is bravely applied to a system of atoms, ions, or molecules pretending these are not quantum objects. For molecules, such classical parametrizations go by the name *force field* methods, and they make up an important class of computational chemistry [58]. Given the knowledge of an atom's mass m and the force $\boldsymbol{F}(t)$ acting on it at some time t, the atom's acceleration $\boldsymbol{a}(t)$ is calculated according to $\boldsymbol{F}(t) = m\boldsymbol{a}(t)$; the force itself is the derivative of the potential energy $V(\boldsymbol{r}(t))$ such that we have

$$-\frac{\partial V(\boldsymbol{r}_i(t))}{\partial \boldsymbol{r}_i} = m_i \frac{d^2 \boldsymbol{r}_i}{dt^2} \tag{2.114}$$

for each atom i, and the deterministic evolution of all atomic positions $\boldsymbol{r}_i$, accelerations $\boldsymbol{a}_i$, and also velocities $\boldsymbol{v}_i$ must be recorded. These equations of motion may be formulated more elegantly by introducing the system's Lagrange function $\mathcal{L}$ which is the difference between the kinetic and the potential energy

$$\mathcal{L} = T - V(\boldsymbol{r}), \tag{2.115}$$

and the classical Euler–Lagrange equations of motion are solved for all particles which are coupled through the assigned interatomic potential $V(\boldsymbol{r})$. Eventually, all equations are integrated and the atomic trajectories – the sequence of states – are followed on the potential-energy surface. Because of the high complexity involved, different *numerical* integration strategies may be used, for example, the *velocity Verlet* algorithm [250]. Here, the new coordinates and velocities at time $t + \delta t$ are

$$\boldsymbol{r}_i(t + \delta t) = \boldsymbol{r}_i(t) + \delta t \boldsymbol{v}_i(t) + \frac{\delta t^2}{2} \boldsymbol{a}_i(t) \tag{2.116}$$

and

$$v_i(t + \delta t) = v_i(t) + \frac{\delta t}{2}[\boldsymbol{a}_i(t) + \boldsymbol{a}_i(t + \delta t)]. \tag{2.117}$$

A molecular dynamics (MD) simulation is usually performed for a fixed number of particles N (typically hundreds or thousands) inside a fixed volume, and the total energy is a constant of the motion. The time evolution of such a *microcanonical ensemble* is accumulated over many time steps (typically millions and more) of which a single time step is extremely small (typically in the femtosecond range). Eventually, structural and thermodynamical properties are extracted via *time averages* which are thought of as being comparable to experimentally observed ensemble averages; this is called the *ergodicity* hypothesis. For example, the absolute temperature of the system is proportional to its kinetic energy, which is given as the sum of all kinetic energies of the particles. Indeed, the advantage of MD simulations results from the access of *thermodynamical properties* through finite-temperature calculations, albeit (mostly) without quantum mechanics. Second, theoretical structural optimizations used to find the global minimum of a system, are less likely to get stuck in a local one because, at finite temperatures, atoms may "hop" out of relative minima. At present, a very wide range of MD methods is available, and computer experiments of this kind have become worthwhile (they are often cheaper) alternatives to real experiments.

Within solid-state chemistry, MD simulations have not been widely used even though there are prominent and impressive counterexamples [251, 252]. Traditionally, the synthetic solid-state chemists have concentrated on the synthetic and crystallographic aspects, but questions concerning chemical reactivity or reaction mechanisms (here, atomic movement comes into play) have been somewhat neglected. The classical approaches to the understanding of solid-state materials (see introductory sections) are *very* static. Also, the attempt to use standard MD-like simulations in modelling solid-state materials is ultimately challenged by the fact that inorganic phases can be composed of atoms from *any* part of the periodic table. Thus, an operational atomistic simulational method for such matter immediately faces an incredibly complex *parametrization problem* [253, 254] which must be tackled for a given class of compounds. Because potentials determine the quality of any kind of atomistic simulation, choosing a set of suitable potentials $V(\boldsymbol{r})$, which are either derived from experimental data, from quantum-mechanical calculations, or ingeniously imagined, is the most critical issue and great care must be taken to verify their validity and also transferability. Nonetheless, it remains almost impossible to *classically* model the interplay between electronic and geometrical structure, which becomes crucially important in electronically induced structure transformations, for example in Jahn–Teller instabilities.

An ingenious method of extending the molecular dynamics approach when incorporating quantum mechanics, is given by the scheme of **Car** and **Parrinello** [255]. Their method combines classical MD with parameter-free quantum mechanics to result in an *ab initio* molecular dynamics method, sometimes also called **dynamical simulated annealing**. In this approach, the nuclei are treated as classical objects (the Born–Oppenheimer approximation is still valid) but the electrons are understood from density-functional theory. As we have seen already, the interactions between the electrons and the nuclei can be described satisfactorily by pseudopotentials (see Section 2.15.2), together with plane-wave basis sets, supercells (see Section 2.18), and periodic boundary conditions.

The essential trick of the Car–Parrinello method is not to parametrize the potential-energy surface, but to calculate it *on the fly* from first principles when generating the nuclear trajectories by Newton's mechanics. The electronic ground state is calculated at the same time. To do so, one starts with an extended Lagrangian which corresponds to a *fictitious* dynamical system

$$\mathcal{L} = T_{\text{nuc}} + T_{\text{ele}} - E[\{\psi_i\}, \boldsymbol{R}] + \sum_{ij} \Lambda_{ij} \left(\int \psi_i^*(\boldsymbol{r},t)\psi_j(\boldsymbol{r},t)d\boldsymbol{r} - \delta_{ij} \right), \quad (2.118)$$

where $\boldsymbol{r}$ and $\boldsymbol{R}$ stand for electronic and nuclear coordinates. The electronic orbitals ψ_i are provided with a classical (!) kinetic energy because of likewise *fictitious* wave-function masses μ_i, that is

$$T_{\text{ele}} = \sum_i \frac{1}{2}\mu_i \int |\dot{\psi}_i(\boldsymbol{r},t)|^2 d\boldsymbol{r}, \quad (2.119)$$

and the sum runs over all occupied levels ψ_i. In the above equations, the subscripts designate to which entities (electrons or nuclei) the energy terms belong. The so-called Lagrange multipliers Λ_{ij} are needed as (holonomic) constraints to ensure that the wave functions remain normalized and orthogonal during the entire process. Eventually, Euler–Lagrange equations of motion

$$\mu\ddot{\psi}_i(\boldsymbol{r},t) = -\frac{\delta E}{\delta\psi_i^*(\boldsymbol{r},t)} + \sum_j \Lambda_{ij}\psi_j(\boldsymbol{r},t) \quad \text{and} \quad M\ddot{\boldsymbol{R}}(t) = -\frac{\partial E}{\partial \boldsymbol{R}(t)} \quad (2.120)$$

are solved, in which the orbitals are *classically* minimized. Nuclear and electronic degrees of freedom are relaxed simultaneously, and the choice of the fictitious mass of the orbitals accounts for different time scales such that the wave functions (the electrons) with high-frequency motion will follow the slow ionic motions adiabatically. Note that this recipe does *not* strive for self-consistent wave functions (orbitals), and the evolution of the latter does not occur on the Born–Oppenheimer surface.[33]

The striking advantage of the Car–Parrinello approach is that finite-temperature simulations, needed in the search for global minima, can be performed with DFT-typical accuracy because the electrons–nuclei interactions are correctly described; therefore, the transferability problem of classical MD does not exist. The time steps are within the femtosecond range, allowing MD simulations up to several picoseconds. During structural optimization, the average interatomic forces are close to the exact Hellmann–Feynman forces because of error cancellations in the (oscillating) orbitals. In most cases (insulators, semiconductors, molecules), an irreversible nuclei–electrons energy transfer (due to the fictitious coupling of both systems) does appear but it is rather small.

For completeness, we mention that other competing approaches targeted at the *direct* minimization of the density-functional (Kohn–Sham) energy functional [256] have also been successfully developed, for example, the **conjugate gradient** method, a second-order search for the energy minimum. Recent Car–Parrinello-related work includes complex simulations using path-integral molecular dynamics for the nuclei and density-functional calculations for the electronic wave function [257]. In addition, work is going on in the realm of parameter-free nonadiabatic molecular dynamics required when moving away from the electronic ground state [258].

2.18 Practical Aspects

Fortunately, there are many excellent approaches which can be used to calculate the electronic structures of solids. All of them have their strengths (and also weaknesses), and the choice of the method depends on the material, the chemical or physical question to be answered, the amount of information (in particular, structural information) already known and, of course, on the amount of computing power (that is, money) which is available. Thus, if qualitative results for structurally well-characterized materials are required,

33) To discriminate Car–Parrinello molecular dynamics more clearly, let us compare it with two other approaches. Ehrenfest molecular dynamics is based on the time-*dependent* Schrödinger equation for which the wave function stays minimized throughout the evolution of time such that the maximum time step is very limited because of the fast dynamics of the electrons. Born–Oppenheimer molecular dynamics utilizes the time-*independent* Schrödinger equation and requires self-consistency at all time steps but the latter are longer as a consequence of the slower nuclear dynamics.

any semiempirical but well-parametrized method is certainly a good choice in order to understand the electronic structure. If numerical results of higher accuracy are needed, parameter-free full-potential methods for all the electrons are justified. Given that sufficient structural information is lacking and structures must be generated at the very beginning, a fast pseudopotential method concentrating on the valence electrons only, is a reasonable alternative. It goes without saying that personal *experience* is very much needed to successfully navigate in the realm of electronic-structure methods, and the question of the "best" method is impossible to answer if the scientific question has not been specified in detail. Also, *people* solve problems, computer programs do *not*. Yet, there are still two more things to mention.

2.18.1 Structural Models

For completeness, let us briefly describe, using a few examples, the kind of structural input which is needed to cover the typical aspects of electronic-structure theory. For simplicity, we may imagine a two-dimensional world in which gallium nitride, GaN, is the object of our study. Figure 2.40(a) shows the unit cell of this two-dimensional GaN with orthorhombic symmetry and planar four-fold coordination for Ga and also N (in three dimensions, the four-fold coordination is tetrahedral, of course). The broken lines indicate the size of the unit cell which is repeated over and over again, to infinity (we are going to use Bloch's theorem) within two dimensions by periodic boundary conditions. Given that only the valence orbitals are needed for Ga ($1 \times 4s, 3 \times 4p$) and N ($1 \times 2s, 3 \times 2p$), the four-atom unit cell contains $2 \times 2 \times 4$ orbitals, and a secular determinant sized 16×16 must be diagonalized (see Equation (2.10)).

Alternatively, one might decide to double the length of the unit cell along both the x and y directions as given in Figure 2.40(b) – a so-called superstructure setting – but the new unit cell (quadruple in volume when compared with the preceding one) is more challenging to calculate in terms of time and computer memory. Because we now have 16 atoms and, therefore, 64 orbitals, a 64 $\times$ 64 secular matrix must be solved. Although there are mathematical tricks to speed up matrix diagonalization, a DFT approach (and also others) typically scales with N^3 where N is the system size (see Section 2.15.5). Thus, the supercell problem needs more time by a factor of $\approx 64^3/16^3 = 64$, a dreadful waste of computing power.

Nonetheless, the supercell approach may have important advantages. Figure 2.40(c) shows a scenario where one N atom has been exchanged for a higher homologue, P (in bold). The resulting composition of the unit cell is now Ga_8N_7P or $GaN_{0.875}P_{0.125}$, and the additional cost in calculating the enlarged structure is willingly paid because it is possible to investigate the

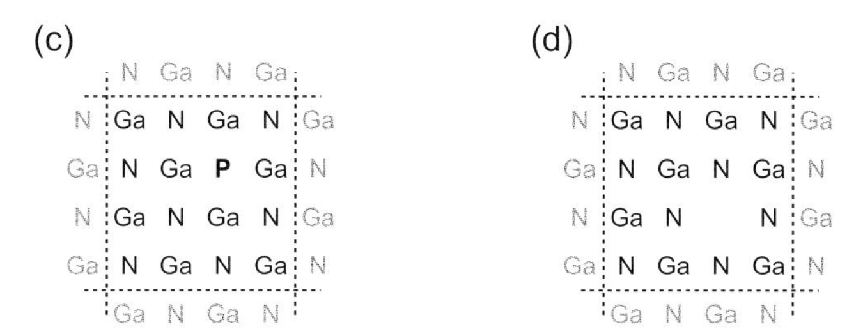

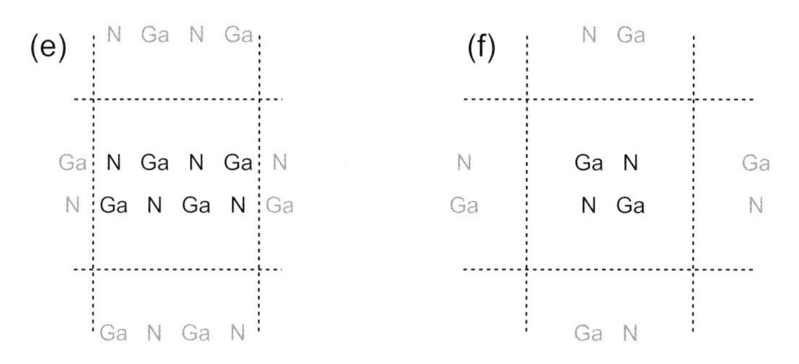

Fig. 2.40 Some geometrical choices on how to set up electronic-structure calculations for solids (see text); the broken lines indicate the size of the chosen unit cell.

changes in electronic structure by means of atomic *substitution*. Note that there are also other, perturbative ways to do this using smaller unit cells. Also, the effects of *vacancies* symbolized in Figure 2.40(d) may be easily studied, in this case corresponding to $Ga_7\square N_8$ or $Ga_{0.875}\square_{0.125}N$ where $\square$ stands for a missing atom. The smaller the percentage of impurities or vacancies, the larger the unit cells that have to be dealt with, and consequently, the more expensive the calculations.[34]

The calculation of the electronic structure of a *surface* – as opposed to the bulk electronic structure – is easily accomplished, namely by a geometrical layout as schematically given in Figure 2.40(e). In this approach, one simply leaves enough empty space in one direction such that the atoms on top cannot interact with the periodically repeated atoms on the bottom. Finally, even a Ga_2N_2 *molecule* – (f) – may be studied if the unit cell is made so large that the central entity is geometrically (and also electronically) isolated. It seems awkward, at best, to calculate molecules using periodic boundary conditions – a straightforward molecular quantum-chemical calculation would be *much* more appropriate – but there are cases when this is indeed a good choice, for example, if the electronic structures of a solid-state material *and* its molecular precursor need to be studied using *exactly* the same computer program characteristic of a particular basis set and quantum-mechanical method. A good example is given by the *ab initio* molecular dynamics approach based on plane waves, which typically treats both extended solids as well as molecules using the related strategies, namely (a) for the bulk and (f) for the molecule.

2.18.2 Energy, Enthalpy, Entropy and Gibbs Energy

The reader will have noticed that both classical and quantum-mechanical theories based on energetic criteria rely on the same energetic principle: a lower energy always refers to a larger stability, and vice versa. This is true for purely electrostatic reasoning (such as that derived from lattice energies) and also for quantum chemistry, simply because of the underlying variational principle from wave mechanics or density-functional theory. Given an arbitrary system of nuclei and electrons (in molecules and solids), there is *one* optimum wave function (or optimum electron density) which may be found by searching for the lowest-energy solution. If this solution is found, energetic rankings are straightforward. For example, we might envisage two structurally different molecules AB and *AB* but with *identical* chemical compositions; chemists would then call AB and *AB* molecular *isomers*. If a variational quantum-chemical calculation yields a lower energy for AB than for *AB*, we will regard molecule AB as the stable isomer adapting the ground-state structure. Likewise, AB is more strongly bonded than *AB* because one needs more

34) This seems counterintuitive in the first place but is easily explained. A ten-atom unit cell with nine correct but one false atom may serve as a computational impurity model for a 10% doping scenario. To get to only 1% of doping, a hundred-atom unit cell with ninety-nine correct atoms and one false atom is needed.

energy to separate it into atoms A and B. The same trivial argument applies to spin isomers like $^3O_2$ and $^1O_2$ (see Section 2.3); the energy is always decisive.

For solid-state materials, however, the above reasoning must be somewhat enlarged in scope. Think of a tiny piece of rock salt which we wish to describe. Despite the fact that the unit cell of NaCl is easily defined and repeated over and over again in all three dimensions by periodic boundary conditions, this purely mathematical trick does not conceal the fact that we are *not* dealing with a single NaCl entity. What is meant, instead, is a *macroscopic* amount of NaCl incorporating a huge number (of the order of 10^{23}) of Na and Cl atoms. Here, the laws of classical thermodynamics become important [259].

One typically starts with an internal energy E of a macroscopic system, expressed as the internal energy of the periodically repeated unit cell. This state function[35] is part of another state function H, the enthalpy, which is a very useful energetic measure for conditions of constant pressure p. For a complete picture, one also needs to know the value of the state function T, the temperature, and that of the state function S, the entropy, a measure of chaos and also probability. These functions may be combined to yield the Gibbs energy (or *free* enthalpy) G, the true and final measure of stability. In its difference form, the so-called Gibbs–Helmholtz formula reads

$$\Delta G = \Delta H - T\Delta S \tag{2.121}$$

$$= \Delta E + p\Delta V - T\Delta S. \tag{2.122}$$

If ΔG corresponds to a chemical reaction, a negative/positive value characterizes the reaction products as *thermodynamically* stable/unstable. In comparison, negative/positive ΔH values indicate *exothermic/endothermic* reactions and the reaction products are considered *enthalpically* stable/unstable. Note that entropic contributions are deliberately neglected in the latter case.

Because the thermodynamic ranking requires a knowledge of the Gibbs energy G, the quantum-mechanically calculated electronic energy E must be converted into G according to the very last equation; a complication arises from the fact that G is temperature-dependent. Unfortunately, the presently available electronic-structure theories do not easily operate at finite temperatures (for example, 300 K) and their results merely refer to T = 0 K. It is extremely difficult to imagine how to calculate $G(T)$ correctly from first principles. Fortunately, an *ad hoc* solution to this problem of simply setting $G(T) \equiv G(0)$ at about room temperature (and doing the same for H and S) is quite tolerable because this is done *both* for educts and products (A + B $\leftrightharpoons$ AB) such that the errors made either left or right largely cancel. In addition, the entropic contribution of ΔG (the last term in Equation (2.121)) can be safely neglected

35) A state function does not depend on the history of the system; see also introductory textbooks on physical chemistry [259].

for the plethora of solid-state materials because all of them are pretty well ordered. Recall that, at room temperature, the very different carbon polymorphs diamond and graphite differ by less than 1 kJ/mol in $T\Delta S$, and this is often smaller than the error of most electronic-structure methods. We may therefore simplify according to

$$\Delta G(T) \approx \Delta G(0) \equiv \Delta G \qquad \text{and} \qquad \Delta G \overset{\text{solids}}{\approx} \Delta H = \Delta E + p\Delta V, \tag{2.123}$$

which also gives the method for converting a given energy–volume diagram (which results from electronic-structure theory, namely E and V) into a much more convenient enthalpy–pressure diagram. First, we restrict ourselves to absolute zero temperature. Second, ΔG is approximated by ΔH because we are dealing with solid-state materials. Third, the theoretically calculated E vs. V data of the solid-state phases are numerically fitted on the basis of some analytical function, such as the formula of Murnaghan [260] or Birch–Murnaghan [261]:

$$\begin{aligned} E(V) &= \frac{9B_0V_0}{16}\left[(B_0'-4)\left(\frac{V_0}{V}\right)^2 + (14-3B_0')\left(\frac{V_0}{V}\right)^{\frac{4}{3}} + (3B_0'-16)\left(\frac{V_0}{V}\right)^{\frac{2}{3}}\right] \\ &\quad + \left(E(V_0) - \frac{9B_0V_0}{16}(B_0'-6)\right). \end{aligned} \tag{2.124}$$

Here, V_0 is the minimum volume and B_0 the bulk modulus in the ground state; B_0 is defined through the curvature of $E(V)$ at the minimum volume,

$$B_0 = B(V_0) = V\left.\frac{\partial^2 E}{\partial V^2}\right|_{V=V_0}, \tag{2.125}$$

and B_0' is the dimensionless derivative of $B(V)$ at V_0. Fourth, the pressure p is extracted by a numerical differentiation of the fitted curve according to $p = -\partial E/\partial V$. Eventually, the enthalpy H serves as an approximate measure of stability. Reiterating, enthalpy differences ΔH can be expressed as differences between calculated electronic energies ΔE for zero pressure. For absolute zero temperature, ΔE also equals ΔG. If gaseous species are not involved, the close correspondence between ΔE and ΔG approximately holds also for finite temperatures.

2.19 Computer Implementations

Even at the simplest theoretical level, band-structure and related calculations are far too complex to be performed by hand; thus, a large number of computer programs have been developed. It is difficult to estimate the differing strengths (and weaknesses) of the various kinds of software, therefore I will not even try to do so. Also, corresponding links to Internet resources are not given on purpose (because they tend to change too often) but any search engine will find them quite easily if the correct names of the computer programs are supplied. An exception is made with respect to the nicely maintained website of the Ψ_k initiative but this solid-state physics resource lacks the proper links to the important quantum-chemical sites. In the following, I will mostly (but not exclusively) concentrate on the nonmolecular computer programs.

With respect to (semiempirical) tight-binding and molecular-orbital calculations, EHMACC (in combination with EHPC and EHPP) written by the Whangbo group at North Carolina State University (USA) is a popular program package based on the extended Hückel approximation. It is available from the Quantum Chemical Program Exchange QCPE. Another extended Hückel program by the Whangbo group goes by the name CAESAR and has been commercialized. The computer program BICON-CEDiT by the Calzaferri group at Bern (Switzerland) is an EHMACC successor and includes a number of extensions, for example, a two-body repulsive energy term for geometry optimization. It is freely available from the authors. Yet Another extended Hückel Molecular Orbital Program, called YAeHMOP (Landrum and Glassey), can also be used for analyzing the electronic structure of molecules and solids, and it is available, for free, at sourceforge.

Upon going *ab initio*, the Dovesi and Pisani groups at Torino (Italy) are distributing their tight-binding CRYSTAL program which is based on the Hartree–Fock approximation but which also contains density-functional routines. Also, the latest version of the quantum-chemical GAUSSIAN software package, very widely used in the molecular communities, now also covers periodic systems. Jug's group at the Technical University of Hannover (Germany) offers its Hartree–Fock-style semiempirical MSINDO program to deal with the electronic structures of molecules, clusters, and solids.

By far the largest group of computer programs for solids follows the pseudopotential/plane waves approach. For example, the Vienna *ab initio* Simulation Package (VASP) by the Hafner group at Vienna University (Austria) is just such a density-functional code, which also allows molecular dynamics calculations. The corresponding Car–Parrinello molecular dynamics code CPMD, coordinated by the CPMD consortium (IBM and Max Planck Institute for Solid State Research, Stuttgart, Germany), is distributed free of charge. An alternative program (dubbed CASTEP) by the Payne group (Cavendish, UK) is commercially avaliable. PWscf is another plane-wave pseudopotential package promoted by the DEMOCRITOS center at Trieste (Italy); the program's development is coordinated by the Baroni group, and the program is distributed under the GNU public license. Also, PEtot by the Wang group at Lawrence Berkeley National Laboratory (USA) is freely distributed for noncommercial purposes, as is the ABINIT program (GNU public license) started by Gonze's group at the Catholic University of Louvain-la-Neuve (Belgium). Another program, PARATEC, has been designed for parallel computing platforms at Lawrence Berkeley National Laboratory. Likewise, Octopus is a pseudopotential program from a Spanish group focusing on time-dependent Kohn–Sham equations, distributed under the GNU public license. In addition, there is a free plane-wave pseudopotential DFT program (FHI98md) available from the Scheffler group at the Fritz Haber Institute of the Max Planck Society (Berlin, Germany), and this institute is also developing a new program, called SFHIngX.

An order-N density-functional method based on local orbitals is developed and distributed under the name SIESTA by the SIESTA group (Spain) under the GNU public license. Another order-N program, CONQUEST, is made by Gillan's group at University College London (UK). The commercially available band version of the original Amsterdam Density-Functional (ADF) program is one of the few DFT codes that is based on Slater-type orbitals. The FPLO package, a full-potential DFT implementation operating with local atomic orbitals, is distributed by Eschrig's group at the Leibniz Institute for Solid State and Materials Research (Dresden, Germany).

With respect to augmented-wave methods, there is the TB-LMTO-ASA program package by the Andersen group, available for free from the Max Planck Institute for Solid State Research (Stuttgart, Germany). Savrasov's FLMTO program can also be downloaded here. A version of the augmented spherical wave method (ASW) is distributed by Eyert's group (University of Augsburg, Germany) including an ASA and a full-potential version. The program WIEN2k is a widely used implementation of the FLAPW approach (also including local orbitals), and it is developed and distributed by the Schwarz group at the Technical University of Vienna (Austria). Alternatively, FLEUR by the Blügel group (Jülich Research Center, Germany) is another implementation of the FLAPW method. Both programs are available from the authors.

The PAW all-electron implementation of *ab initio* molecular dynamics has been developed by the Blöchl group (Clausthal, Germany) and can be received from there but is licensed from IBM (Rüschlikon, Switzerland). The PAW method is also part of other plane-wave packages such as VASP or ABINIT (see above).

In addition, there is the CASINO program based upon the quantum Monte Carlo approach by the Needs and Towler groups from the Cavendish Laboratory (UK), and the code is available from the authors.

In the case of the classical molecular-dynamics type of calculations, the GULP program package by Gale's group at the Curtin University of Technology (Perth, Australia) is freely available, to academics, from the author. The alternative program DL-POLY, originally developed by Smith and Forester at Daresbury Laboratory (UK), may be obtained from there. My own group at Aachen offers, free of charge, the bond-valence-based aixCCAD program which is especially useful for teaching purposes.

3
The Theoretical Machinery at Work

Practice is the best of all instructors.
PUBLILIUS SYRUS

However beautiful the strategy, you should occasionally look at the results.
WINSTON S. CHURCHILL

Most solid-state chemists will probably agree that the important goals of theoretical solid-state research are, first, correctly describing the experimental facts, second, understanding them in one way or another, and, eventually, predicting how to make new materials with the desired properties. The last goal – theoretically paving the way towards the new – is probably the most challenging one. Note that Niels Bohr has already said that predictions are difficult, especially for the future, and there is a lot of truth in this strange remark.

In addition, scientific practice shows that there is no clear distinction between describing/understanding/predicting various chemical and physical properties because, all too often, the process of adequately describing a phenomenon already involves a certain kind of "understanding". It is also puzzling to note that sometimes the same phenomenon can be nicely described and also explained from two greatly differing perspectives, and each side will surely claim that their own way of seeing things is exactly right, whereas the others simply "do not understand". Finally, some computational approaches have already become so accurate that they are able to deliver perfectly reliable results *without* human intervention. In other words, the correct numerical answer is produced without any understanding (although some scientist is still commanding the computer program) but the consequences for the laboratory work must be formulated by a human being. Whether the computational tools are of traditional, classical origin or are quantum-mechanical in nature, should not be considered, uncritically, as being characteristic of their validity. Even

Computational Chemistry of Solid State Materials. Richard Dronskowski

ISBN: 3-527-31410-5

within the quantum-mechanical field, differing labels (in terms of *semiempirical, approximate, ab initio* or *first principles*) do not automatically characterize quality. In an ideal world, one would simply run the most advanced computer program and it would provide us with all the information we are looking for. Because the world is real, there are satisfactory semiempirical results in cases where more "sophisticated" approaches encounter problems.

In the following, we will deal with fourteen solid-state chemical problems and apply various computational tools in order to understand, first, the structural information and then, second, the chemical and physical (other than structural) properties where the electronic structure fully comes into play. There will also be examples for predicting new compounds and the conditions of their synthesis when the theory proves to be so trustworthy that this may be tried fairly painlessly. We will start with very simple structures/compounds but the complexity will slightly increase in going from section to section, in particular with respect to physical properties. Admittedly, the selection is heavily biased by what has been accomplished by my own group within recent years (these problems have been important to *us*) but, on the other hand, there is no better way for me to acknowledge the contributions of former and present coworkers of mine. Also, the breadth of the materials under study makes me hope that the selection offers a rather typical overview for current solid-state chemical problems. If possible and justified, the connection with the related work of other groups will also be referenced.

3.1 Structure and Energetics: Calcium Oxide

As indicated before, the *ab initio* electronic-structure theory of solid-state materials has largely profited from density-functional theory (DFT), and the performance of DFT has turned out well even when the one of its molecular quantum-chemical competitors – Hartree–Fock theory – has been weakest, namely for *metallic* materials. For these, and also for covalent materials, DFT is a very reasonable choice. On the other hand, *ionic* compounds (with both metals and nonmetals present) are often discussed using only the ionic model, on which most of Section 1.2 was based, and the quantum-mechanical approach is not considered at all, at least in introductory textbooks. Nonetheless, let us see, as a first instructive example, how a typical *ionic* material can be described and understood by the ionic *and* the quantum-chemical (DFT and HF) approaches, and let us also analyze the strengths and weaknesses.

Calcium oxide, CaO, or simply *lime* is a white, caustic, alkaline chemical which has found applications in a number of fields. CaO can be made by heating calcium carbonate, $CaCO_3$, to several hundred degrees Celsius, and CaO rapidly reacts with water to form Ca^{2+} and OH^- ions. At room temperature, it also absorbs water and carbon dioxide. The use of CaO in the fabrication of glasses and metals, in paper making, in water processing, in pottery, and in pollution control (binding gaseous sulphur trioxide, SO_3, in the form of calcium sulfate, $CaSO_4$) is certainly worth mentioning. CaO is a ubiquitous chemical, for good reasons.

CaO adopts the sodium chloride structure type (Figure 1.1(b)) with an experimental lattice parameter of $a = 4.811$ Å. Given some freshmen-type chemistry, the typical starting point for describing the solid would be an ionically bonded material made from Ca^{2+} and O^{2-} ions. The latter ion is entirely unstable (see Section 1.2) in the gas phase and is only existent within an ionic solid. The very high melting point of CaO (2927 °C) also points towards the idea of a strongly bonded ionic structure with a large lattice energy.

If we do *not* know which crystal structure (and associated lattice parameter) is adopted by CaO, how can one proceed before performing the diffraction experiment or running the quantum-chemical calculation? Certainly, the [ZnS], [NaCl], and [CsCl] structure types are potential candidates, so starting with the (additive) ionic radii by Shannon (Table 1.1) is a perfectly reasonable way to go. The values for Ca^{2+} are ca. 0.95, 1.00, and 1.12 Å for four-, six-, and eight-fold coordination. Here we have estimated the value of ${}^{(4)}r(Ca^{2+})$ from ${}^{(6)}r(Ca^{2+})$ by the use of Equation (1.1). For O^{2-}, the corresponding anionic radii are tabulated as 1.38, 1.40, and 1.42 Å such that the Ca^{2+}–O^{2-} interionic distances r_{ij} are classically *predicted* to be 2.33, 2.40, and 2.54 Å for the above three CaO structure types. For the corresponding lattice parameters a, this is equivalent to 5.38, 4.80, and 2.93 Å, because the lattice parameters a and the interionic distances r_{ij} relate to each other according to $r_{ij} = a\sqrt{3}/4$ in the [ZnS] type, $r_{ij} = a/2$ in the [NaCl] type, and $r_{ij} = a\sqrt{3}/2$ in the [CsCl] type, as depicted in Figure 1.1.

In order to predict which structure type is actually found, one often reads (see Section 1.1) that the quotient of the cation and anion radii, r_c/r_a, needs to be calculated, and the critical value is 0.732 for the change from [CsCl] to [NaCl] and 0.414 from [NaCl] to [ZnS]. Indeed, r_c/r_a is found to be 0.714 for the [NaCl] type but *only* if the values for six-fold coordination of Ca^{2+} and O^{2-} are used. Taking, for example, the ${}^{(8)}r$ radii, the ratio is 0.789, a self-fulfilling prophecy in favor of the [CsCl] type with eight-fold coordination. Clearly, this classical criterion is unfortunately unreliable.

We may derive another prediction for the lattice parameters of CaO in the three different crystal structures, simply by considering the volume increments of Ca^{2+} (6.5 cm^3/mol) and O^{2-} (11 cm^3/mol), and the total 17.5 cm^3/mol which corresponds to 29.06 Å^3 per ion pair. Given that the unit cells of the [ZnS]/[NaCl] and [CsCl] structures contain four and one formula unit per cell, respectively, the lattice parameters of CaO are predicted to be 4.88 Å for *both* the [ZnS] and [NaCl] type and 3.07 Å for [CsCl]. The related Ca^{2+}–O^{2-} distances (using the above r_{ij}/a conversion) are 2.11 Å for [ZnS], 2.44 Å for [NaCl], and 2.66 Å for [CsCl] based on volume increments.

There is yet another classical method used to predict these lattice parameters; namely by the bond-valence method. In the [ZnS] type, the +2 bond-valence sum of Ca^{2+} results from four Ca^{2+}–O^{2-} bonds of bond valence $2/4 = 0.5$. For the [NaCl] and [CsCl] types, there are six/eight corresponding bonds with bond valences of $2/6 = 0.3333$ and of $2/8 = 0.125$. Together with the bond-valence parameter of 1.967 Å from Table 1.3 and using Equation (1.15), the interionic Ca^{2+}–O^{2-} distances are 2.22, 2.37, and 2.48 Å which corresponds to (see above conversion) lattice parameters of 5.14, 4.75, and 2.86 Å for the [ZnS], [NaCl], and [CsCl] types.

These back-of-the-envelope calculations may be compared with the results of full-potential electronic-structure calculations of the FLAPW type (WIEN2k) based on the generalized gradient approximation (GGA) for the exchange–correlation potential. On purpose, we use a very accurate DFT method. Figure 3.1 shows the results of these quantum-mechanical calculations as a plot of the total energy as a function of the volume per formula unit for all three structure types. The calculations are based on 482/266/289 plane waves for

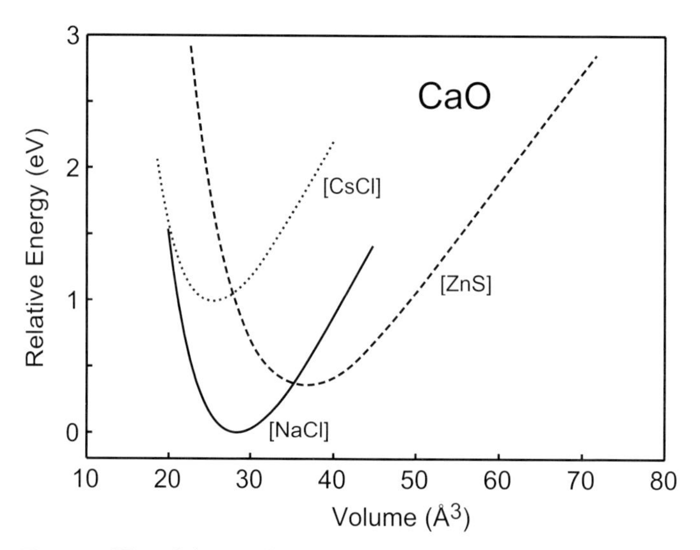

Fig. 3.1 Plot of the total electronic energy of calcium oxide, based on the full-potential LAPW method and the GGA, as a function of the volume per formula unit for three different structure types.

[ZnS]/[NaCl]/[CsCl]-like CaO, and the diagonalization of the Hamiltonian was performed at a total of 1496/816/816 $\boldsymbol{k}$ points for which the calculations appeared to be fully converged. Note that *no* specification of the atomic "nature" (in particular, charge) needs to be made at the beginning of such a first-principles calculation. It is a good idea, however, to approximately guess the interatomic distances (lattice parameters) in order to begin close to the minimum geometry and thus save precious CPU time. The total energy is calculated for the given structure type and lattice parameter and then repeated until a smooth curve can be drawn in the proximity of the lowest energy.

The DFT calculation correctly reproduces the favored structure type, [NaCl], found for calcium oxide because, for this one, the lowest total energy is achieved. Also, the GGA approach almost exactly hits the experimental lattice parameter, and the lowest energy is found for $a = 4.83$ Å. The slight overestimation of only 0.02 Å (i.e., a 0.4% error) is a common feature of the GGA functional (see Section 2.12.1), whereas the LDA functional would have yielded a small underestimation (and also a too-large bonding energy, the LDA "overbinding" problem). This is again looked at in later sections.

To compare the experimental lattice parameter more conveniently with the quantum-mechanical and also the classically derived ones, these are all found together in Table 3.1. Indeed, all of them agree, with respect to the [NaCl]-type lattice parameter, to lie around 4.81 Å. The best value is the one derived from the empirical ionic radii but it may be argued that the latter data were, in fact, *fitted* to match the experimental crystal-structure results. The GGA result further allows us to estimate the classical predictions for the hypothetical lattice parameters of the [ZnS]/[CsCl] types, and here the fit is acceptable, too, for [ZnS] ($a \approx 5.2$ Å) and [CsCl] ($a \approx 3.0$ Å), with the only exception for the volume-increment method, in particular the [ZnS] entry. Obviously, we have already gone beyond the limits of this simplistic method and will therefore not consider it any further for the present system.

Tab. 3.1 Experimental and theoretically predicted lattice parameters (Å) for CaO in the [ZnS]/[NaCl]/[CsCl] structure types derived from GGA calculations, ionic radii tables, bond-valence calculations, and volume increments.

CaO	a_{exp}	a_{GGA}	a_{radii}	a_{BV}	a_{incr}
[ZnS]		5.21	5.38	5.14	4.88
[NaCl]	4.811	4.83	4.80	4.75	4.88
[CsCl]		2.96	2.93	2.86	3.07

When it comes to energetics, we may use the *geometrical* information from the classical calculations and convert them into *lattice energies* by assuming ideal ionic charges (±2) and an average repulsion parameter of $\rho = 0.345$ Å. Then, the calculation by means of Equation (1.11) using Madelung constants of 1.6381 for [ZnS], 1.7476 for [NaCl], and 1.7627 for [CsCl] yields lattice energies U_{L} of 3329/3465/3333 kJ/mol for the [ZnS]/[NaCl]/[CsCl] types, if interatomic distances based on ionic radii are used. This looks fine since the correct type, [NaCl], is most stable. Performing the same calculations of lattice energies based on the interatomic distances from the bond-valence approach, the lattice energies are 3460/3498/3401 kJ/mol, which again favors the correct [NaCl] type and disfavors [ZnS]/[CsCl]. For another convenient comparison of the relative CaO energies in the three structure types, quantum-mechanical (GGA) and classical results are listed together in Table 3.2, but now expressed in eV.

Tab. 3.2 Relative energies (eV) of CaO in the [ZnS]/[NaCl]/[CsCl] structure types derived from GGA calculations, ionic radii tables, and bond-valence calculations.

CaO	E_{GGA}	E_{radii}	E_{BV}
[ZnS]	0.35	1.41	0.39
[NaCl]	0	0	0
[CsCl]	0.99	1.37	1.01

According to density-functional (GGA) theory, [NaCl] is the ground-state structure of CaO and [ZnS]/[CsCl] lie 0.35/0.99 eV higher in energy (see also Figure 3.1). The relative energies, on the basis of ionic radii, largely overestimate the energy differences, and they also give the [CsCl] type as more stable than the [ZnS] type. The energetics derived from the bond-valence (BV) approach is remarkably close to the GGA result, a surprising finding taking into account that the BV calculations were based on only *one* (r_0) parameter for the Ca^{2+}–O^{2-} combination, in contrast to a set of *six* ionic radii. We conclude that the density-functional approach works nicely for the simple case of CaO in terms of *structure* although the (extremely simple) classical estimates – less expensive by orders of magnitude – are also acceptable. Nonetheless, the quantum-mechanical theory allows deeper insights because we may proceed even further.

Given the self-consistent density-functional calculation, let us now visualize the electronic structure of CaO in real space. Because the rock-salt structure is very regular, even a simpler DFT calculation (in terms of atomic potentials) will now suffice, and we therefore recalculate the electronic structure by means of the, much faster, TB-LMTO-ASA method. Since we also know the experimental lattice parameter a with absolute certainty, the less accurate LDA functional will perform nicely for that a. Figure 3.2 shows the theoretical electron density $\rho(\boldsymbol{r})$ in the (100) plane, with the Ca atoms in the corners/center and the O atoms lying inbetween.

For the chosen density window, it is obvious that the oxygen atoms appear larger than the calcium atoms, in accordance with the expectations based on ionic radii. We note that this finding should not be overinterpreted, however,

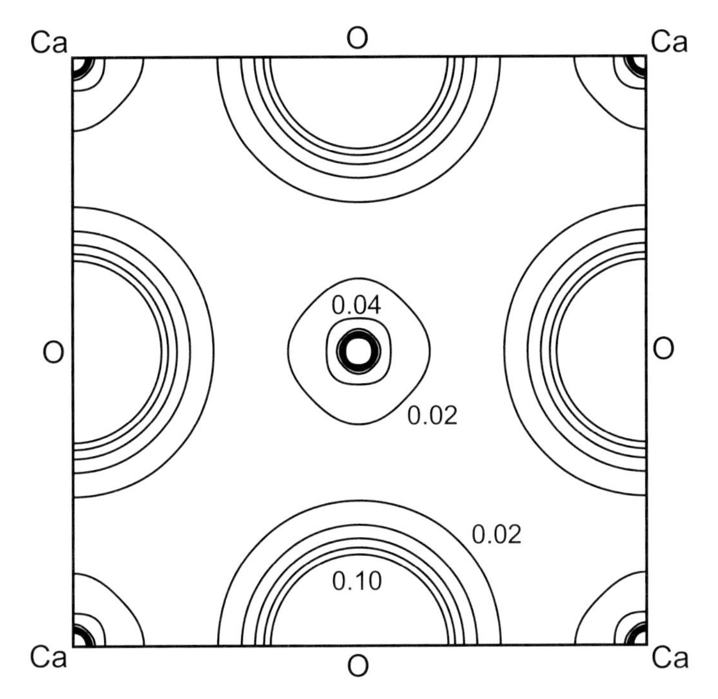

Fig. 3.2 Plot of the calculated electronic density in the (100) plane of CaO according to a TB-LMTO-ASA calculation and the LDA. Isodensity lines are indicated in steps of 0.02 e/a_0^3 where a_0 is the Bohr radius of ca. 0.529 Å.

because a larger density window would yield quantitatively different results and show Ca and O to be of similar size. Nonetheless, all atoms look spherical, the O atoms even more than the Ca atoms. Between Ca and O, there are regions of *minimum* electron density, and this is what we expect to recognize, being biased by the ionic model.

In addition, let us quantum-chemically analyze the electronic structure of this simple material in terms of chemical bonding. This is very easy to do because the above-mentioned TB-LMTO-ASA method operates with an extremely short-ranged basis set, such that electron and energy-partitionings are straightforward. Figure 3.3 shows the results, namely the band structure, the density-of-states, and the crystal orbital Hamilton population analysis.

As a good rule of thumb, the more ionic the material, the less spectacular the band structure. This is particularly true for the band structure of CaO, thereby supporting the ionic model. To a very good approximation, the occupied valence region (a) is represented by the $2p$ orbitals of oxygen, and their $2s$ orbitals are so deep in energy that they have not been included in the figure. The empty conduction bands are given by the $4s$ and $4p$ orbitals of calcium. The amount of covalent interaction looks small, which may be deduced from

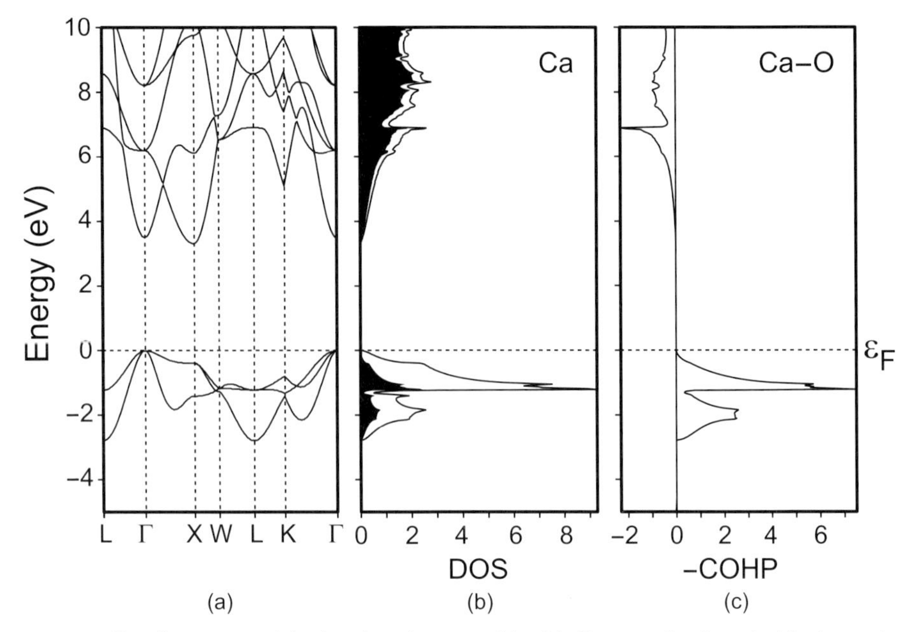

Fig. 3.3 Band structure (a), density-of-states (b) with Ca contributions in black, and Ca–O crystal orbital Hamilton population (c) of CaO in the [NaCl] structure type, according to a TB-LMTO-ASA calculation and the LDA.

the, likewise small, dispersion of the oxygen $2p$ bands which is less than 3 eV. In simple chemical language, the oxygen $2p$ orbitals merely act as charge sinks into which the former $4s$ electrons of calcium are dumped such that the ionic description is recovered, very simple (and boring) indeed.

A more appropriate conclusion can be drawn from the $\boldsymbol{k}$-averaged DOS shown in Figure 3.3(b) because the black projection of the DOS – corresponding to the covalent Ca contribution – is nonzero below the Fermi level such that the chemical bonding cannot be entirely ionic; the ionic model *is* an oversimplification. Although one would expect a large band gap between valence and conduction bands, which is so typical of an electric insulator, the DFT result for an indirect gap[1] (less than 4 eV) between $\mathbf{\Gamma}$ and $\mathbf{X}$ very much underestimates the band gap, which is around 7 eV experimentally. This is a normal DFT weakness because density-functional theory only relates to ground-state properties (see Section 2.12), and the band gap itself refers to an excited-state property which standard DFT is unable to describe correctly. With respect to an energy-resolved analysis of chemical bonding, let us look at the COHP plot in Figure 3.3(c). Although there is significant charge transfer (thus, elec-

1) One refers to an *indirect* band gap if the top of the valence band and the bottom of the conduction band do *not* occur for the same $\boldsymbol{k}$ point. For a *direct* band gap, however, a vertical excitation from the top of the valence band leads directly into the lowest part of the conduction band.

trostatic energy) contained in the structure of CaO, even the COHP analysis of the orbital interactions is indicative of some covalent calcium–oxygen bonding interaction. The states up to the Fermi level are purely bonding, and antibonding states are found way up in the unoccupied part. The Fermi level, drawn in the "chemical" style of being equivalent with the highest occupied level, perfectly fills up all bonding states, and this is because the structure (lattice parameter, interatomic distances) is *optimized* for the electron count of CaO.

Are there other quantum-chemical alternatives to the density-functional approach used here? Yes, another choice for generating a reliable electronic-structure calculation for CaO would have been to use a Hartree–Fock-based strategy. Because CaO is nonmetallic, the method works well [262], and it yields theoretical lattice parameters of 4.864/4.940 Å – depending on the inclusion/exclusion of virtual 3*d* orbitals for Ca – which stand for a 1.1/2.7% overestimation of the experimental value, almost as good as the GGA result.

3.2 Structural Alternatives: Transition-metal Nitrides

Because CaO is not extraordinarily fascinating in terms of electronic structure, let us move on to other simple binary compounds which are electronically more challenging. We will replace the oxygen atom by nitrogen where the smaller electronegativity of N compared with O (see Table 2.2) is indicative of a higher tendency to covalent bonding. Also, the alkaline-earth metal Ca is replaced against its immediate successors in the periodic table, that is, the 3*d* transition metals. Let us therefore discuss the binary transition-metal nitrides of general composition MN, M starting with Sc and beyond.

To the delight of all solid-state chemists, these transition-metal nitrides have recently caused a lot of interest in the materials community. Among possible applications, transition-metal nitrides might serve as materials for (to give but one example) magnetic storage devices as an alternative to the established oxides.[2] At the present time, the chemistry and physics of the *iron* nitrides, in particular, are of fundamental importance for both basic research and technology, the latter due to steel production and hardening. In fact, iron nitrides already possess a proud history as corrosion-protective surface materials, and the definite phase diagram for the iron nitrides [263] is more than half a century old. One has to keep in mind, though, that the number of well-characterized metal nitrides is generally much smaller than that of the corresponding oxides, and part of the explanation goes back to the synthetically

2) Despite the fact that this section deals with transition-metal nitrides, we should also mention that *rare-earth* nitrides emerge as superior luminescent materials because of their very special properties; see also Section 3.3.

more challenging techniques required to prepare nitride phases, especially when it comes to the activation of nitrogen. If we were to start with molecular nitrogen, the N≡N molecule (bond energy about 945 kJ/mol) must be broken to yield two N atoms, and this requires almost twice as much energy as for the O=O molecule (bond energy about 498 kJ/mol). Also, as has been said at the beginning, N is less electronegative than O, thus its oxidizing power is weaker. Despite these synthetic difficulties, nitride chemistry is blossoming [264].

For the entire group of 3*d* transition-metal nitrides, it is probably a good guess to assume the [NaCl] structure type throughout, not simply because CaO adopts it, but also because no other AB structure type is more common than the [NaCl] type. If we deliberately neglect small deviations from the ideal geometry (that is, small tetragonal distortions and such-like), then, in fact, many MN compounds adopt the sodium chloride structure type, for example, the earlier members (lattice parameter in parentheses) ScN (4.44 Å), TiN (4.24 Å), VN (4.10 Å), CrN (4.14 Å), and MnN (4.23 Å) [265]. Surprisingly, FeN and CoN both crystallize in the [ZnS] structure [266], and the lattice parameters are 4.31 and 4.30 Å. Nitride structural chemistry *is* interesting.

How can the structural change from [NaCl] to [ZnS] be explained? A quick look into the list of ionic radii (Table 1.1) shows that the cationic radii for the trivalent state (assuming M^{3+} and N^{3-} throughout) slowly *decrease* on going from Sc^{3+} to Co^{3+} because of the continuously increasing nuclear charge. On the other hand, the size of the nitride anion is 1.46 Å for tetrahedral coordination. It seems that the (numerically unreliable) criterion of the radius ratio is supportive of FeN and CoN to adopt the zinc-blende type. The situation is much more difficult, however, since it is unclear, at the outset, which cationic radius should be used as we do not know whether high-spin or low-spin cations are present. Let us focus on FeN. The rather primitive ionic picture needs to be augmented by a little theory, and this we attempt visually in Figure 3.4, schematically showing the crystal-field splittings of an Fe^{3+} cation, assuming low-spin configurations at the outset.

For octahedral coordination in the [NaCl] type and a strong ligand field (a good approximation for the N^{3-} anion), the five electrons will go into the lower-lying t_{2g} set, and none of the strongly Fe–N antibonding e_g levels will be filled.[3] For a tetrahedral coordination in the [ZnS] type and a correspondingly smaller splitting, the two low-lying e levels are already filled such that at least one electron must enter one of the three Fe–N antibonding t_2 levels.

3) Because of the crystal field around the metal cation and the latter's *antibonding* interactions with the ligands, *all* five formerly degenerate *d* orbitals become energetically destabilized within the octahedral environment, but the t_{2g} set is *less* destabilized than the e_g set. One therefore sometimes (falsely) assumes that the t_{2g} set is "stabilized". The situation within the tetrahedron is similar except that the two-above-three splitting changes into a three-above-two splitting.

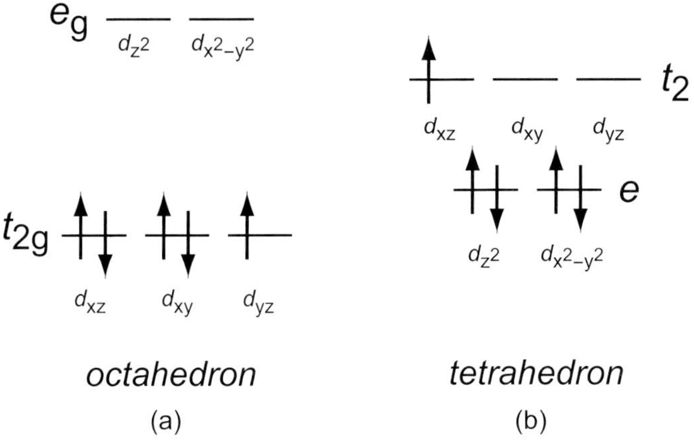

Fig. 3.4 Schematic crystal-field splitting for an octahedrally (a) and tetrahedrally (b) coordinated metal cation. The electron filling corresponds to a d^5 count, appropriate for a formal Fe^{3+}, and a low-spin scenario.

The small splitting furthermore suggests it to be energetically advantageous to adopt a high-spin configuration for the tetrahedral geometry, such that the five electrons are equally distributed over the e and t_2 sets to achieve maximum spin. Because all five electrons are filled into lower-lying levels for an octahedral field, it really looks as if FeN should crystallize in the sodium chloride and not in the zinc-blende structure, in conflict with the experimental result.

The question of alternative structure can be answered by electronic-structure theory, and it turns out that a quantitative answer is slightly more complicated because different *magnetic* properties are calculated for the [NaCl] and [ZnS] types. Nonetheless, non-spin-polarized band-structure calculations are quite sufficient to supply us with a correct *qualitative* picture. This has been derived using the TB-LMTO-ASA method and the LDA functional, and they give the correct lattice parameters with lowest energies for both structure types [267], just as for the case of CaO.

Upon performing these calculations for the entire series of 3*d* metal nitrides, that is, from ScN to NiN, the density-of-states (DOS) curves and also the crystal orbital Hamilton population (COHP) curves appear to be quite similar to those of FeN. Therefore, we adopt a *rigid-band model* in which we assume the electronic structure to be *fixed* but we allow different degrees of electron filling in the bands, DOS, and COHP curves. So, let us schematically show the DOS and also M–N and M–M COHPs of all MN phases in the sodium chloride (Figure 3.5) and zinc-blende (Figure 3.6) structures along with arrows indicating the positions of the Fermi energy ε_F for the isostructural compounds. The behavior of the Fermi level across the series is what we would expect because it is moving up through the bands as the atomic number (and thus the number of valence electrons) increases.

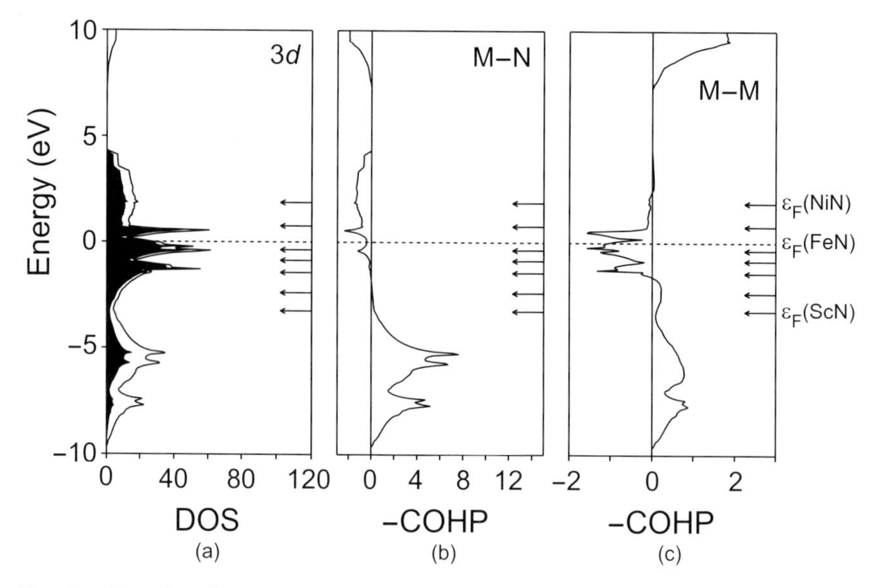

Fig. 3.5 Density-of-states (DOS) including local $3d$ projections (a) and the crystal orbital Hamilton populations (COHP) for the M–N interactions (b) and the M–M interactions (c) in MN, adopting the sodium chloride structure. The original curves go back to a calculation performed for FeN, and the arrows denote the Fermi levels in the other isostructural compounds.

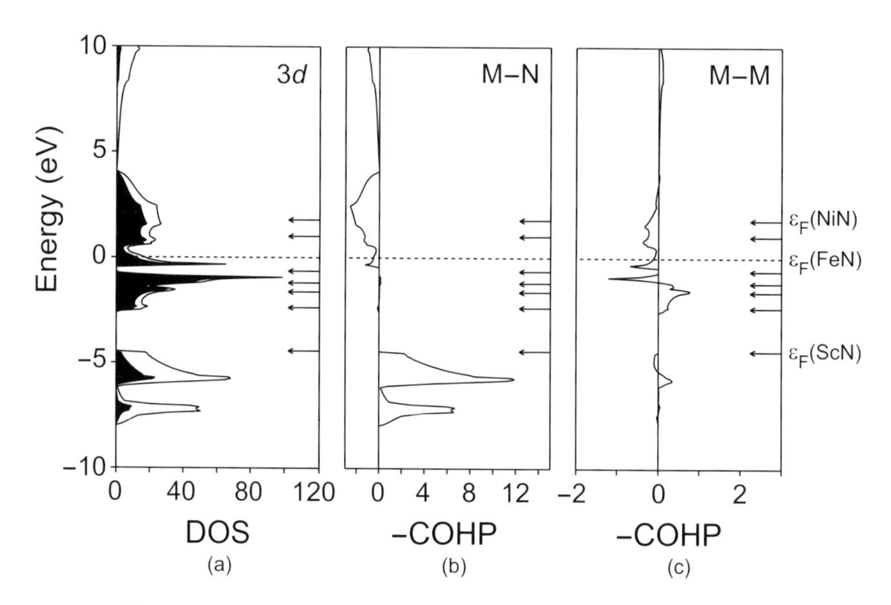

Fig. 3.6 The same as above, but for the zinc-blende structure.

If we first concentrate on an electron filling which corresponds to FeN, then this phase is predicted, for both structure types, to be a metal because of the nonvanishing DOS at the Fermi level (dashed lines). In the occupied region of the two DOS, there are two sets of peaks. The first (around −6 eV) is primarily composed of N $2p$ orbitals, with a bonding admixture of Fe $3d$. The familiar three-below-two (t_{2g} below e_g) crystal-field splitting pattern for octahedrally coordinated Fe (Figure 3.4) is obscured by the large dispersion of the $3d$ peaks in the DOS, indicative of strong covalency. In the [ZnS] structure, however, the $3d$ peaks are narrow enough for the crystal field splitting (e below t_2, tetrahedral Fe coordination) to be discerned (Figure 3.6). Also, the Fe–N π^* interactions in the e set (ca. −2 eV) are considerably weaker than the σ^* interactions in the t states, at about +2.5 eV.

Direct analysis of the chemical bonding between Fe and N in the COHP plots shows that, for both structures, all bonding levels have been filled with electrons, and there is only a small Fe–N antibonding peak visible for the highest occupied bands. With respect to Fe–N bonding, both structure types, [NaCl] and [ZnS], are ideal. To understand the difference in stability between them, however, the Fe–Fe interactions must be considered. Obviously, there are strong *antibonding* Fe–Fe interactions close to the Fermi level (Figure 3.5(c)), and these destabilizing effects are more pronounced in the NaCl structure than in the ZnS structure (Figure 3.6(c)). This finding also results from an energy integration of the Fe–Fe COHPs, showing that the band-structure energy is weakened by 0.21 eV per Fe–Fe contact in the [NaCl] structure, but only by 0.03 eV per Fe–Fe contact in the [ZnS] structure. This result is a simple consequence of the [ZnS] structure allowing *wider* Fe–Fe distances (2.97 Å) than the [NaCl] structure (2.79 Å), given that the stronger (therefore, energetically more important) Fe–N bonds are optimally adjusted in terms of geometry.

Transformations to the [ZnS] structure type considerably lessen the strength of these unfavorable Fe–Fe interactions. In the earlier members of the MN series (ScN, TiN, VN, ...), the Fermi level lies low in the M–M COHP curve, not yet sampling so many antibonding states. For these electron counts, the sodium chloride structure, with much shorter M–M contacts, is favored. For higher electron counts, switching to the [ZnS] structure type reduces the antibonding M–M interactions by a significant amount. It also can be shown, numerically, that the Fe–N bond for the tetrahedral coordination is more covalent than for the octahedral coordination (see also below).

Consequently, a total-energy calculation of the structural alternatives for ScN and FeN based on GGA pseudopotential calculations (VASP), given in Figure 3.7, yields the preference of the [NaCl] type for ScN and the [ZnS] type for FeN, and we have just given a qualitative explanation for this finding in terms of metal–metal *antibonding*.

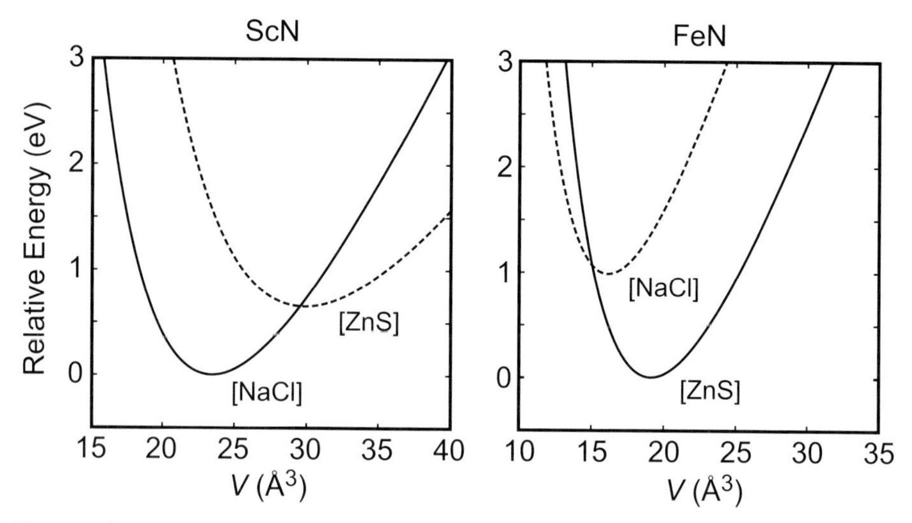

Fig. 3.7 Energy–volume diagram for the [NaCl] and [ZnS] structural alternatives of ScN and FeN using GGA pseudopotential calculations including spin-polarization.

Although the latter explanation does not depend on the magnetic state, we note, for completeness, that FeN would exhibit about 1.3 unpaired electrons in the less stable [NaCl] polymorph. [ZnS]-type FeN, however, is almost nonmagnetic (Pauli paramagnetic, without local moments). ScN is magnetically inactive throughout because there are no leftover *d* electrons for its trivalent state.

Let us be creative and present another way of looking at the structural change from the [NaCl] type to the [ZnS] type, not based on any quantum-mechanical calculation but focusing on the strength of the M–N bonds in both structures, qualitatively. If we stick to the oversimplified notion that the chemical bonding nature in the early transition-metal nitrides (such as ScN) and the later ones (such as FeN) is identically ionic, then their lattice energies determining the structural stabilities are simply given by Equation (1.11), and the details of the structure only enter through the Madelung constant and the shortest interionic distance. Because the Madelung constant $\mathcal{M}$ for the [ZnS] type is *smaller* than for the [NaCl] type, a shorter M–N distance r_{MN} is needed to ensure the same lattice energy, namely by

$$\frac{\mathcal{M}^{\text{NaCl}}}{r_{\text{MN}}^{\text{NaCl}}} = \frac{\mathcal{M}^{\text{Zns}}}{r_{\text{MN}}^{\text{ZnS}}} \quad \text{or} \quad \frac{r_{\text{MN}}^{\text{ZnS}}}{r_{\text{MN}}^{\text{NaCl}}} = \frac{\mathcal{M}^{\text{ZnS}}}{\mathcal{M}^{\text{NaCl}}} = 1.6381/1.7476 \approx 0.937, \tag{3.1}$$

which corresponds to a 7% shortening for the [ZnS] type. We note that the quantum-mechanical optimization of the lattice parameters of FeN yields Fe–N distances of 1.98 Å for the [NaCl] type and 1.82 Å for the [ZnS] type, which indeed reflects an 8% shortening, nicely fitting into the above picture! Admittedly, such an electrostatic estimate does not take the chemical nature of the atoms into account at all, it merely reflects that the Coulomb interaction is infinite in range, decaying with the reciprocal interionic distance. If ionic interactions alone determined the system, one could equally ask for a [CsCl] polymorph of FeN, with a Madelung constant of 1.7627 and slightly *widened* Fe–N distances. This cannot be correct, but why?

As soon as covalency comes into play, small interatomic distances with small coordination numbers are preferred. This is because covalent interactions do *not* fall off as r^{-1} but much faster. For example, the overlap integral S_{ij} between two atomic orbitals goes to zero quite rapidly because the atomic valence functions themselves scale as $e^{-\zeta r}$, as has been indicated in Section 2.2. Thus, if the *covalent* contribution to the cohesive energy increases, smaller coordination numbers (4) are favored over larger coordination numbers (6) although both would be equally suitable for a purely Coulomb interaction. A more detailed discussion has been given by Pearson [201].

To further support this argument, we only have to demonstrate that covalency increases in going from ScN to CoN, and this is easily accomplished. Taking a look at the absolute electronegativities in Table 2.2, the early transition metals (Sc–Mn) exhibit χ values between 3.34 and 3.72, eV whereas the electronegativity of N is 7.30 eV. Also taking into account their various absolute hardnesses, the electron transfer between M and N (calculated by using Equation (2.101)) starts with approximately 0.190 for ScN and falls off continuously to arrive at 0.139 for CoN. All values are plotted in Figure 3.8, showing that ionicity decreases, whereas covalency increases upon going to the right side of the 3d row, and the largest "jump" occurs between MnN and FeN. Because covalent interactions are bound to shorter interatomic distances, the coordination number must decrease such that FeN (and also CoN) crystallizes in the [ZnS] structure type.

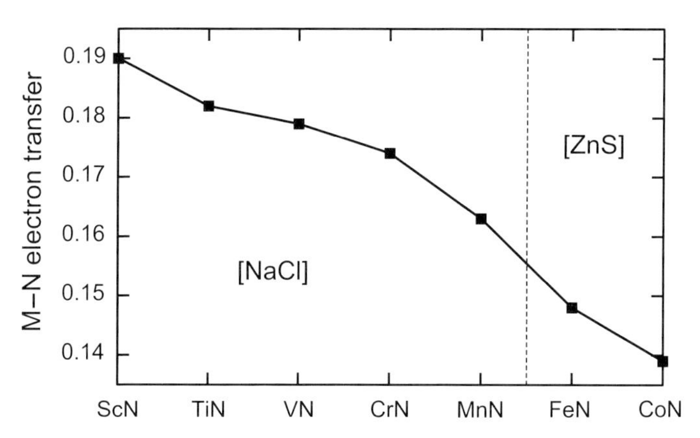

Fig. 3.8 Approximate charge transfer (Equation (2.101)) in the 3d transition-metal nitrides.

3.3
Structure and Physical Properties: Cerium Pnictides

In the last chapter, we observed and explained a structural change on moving horizontally through the periodic table. Let us now move *vertically* and investigate the change in another important physical property, electrical conductivity, while still keeping the structure unchanged. Because simple 1:1 nitrides are also formed by the rare-earth elements, cerium nitride, CeN, will be the starting point and also the major focus of this section.

Cerium is one of the rare-earth or $4f$ elements (according to the valence shell that is gradually filled), and this particular class of metallic elements may be viewed as truly technology-decisive materials. The partially filled $4f$ levels of the lanthanoids (another name for the rare earths) manifest themselves in exciting physical properties, and large parts of solid-state physics deal with the luminescent and also magnetic properties of these metals and their compounds. Within the field of chemistry, important catalytic processes as well as anticorrosion technologies depend on the presence of the rare earths. The element cerium, in particular, is mainly used in automotive antipollution catalysts (CeO_2), in optical glass fabrication and also in various oxide pigments contained in (soon to be outdated) color television tubes, in fluorescent lamps and in novel white light-emitting diodes, *the* optical materials of the 21st century. The ease with which cerium can switch between the oxidation states +3 and +4 is so fundamental that a classic analytical technique, called cerometric titration, depends on it. This is taught to freshmen chemists from the very beginning, at least at RWTH Aachen.

For the convenience of the reader (here I am joking), CeN also crystallizes in the sodium chloride structure type with a lattice parameter of 5.01 Å. The other cerium pnictides incorporating the higher homologues of nitrogen also adopt this structure, and the lattice parameters are 5.91 Å (CeP), 6.07 Å (CeAs), 6.41 Å (CeSb), and finally 6.50 Å (CeBi). The large difference between the lattice parameters of cerium nitride and cerium phosphide is due to the special situation of the elements in the first long period. Because of their "missing" $1p$ orbital, the $2p$ atomic orbital experiences a relatively high nuclear charge and is strongly contracted. In Section 2.9, we have already mentioned the consequences of this, namely high electronegativities for these elements and their ability to easily mix $2s$ and $2p$ orbitals, allowing multiple bonding (N≡N, C=C, etc.). In addition, the course of the pnictide lattice parameters reflects that the Bi atom is only slightly larger than its lighter homologue Sb, and this is due to the lanthanoid contraction [63] plus, to a lesser extent, the relativistic shrinkage of the heavy atom Bi [118].

If one were to assign oxidation states to the cerium pnictides, a formulation of trivalent cerium and also N^{3-}, P^{3-}, As^{3-}, Sb^{3-}, and Bi^{3-} would be straightforward, and the ionic radius for Ce^{3+} (1.01 Å for octahedral coordination, Table 1.1) in combination with the experimental lattice parameter, translates into an N^{3-} radius of 1.50 Å, somewhat enlarged when compared with that for four-fold coordination (1.46 Å). Unfortunately, the other anionic radii are unavailable from the Shannon tabulation. Still assuming Ce^{3+} throughout, its valence electron configuration should be $4f^1 5d^0 6s^0 6p^0$ or, in short, [Xe]$4f^1$, with one unpaired electron in a tightly bound, low-lying, strongly contracted $4f$ level which can naively be expected not to be heavily involved in chemical bonding. If that is so, the cerium pnictides should all be salt-like materials, similar to main-group nitrides such as AlN and GaN when it comes to transport properties. Thus, we expect the whole CeX series to be semiconducting. On the other hand, the single unpaired electron which we believe to be buried in a $4f$ atomic orbital should clearly manifest itself from a paramagnetic signal in any magnetic measurement.

This prediction is a reasonable one for most cerium pnictides, namely CeP, CeAs, CeSb, and CeBi which, in fact, exhibit localized spin moments with an antiferromagnetic ordering of the $4f$ electron remaining on each Ce [268]. CeN, however, is a *metallic conductor* with the corresponding magnetic properties and it only shows Pauli paramagnetism of the metallic electron gas, such that *no* local spin moment, characteristic for an unpaired electron, can be detected. This behavior leads to the possibility of an electronic formulation according to $Ce^{4+}N^{3-}e^{-}$, with one electron left in the conduction band, but the smaller cationic radius of Ce^{4+} (0.87 Å) would require a super-enlarged N^{3-} anion of 1.64 Å which is very hard to believe. The interesting puzzle of the peculiar electronic (and nonmagnetic) properties of CeN has been extensively debated in the literature, and the most prevalent explanations are mixed-valency for cerium [269] and itinerant (delocalized) $4f$ electrons [270] that are involved in the chemical bonding.

If cerium in CeN is trivalent, what happened to the leftover electron? If cerium is tetravalent, why do the interatomic distances better match the trivalent state? Why is CeN metallic although the other cerium pnictides are not? A first insight is given by a density-functional band-structure calculation on CeN [271] using TB-LMTO-ASA theory and the spin-unrestricted variant of the local-density approximation. When a spin-polarized calculation is started with a local moment on cerium, this given moment immediately vanishes upon iterating towards self-consistency, in harmony with the experiment. This is a very good sign for the density-functional approach using the LDA. Also, the Fermi level cuts through a bunch of valence levels, indicative of the metallic properties of CeN. To begin the closer analysis, Figure 3.9 shows a *fatband* plot for CeN, a band-structure equivalent of projected DOS plots. Not only is the band structure shown, but the *widths* of all bands (therefore *fat*band) are proportional to the contribution of a given set of atomic orbitals to the crystal orbitals.

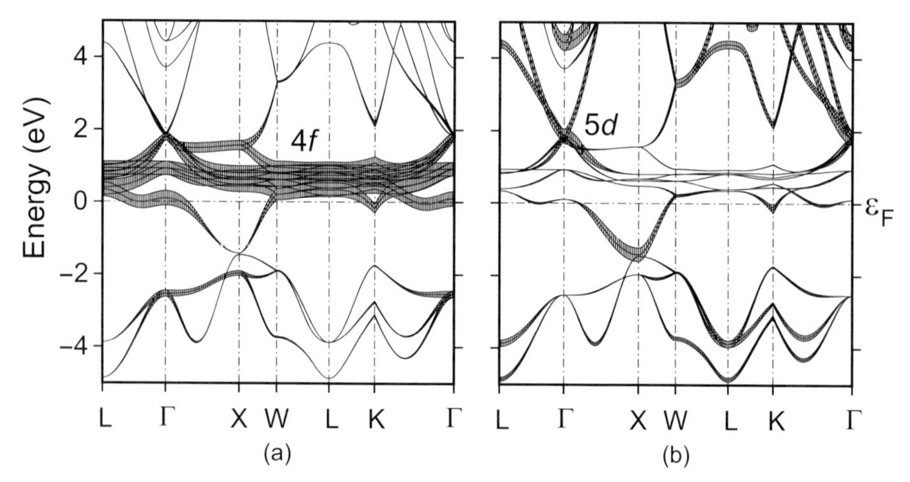

Fig. 3.9 Fatband plots for CeN; the widths of the bands are proportional to the contribution of the Ce $4f$ (a) and Ce $5d$ (b) atomic orbitals. The bands are shifted so that the Fermi level (indicated by a horizontal dotted line) lies at the energy zero.

Clearly, the cerium $4f$ levels are very localized in energy and do *not* show a large dispersion (flat bands), thereby indicating strong spatial localization. It is also somewhat surprising that the primarily $4f$ states of CeN lie *above* the Fermi level such that they are not occupied, and the admixture of Ce $4f$ orbitals into occupied valence states is quite small. Remember that, at the outset, we had assumed the $4f$ levels to be lying very low in energy, resembling core-like functions.[4] The behavior of the Ce $5d$ orbitals in CeN is also very interesting. The occupied band immediately below the Fermi level, clearly visible along the symmetry lines $\mathbf{\Gamma} \rightarrow \mathbf{X}$ and $\mathbf{X} \rightarrow \mathbf{W}$, must be the one which holds the remaining Ce-centered electron, and it has quite a large contribution from the Ce $5d$ orbitals. Averaged over the entire Brillouin zone, this situation is also visible from the DOS plots of CeN given in Figure 3.10(a) and (b).

As seen from the band structure, the cerium $4f$ contribution is mostly restricted to the empty conduction band, but the valence band exhibits a significant $5d$ character. The corresponding COHP curves in Figure 3.10(c) and (d) show, besides the expected strong Ce–N bonding in the occupied region, that the states just below the Fermi level (between 0 and −2 eV) generate *bonding* Ce–Ce interactions but they do *not* contribute to the Ce–N bonding. Indeed, a numerical integration of the Ce–Ce COHPs yields −1.14 eV, that is, −0.10 eV for each of the twelve 3.57 Å contacts of formally Ce^{3+}–Ce^{3+} type. This is one-

4) On the other hand, pure LDA calculations are known to typically overestimate the itinerancy (that is, the delocalized behavior) of the $4f$ levels, such that their exact energetic positioning must also be considered with a little caution.

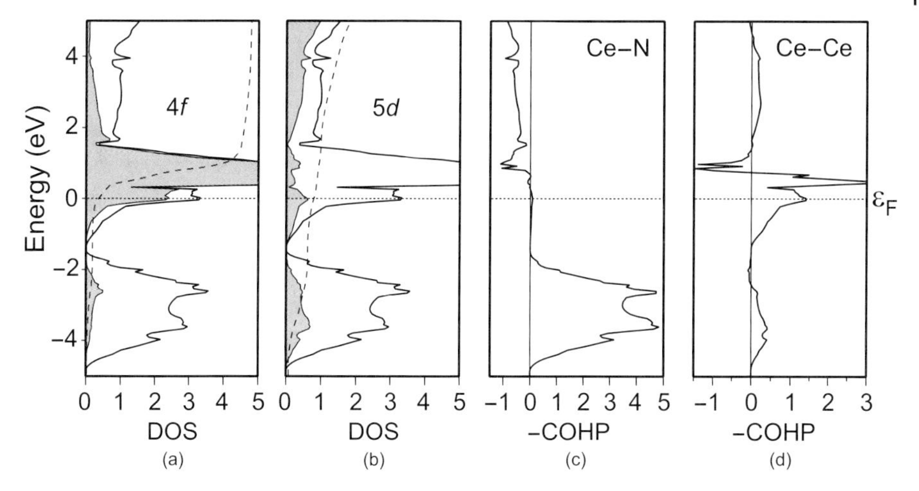

Fig. 3.10 Total DOS, Ce $4f$ and $5d$ projected DOS, and COHP curves for the Ce–N and Ce–Ce interactions in CeN. The dashed line in each DOS curve shows the integration of the projected states. To make the projected DOS curves more visible, the DOS is magnified. The Ce–N and Ce–Ce COHPs include all nearest atomic neighbors.

sixth of the value in face-centered cubic Ce in which each of the twelve 3.64 Å Ce–Ce bonds contribute -0.59 eV to the integrated Ce–Ce COHP.

Curiously, the reliance upon ionic radii can get us into trouble when trying to understand CeN, and *attractive* cation–cation interactions are *very* surprising for the [NaCl] type which alludes to ionicity. The Ce^{3+}–Ce^{3+} distance, 3.57 Å, is *much* larger than twice the ionic radii for six-fold coordination (2×1.01 Å = 2.02 Å) and also definitely larger than the one for twelve-fold coordination (2×1.34 Å = 2.68 Å; remember that each Ce^{3+} has twelve cerium neighbors) but bonding cerium–cerium interactions arrive unexpectedly for the "ionic believer". Let us keep in mind, though, that the above distances are about the same length as those in metallic Ce. Obviously, ionic radii cannot be used to *a priori* rule out covalent interactions. The COHP and band structure results suggest the reason for the metallic behavior of CeN and complete spin pairing, namely that the remaining electron on the Ce^{3+} is located in a spatially diffuse $5d/4f$ hybrid orbital instead of in a highly localized $4f$ orbital. The spatial extent of the electrons in the band just below the Fermi level can be quite clearly seen in a plot of the electron density of CeN, given in Figure 3.11. In (a), all occupied bands are used for that purpose whereas (b) corresponds to an energy slice just below the Fermi level.

The electron density $\rho(\boldsymbol{r})$ which emphasizes the frontier levels (b) clearly shows that the states just below ε_F are centered along the directions of the Ce^{3+}–Ce^{3+} interactions, and the remaining electron on cerium is localized in a cerium–cerium (partial) bond instead of being localized on the ion it-

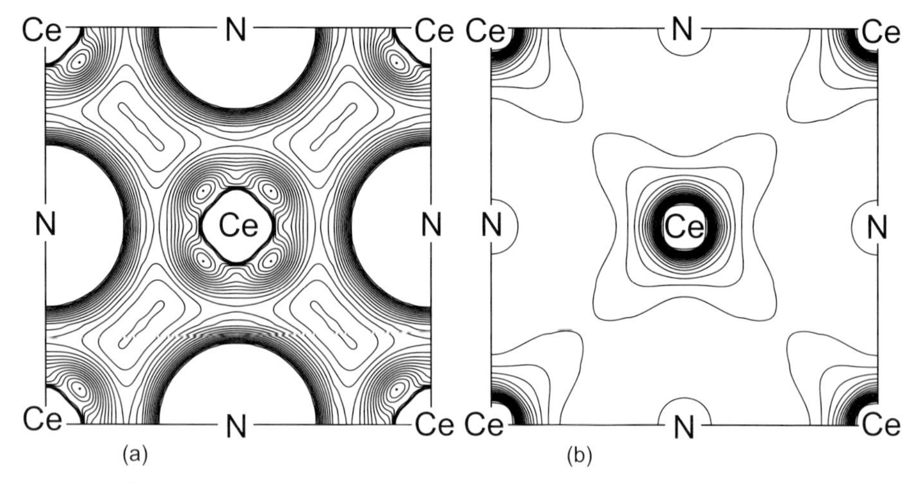

Fig. 3.11 Contour plots of the total electron density (a) in the (100) plane of CeN and an energy slice containing only the bands between the Fermi level and approximately 1.5 eV below (b). 20 contours equally spaced between 0 and 0.05 electron/a_0^3 are shown.

self. The spatial character of the wave function is *d*-like (see central Ce atom), and the nonzero DOS at the Fermi level explains the metallic properties. The leftover electron, however, is associated with the Ce bonding, not delocalized in some free-electron-like band. This rules out the crystal composition $Ce^{4+}N^{3-}e^{-}$ mentioned above. Summarizing, the lack of spin-polarization is due to unexpected Ce^{3+}–Ce^{3+} interactions which cause occupation of Ce 5*d* orbitals instead of 4*f* orbitals, and the valence electron configuration of Ce in CeN is perhaps best described as somewhere inbetween $4f^0 5d^1 6s^0 6p^0$ and $4f^1 5d^0 6s^0 6p^0$. This is in accordance with independent (earlier, in fact) calculations by others [270]. There are many other examples of such metal–metal bonding in metal-rich solid-state materials [272].

The above reasoning is also supported by a numerical experiment in which the cerium–cerium interactions in CeN are weakened by artificially increasing the lattice constant in the calculation. This experiment cannot, unfortunately, be experimentally realized but the theoretical calculation shows that for a lattice parameter larger than 5.17 Å – that is, a more than 3% widening of the experimental CeN structure – the leftover electron moves away from the bonding region and localizes at the Ce^{3+} atom. At the same time, a spin moment pops up at Ce^{3+}, and a band gap opens at the Fermi level such that CeN becomes a semiconductor. This is close to the situation found in cerium phosphide, CeP, with a lattice parameter of 5.91 Å. Because of the larger size of P^{3-}, that leftover electron on Ce^{3+} *cannot* acquire substantial *d* character because of insufficient *d*–*d* overlap that goes back to too large cerium–cerium distances. Instead, the electron resides in a highly localized Ce $4f$ level. The opposite is found in CeN simply because N^{3-} is much smaller.

Fortunately, the DFT approach (LDA) used here is able to deal with that situation quantitatively [271]. In general, some caution is needed when $4f$ levels are involved because of the severe problems mentioned before in the context of strong electronic correlation (see Section 2.13) which may render LDA-like calculations entirely useless. This was not the case for CeN, luckily.

3.4
Structures by Peierls Distortions: Tellurium

It goes without saying that questions touching upon structure–property relationships are not restricted to binary (or higher) chemical compounds, they are already there for the elements. Before we address these for the case of the transition metals, we will start with something quite simple (but only at first sight!), namely the structural behavior of the elements of the main group VI,[5] the chalcogenides. This group contains true nonmetallic elements such as oxygen and sulfur, followed by half-metals like selenium and tellurium, and is completed by a typical metal, polonium. The gradual increase in metallicity is reflected by a change from molecular, covalently bonded species (O_2 molecule, S_8 molecule) over extended but still covalently bonded ones (Se_∞ and Te_∞ chains) to a metallic system with fairly delocalized electrons and almost undirectional bonding (Po).

From the perspective of a materials scientist, tellurium is certainly the most attractive element, and it has found applications mainly in the electronics industry because of its p-type semiconducting properties. It can also be used for the synthesis of so-called II-VI semiconductors (such as MgTe) or may be alloyed to yield superior intermetallic phases. In addition, some Te alloys are under investigation in the realm of phase-change media for rapid data storage [273].

Under standard conditions, elemental tellurium crystallizes in the trigonal [α-Se] type with space group $P3_121$ and lattice parameters of $a = 4.46$ Å and $c = 5.92$ Å. In this structure, the Te atoms form infinite helical chains with three atoms per turn, and the chains run parallel to the crystallographic c axis, depicted in Figure 3.12.

As we will show in the sequel, this very structure may be understood to arise from a *three*-dimensional Peierls distortion away from a simple cubic structure. Although Peierls theory (see Section 2.6.2) is one-dimensional in its original formulation, there have been extensions to real systems in which two- and three-dimensional cases of Peierls-like distortion were observed. For ex-

5) IUPAC recommends to call it group 16; whether this is easier to memorize is left to the reader.

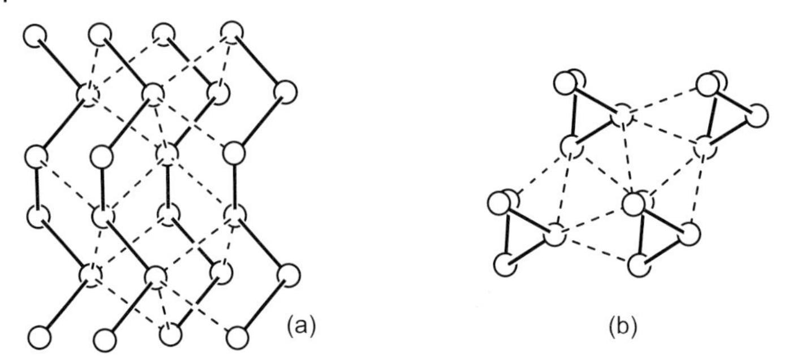

Fig. 3.12 Two views of the normal tellurium crystal structure, side view (a) and view along the chain direction (b).

ample, Burdett and coworkers [274] have shown that the structures of black phosphorus and arsenic can be regarded as distortions of a simple cubic structure due to a half-filled p valence band of these main group V elements, and black phosphorus can, indeed, be forced into a simple cubic phase with metallic properties by using very high pressure. Other examples of Peierls distortions have also been extensively reviewed [275].

Let us now return to tellurium under standard conditions in which the Te atom has a two-plus-four coordination: two intrachain nearest neighbors at a distance of 2.83 Å and four interchain second-nearest neighbors at 3.49 Å (dotted contacts in Figure 3.12). These distances to the second-nearest neighbors are much shorter than the van der Waals contact of 4.12 Å, indicating a little interchain covalent bonding. Bond-valence calculations (see Section 1.5) show that, if 2.83 Å is considered a single bond (i.e., $r_0 \equiv 2.83$ Å), the four bond orders between the second-nearest neighbors sum up to 0.67. The geometrical similarity between this structure, [α-Se] type, and a simple cubic structure, [α-Po], is given in Figure 3.13. By dropping four of the six Te–Te bonds in the [α-Po] structure, the [α-Se] structure emerges. It really seems that both structures should also be electronically related to each other.

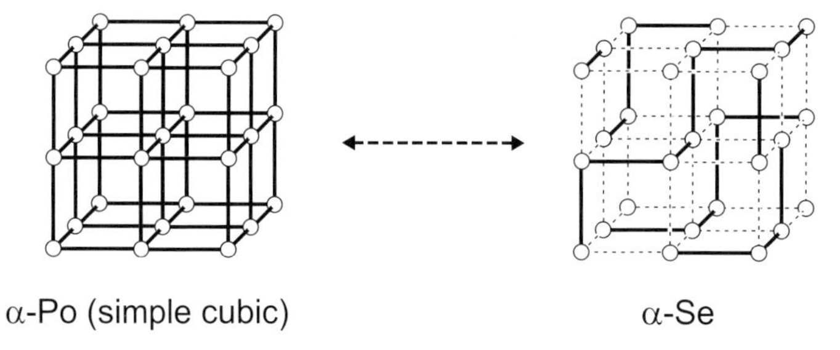

Fig. 3.13 Geometrical relationship between the α-Po (simple cubic) and the α-Se structure.

Although this simple cubic phase is unknown for Te, the structurally closely related rhombohedral Te(IV) phase with the [β-Po] structure (stable between 11 and 27 GPa) exhibits just such an octahedral Te coordination where all six nearest-neighbor distances are equal [276]. This finding may be looked upon as a reminiscence of the pressure–homologue rule of crystal chemistry [5], namely that, under pressure, an element tends to adopt the structure of its higher homologues.

We will first analyze the electronic structure of a *hypothetical* Te adopting the simple cubic structure with space group $Pm\bar{3}m$, and this kind of a cubic tellurium will serve as a Peierls-*undistorted* Te. Its total energy is lowest for a Te–Te distance (= lattice parameter) of 3.16 Å using TB-LMTO-ASA theory and the LDA. Interestingly, these 3.16 Å equal the arithmetic average of the nearest and second-nearest neighbor distances in the real ground-state Te structure with Te–Te chains. Figure 3.14 shows the band structure, DOS and COHP curves for the simple cubic tellurium structure.

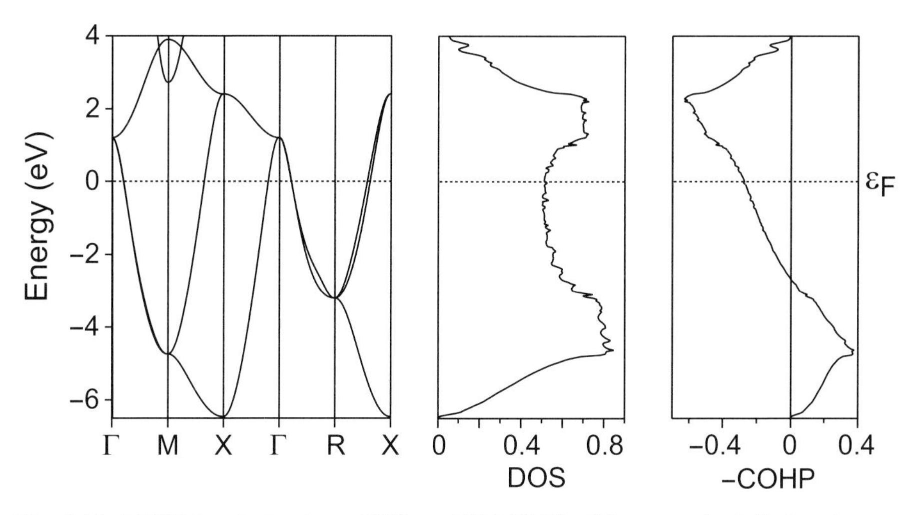

Fig. 3.14 LMTO band structure, DOS and COHP (Te–Te) curves for tellurium in a simple cubic structure with a = 3.16 Å. All curves are shifted so that the Fermi level lies at the energy zero.

The band structure is dominated by the Te $5p$ orbitals in the valence region, and it reflects the typical course of bands in a simple cubic lattice, including the three-fold degeneracy at the high symmetry points **Γ** and **R**. Some bands cross the Fermi level such that there results a significant density of states. Simple cubic Te is thus predicted to be an electrical conductor, just like the higher homologue Po. As can be observed from the COHP curve, the Fermi level lies in a region of Te–Te *antibonding*, thereby destabilizing the structure, making it unfavorable when compared with the ground-state structure. This is the reason why the simple cubic structure is an energetically inappropriate alternative to the ground state. Since the $5p$ bands are only 2/3 occupied, the filling is insufficient to populate the extremely antibonding states at the top.

We note that the simple cubic structure of Te contains a single atom per unit cell. In preparation for a deformation, the system needs to be described using a lowered symmetry which is adopted by the real, Peierls-*distorted* hexagonal Te structure. In the latter, there are three Te atoms per unit cell. Because the valence bands are 2/3 filled, a tripling of the unit cell can remove the degeneracies at the Fermi level (analogous to the doubling for the half-filled bands of the H chain, see Section 2.6.2), but tripling the cell will cause the bands to be folded back twice, not once. Therefore, to triple the cubic unit cell, it must be transformed from a one-atom cell in $Pm\bar{3}m$ to a three-atom cell in $P\bar{3}m1$ (with $a = 4.47$ Å and $c = 5.47$ Å). Later, the atoms in the three-atom cell only have to be slightly shifted such that the ground-state structure of Te (space group $P3_121$) is achieved. The recalculated band structure of cubic Te, based on a *three-atom* unit cell, is presented in Figure 3.15(b) and, for ease of comparison, it is shown together (a) with the preceding band structure, based on a *one-atom* unit cell. Note that both reflect *identical* structures, i.e., the simple cubic one, but using smaller (a) and larger (b) unit cells.

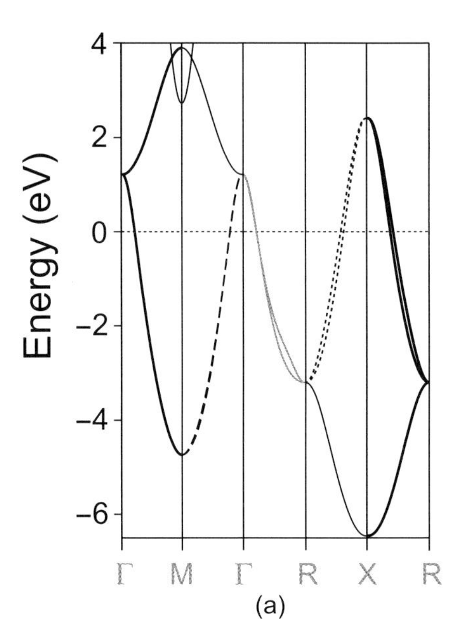

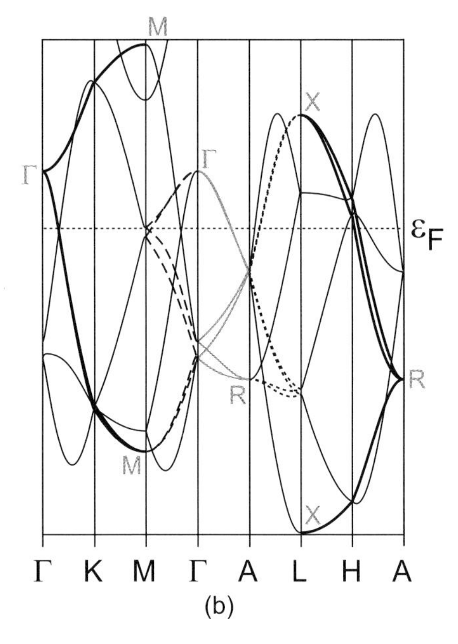

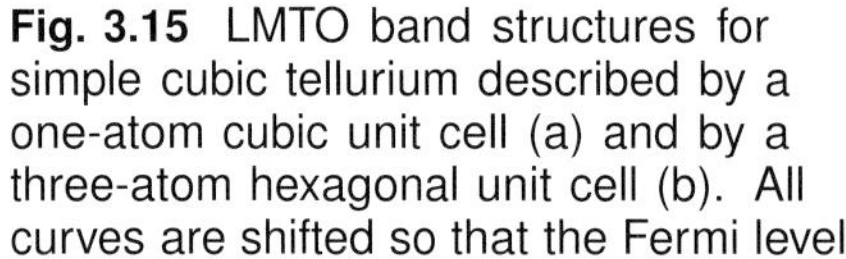

Fig. 3.15 LMTO band structures for simple cubic tellurium described by a one-atom cubic unit cell (a) and by a three-atom hexagonal unit cell (b). All curves are shifted so that the Fermi level lies at the energy zero. In (b) the connections to the k points of (a) are labeled by the plotting style of the bands (also see text).

The real-space tripling of the unit cell also affects reciprocal space, and the first Brillouin zone and the high-symmetry lines along which the band structure is calculated must be changed. Thus, the cubic Brillouin zone of a simple cubic lattice is replaced by a smaller regular hexagonal prism which is the Brillouin zone of a hexagonal lattice, and their relationships, as well as those of their special points, are depicted in Figure 3.16 for the convenience of the reader. In real space, the size of the unit cell has to be tripled for a change from a cubic to a hexagonal lattice. In reciprocal space, there is a *decrease* in size such that the six-sided, regular prism is inscribed in the cube. The centers of both Brillouin zones coincide, and points **R** and **A** correspond to each other; also **L** is equivalent to **X**.

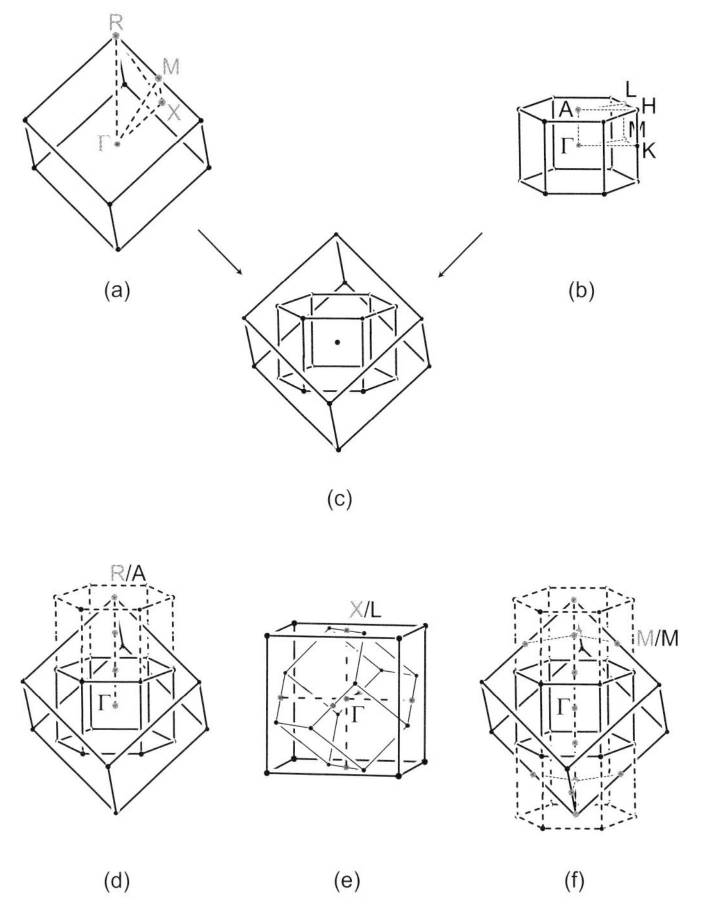

Fig. 3.16 First Brillouin zone for a simple cubic lattice (a), for a hexagonal lattice (b), and their relationships (c)–(f).

Coming back to the band structure of a cubic Te described by a three-atom unit cell, the differences from the former one-atom representation provide a lot of insight. Because the present cell is tripled, there are three times as many bands, so the band structure has become fairly complicated. There are degeneracies at high symmetry points near the Fermi level, easily seen at the (hexagonal) points **M** and **H**. In addition, there are intersecting bands, and some of the crossings lie directly at the Fermi level, that is, along the directions **Γ** → **K** and **Γ** → **M**. The relationships with the bands of the smaller unit cell are also simple to spot (see Figure 3.15). The central grey bands along **Γ** → **R** (a) match the grey bands along **Γ** → **A** (b); they have been folded back twice. The dashed band between **M** → **Γ** (a) is again found along **M** → **Γ** (b). The dotted bands along **R** → **X** are folded back and now lie along **A** → **L**. In addition, the bold bands along **Γ** → **M** and **X** → **R** are stretched across two zones without being folded, and they can be seen (b) running along **Γ** → **K** → **M** and **L** → **H** → **A**. Many band-crossings are created by the backfolding and superpositioning of the bands which result from enlarging the unit cell. Degeneracies at high symmetry points occurring close to the Fermi level are "pure" folding points – the bands used to lie between special $\boldsymbol{k}$ points.

Because of these degeneracies and band-crossings at the Fermi level in Figure 3.15(b), the system is susceptible to a Peierls distortion which splits the degeneracies and creates a band gap. This is immediately obvious from a new band-structure calculation based on the correct ground-state structure of Te of the [α-Se] type, and the bands, DOS and COHP curves are shown in Figure 3.17.

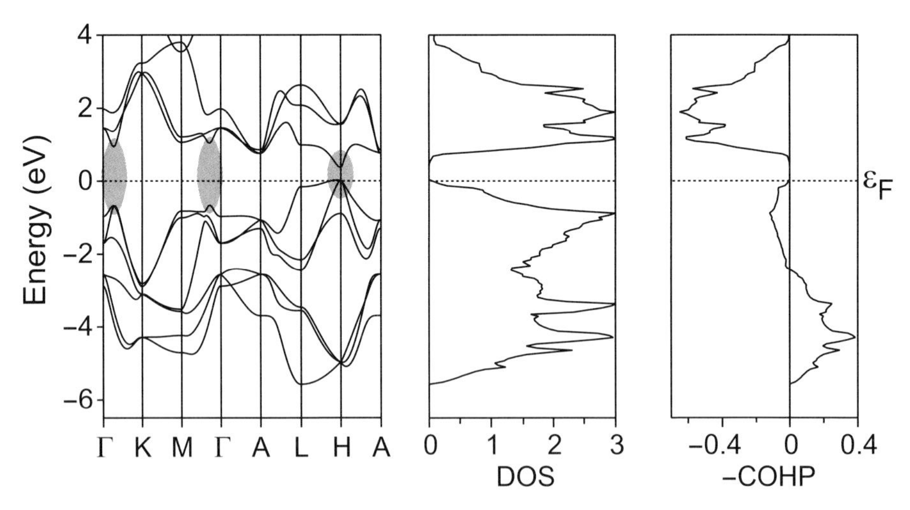

Fig. 3.17 LMTO band structure, DOS and COHP curves for tellurium in the experimental [α-Se] structure. All curves are shifted so that the Fermi level lies at the energy zero.

As expected, the distortion leading away from the cubic symmetry is reflected by the opening of a band gap at the Fermi level, separating the lower-lying occupied bands (2/3 of the total) from the unoccupied bands. This is what the shading in Figure 3.17 indicates. Former band-crossings (see Figure 3.15(b)) split, such that one part of the band moves up in energy, becoming destabilized, while the other part is stabilized, dropping below the Fermi level, and a couple of "avoided crossings" are obvious. In addition, some degeneracies at high symmetry points have been removed. The presence of a band gap is also seen in the newly calculated DOS curve and, accidentally, the prediction of a small band-gap semiconductor is in good agreement with the experimental gap of 0.33 eV. We already know that DFT typically underestimates this excited-state property (see Section 2.12).

With respect to chemical bonding, the COHP curve shows that the formerly antibonding Te–Te interactions in the cubic structure have now been significantly reduced by the Peierls distortion, and we may interpret this behavior as the *driving force* for the symmetry lowering [277]. Obviously, nature avoids antibonding interactions (such as in cubic Te, leading to a high internal stress) whenever possible and thus strives for the hexagonal structure with infinite tellurium helices. The exact size of the energy difference between cubic and hexagonal Te cannot be determined by the too-primitive ASA Hamiltonian [277], but more accurate pseudopotential calculations with the full potential show it to be quite small [278].

Summarizing, antibonding interactions at the Fermi level destabilize a high-symmetry structure and lead to a lowering of structural symmetry. This is what we meant by quantum effects striving against Pauling's fifth rule, in Section 1.3. In the language of band structures, a three-dimensional Peierls distortion of cubic Te sets in to make this process happen whereas, in an alternative chemical phrasing, the principle of maximum bonding lets the Te atom readjust the atomic position to adopt the [α-Se] structure. Other allotropes of Te have also been computationally studied [279].

It goes without saying that similar symmetry-lowering effects may be found virtually everywhere in the periodic table, not just for Te. For example, the main group III element, In, crystallizes with a body-centered tetragonal unit cell which may be thought of as a Peierls-distorted fcc unit cell, adopted by its lighter homologue Al. The quantum-chemical analysis indeed shows that the tetragonal distortion occurs because it increases $5s$–$5p$ mixing and, thereby, strengthens In–In bonding [280]. Likewise, there is good reason to assume that the *very* complicated crystal structure of α-Mn (with 58 atoms in the magnetic unit cell) is the product of a related symmetry-lowering process, thereby optimizing the bonding between certain manganese atoms which are only 2.24 and 2.38 Å apart [281]; antiferromagnetism also plays an important role [282]. In addition, structural distortions as a function of the electron count are present in the intermetallic series MTSi (M = Sc–Ni, T = Co, Ni), an extraordinarily rich group of about 500 (!) representatives. On going from MnNiSi to FeNiSi, a structural change occurs which populates bonding and depopulates antibonding states [283].

This little compilation could probably continue forever. However, we can summarize by noting that, whenever possible, nature tries to get rid of antibonding states, for example, by distorting high-symmetry atomic configurations. If that does not work, then there is a problem.

3.5
Itinerant Magnetism: The Transition Metals

It probably will not be too surprising for the reader that it is not only structural properties which result from symmetry-dependent electronic instabilities, but some of the most exciting physical properties of solid-state materials may also be understood in a very similar way. In this chapter, the macroscopic quantum phenomena of ferro- and antiferromagnetism will be investigated, and we will concentrate on the magnetic properties of the transition metals. We have already touched upon an elemental metallic system (Po) in the last section and will concentrate on the metallic state in greater detail. A special focus will lie on the bonding properties of the transition metals in order to come up with a chemical view of *itinerant* magnetism, the physicist's word for the magnetism of these $3d$ metals.

Although cooperative magnetic phenomena have produced a rich synthetic and theoretical playground for generations of material scientists [284, 285], it is still difficult to understand why some materials are ferromagnetic and others are not. This is especially puzzling taking into account the paramount importance of magnetic compounds for our information society in data storage and retrieval, and it would be highly desirable to derive semiquantitative signposts for the rational synthesis of ferro- and antiferromagnetic materials. Before that, however, we must briefly review the impressive contributions that have been made in the solid-state physics field.

The reader is certainly aware of Curie's heuristic distinction between diamagnetic materials (with small, negative, temperature-independent susceptibilities), paramagnetic materials (with small, positive susceptibilities which typically vary inversely with temperature) and ferromagnetic materials (substances which exhibit magnetism without an applied field below a critical temperature and which behave as strong paramagnets at higher temperatures), to be found in all corresponding textbooks [53]. Using the concept of a magnetic moment by Langevin [286], Weiss explained ferromagnetism as depending on the existence of an extraordinarily large internal molecular field aligning neighboring magnetic moments [287]. The physical origin of the molecular field central to Weiss' theory remained a mystery for years but it was experimentally shown, by a number of ingenious intermetallic dilution experiments, that the molecular field in ferromagnetic iron decays with the sixth power of the interatomic distance.

The puzzle concerning the above classical field was independently solved – by discarding it! – by Heisenberg [288] and Dirac [289] who formulated quantum-mechanical exchange interactions between localized electrons. This approach is somewhat similar to the valence-bond treatment of the hydrogen molecule (see Section 2.11.3), assuming *full correlation*. Within the archetypal ferromagnets (Fe, Co, Ni), however, the conduction electrons are strongly delocalized (itinerant) such that the assumption of localized spins is far from reality. Also, the experimental magnetic moments (numbers of unpaired electrons) are all nonintegral for Fe, Co, and Ni (2.21, 1.60, 0.62) which is difficult to explain assuming localized spins. Therefore, a very influential delocalized theory was introduced by Stoner [290], similar in spirit to the molecular-orbital idea by Hund and Mulliken, and Stoner's criterion for the presence of ferromagnetism is part of many solid-state physics monographs [291]. If the strength of the exchange interactions I times the density-of-states at the Fermi level exceeds unity ($I \cdot \mathrm{DOS}(\varepsilon_\mathrm{F}) > 1$), ferromagnetism sets in.

Like in molecular quantum chemistry, the localized–delocalized antagonism is omnipresent in the theoretical literature on itinerant magnetism. On the one hand, the Hubbard model [292] and related theories for strongly correlated systems have been employed to study rare-earth and also transition metals. Since the latter do not have flat bands, extensions to the Hubbard theory are required [293–295]; also, to make the model Hamiltonians (almost) exactly solvable, simplifications are introduced. On the other hand, density-functional theory is able to extract Stoner's parameters [296, 297] for a self-consistent description of itinerant magnetism [298]. As has been illustrated before, the theoretical limits of the LDA became apparent from Fe phase stability problems (see Section 2.12.1) and were solved by using gradient corrections. The present status of DFT in the treatment of cooperative magnetism has also been reviewed [299].

In general, metallic ferro- and antiferromagnetism is a rare phenomenon because, among the transition metals,[6] only Fe, Co, and Ni are ferromagnetic,

6) The rare-earth metals also show (highly complex) magnetic phenomena but their magnetism is a small-bandwidth magnetism due to electrons that are localized in spatially highly contracted $4f$ orbitals, whereas the metal–metal bonding is mediated through the $5d$ and $6s$ electrons. The $4f$ orbitals are only slightly perturbed by the neighboring atoms and give rise to very narrow bands. Thus, the bands are filled with electrons in a manner analogous to Hund's rules for the fillings of atomic energy levels. This results in unpaired spins and magnetic moments.

while Cr and Mn have antiferromagnetic ground states. The magnetic properties are not solely a consequence of structure or electron count (the higher homologues are *not* magnetic) and the crystal structure is also important because Fe must crystallize in the body-centered cubic structure (α-Fe) to have ferromagnetic properties. It then exhibits a strong saturation moment (2.21 Bohr magnetons) below its Curie temperature (1044 K). As mentioned, the complex structure of α-Mn (58 atoms per unit cell) reverts, in part, to an antiferromagnetic state.

To understand why the ferromagnetic archetype, α-Fe, is ferromagnetic, we will analyze a hypothetical *nonmagnetic* bcc-Fe in which the structure is identical (eight nearest Fe neighbors at 2.485 Å and six second-nearest neighbors at 2.870 Å; see Figure 1.2(c)) but which we do not allow to possess spin moments because all electrons are deliberately *forced* to pair their spins [300]; eventually, we will continue this idea *ad absurdum*. The calculated TB-LMTO-ASA band structure, DOS and Fe–Fe COHP curves for this phase are presented in Figure 3.18, based on the local-density approximation (LDA). Note that we use the LDA on purpose, although it is inferior to the GGA which is needed if delicate phase-stability problems (fcc vs. bcc, see Section 2.12.1) are investigated; our intention here, in *not* focusing on phase stability, is to show that the clue

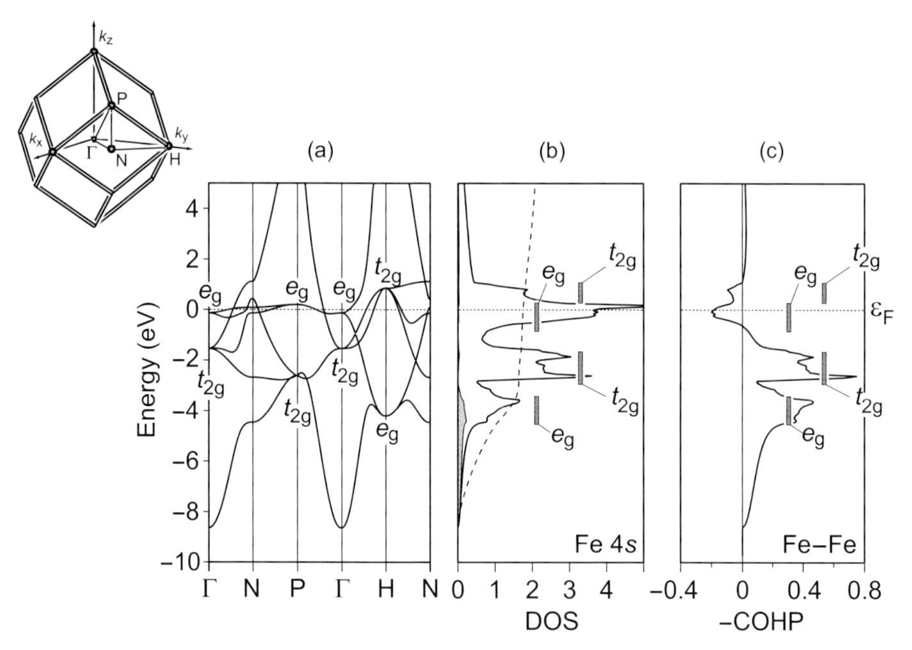

Fig. 3.18 Non-spin-polarized band structure (a), DOS (b) and Fe–Fe COHP curves (c) of α-Fe based on a TB-LMTO-ASA calculation and the LDA. The shaded region and dashed line in the DOS curve corresponds to the projected DOS of the Fe 4*s* orbitals and its integration, respectively. The Brillouin zone is also given.

to our question is so fundamental that it is already apparent from the simple LDA.

In order to ease interpretation, the corresponding Brillouin zone is also included, and a sketch of the associated crystal orbitals at the special points **Γ** and **N** is presented in Figure 3.19. These crystal orbitals can be roughly grouped into the t_{2g} and e_g sets. The, primarily, e_g bands are mostly flat except from **Γ** → **H** and **H** → **N** because they are involved in σ interactions with the second-nearest neighbors. The t_{2g} crystal orbitals, however, experience π-like interactions with their nearest neighbors, thus there is reasonable dispersion along every symmetry direction.

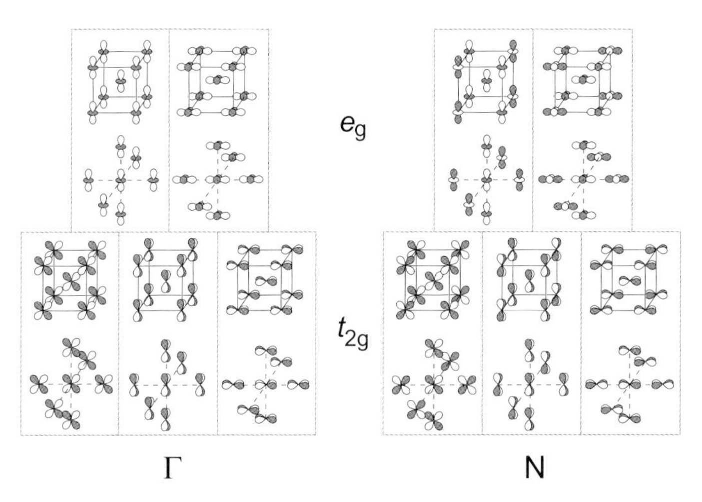

Fig. 3.19 Iconic representations of the crystal orbitals of α-Fe at the high-symmetry points **Γ** and **N**; each orbital is drawn in two parts: one showing nearest-neighbor contacts, the other second-nearest-neighbor contacts. The t_{2g} and e_g labels are used only *qualitatively* to group the orbitals at **N**.

The DOS of nonmagnetic α-Fe (Figure 3.18(b)) is characteristic of a bcc metal because it shows the typical "three-peaked" shape. In the COHP analysis, there is significant Fe–Fe bonding within the valence bands but the curious thing about nonmagnetic Fe is that the Fermi level falls in a strong Fe–Fe *antibonding* region. Consequently, one would expect some kind of instability and an associated structural change, just as in the preceding section dealing with elemental tellurium. A structural change, however, does *not* take place: bcc-Fe stays bcc-Fe. Nonetheless, nonmagnetic iron does undergo a distortion, but instead of the atoms rearranging themselves, *the electrons do so.* Nonmagnetic α-Fe is unstable with respect to an electronic-structure distortion which makes the two spin sublattices inequivalent, thereby reducing the electronic symmetry, lowering the energy, and eventually giving rise to magnetism. For comparison, the spin-polarized DOS and COHP curves are given in Figure 3.20.

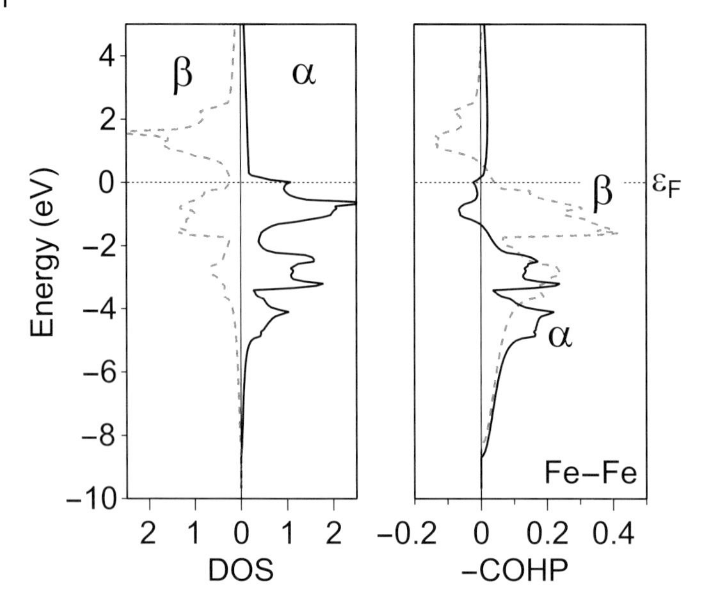

Fig. 3.20 DOS and Fe–Fe COHP curves for ferromagnetic α-Fe. The solid and dashed lines correspond to the α and β spins, respectively.

In the spin-polarized calculation, the two spin sublattices shift in energy because of the exchange hole (see Section 2.9); the shapes of the two DOS do not change very much but the majority spins decrease in energy while the minority spins increase in energy. The resulting theoretical magnetic moment (2.27 Bohr magnetons) is very close to the experimental one (2.21). A closer look at the magnetic COHP curve reveals the reason for the lowering of the total energy on spin-polarization. The shifts in the majority (solid) and minority (dashed) spin sublattices, which mostly take place for the 3d states, have *removed* the antibonding states at the Fermi level, thereby maximizing Fe–Fe bonding as far as possible within this particular structure.

Interestingly, there is a large difference in Fe–Fe bonding due to either α or β spins, and a numerical analysis shows that the minority (β) spins contribute 64% to the total Fe–Fe bonding, whereas the majority α spins are responsible for the remaining 36% (see below). In total, the Fe–Fe bonds become strengthened by roughly 5% upon spin-polarization, and the bond strength itself decays with the sixth power of the interatomic distance, just as the classic molecular field. A further breakdown of the total energy reveals that, similar to the case of H_2^+ (see Section 2.3), the kinetic energy increases on spin-polarization (third chemical bonding myth) but this is more than counterbalanced by a larger decrease in the potential energy such that the total energy is lower.

The spontaneous spin-polarization also results in a rearrangement of the electron density in α-Fe, and the different energetic shifts of the α and β spin sublattices are reflected by their different spatial extents. The majority (α) spin species is not as effectively shielded from the nucleus, goes down in energy and contracts spatially, such that it cannot contribute as strongly to the Fe–Fe bonding as the β species. Those spins, in fact, see a lowered nuclear charge, go up in energy and are more diffuse. Thus, the β wave functions are more involved in Fe–Fe bonding.[7] The difference in spatial extents is visualized in Figure 3.21(a) through a scaled electron-density plot (equalizing the different numbers of α and β spins) evaluated in the (011) plane of bcc-Fe.

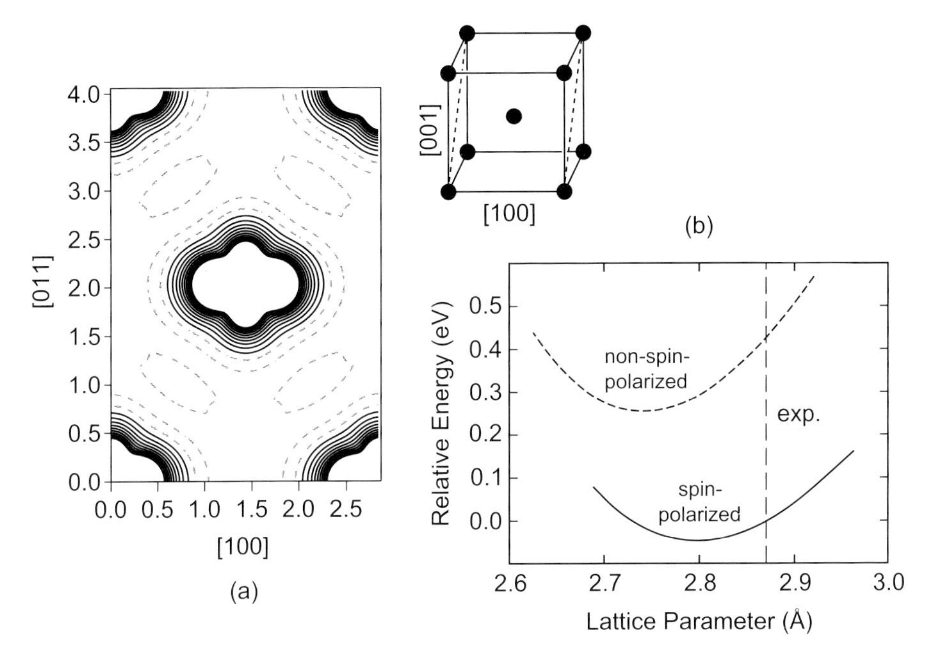

Fig. 3.21 Scaled electron-density difference in α-Fe evaluated in the (011) plane of the cubic unit cell (a); the solid curves indicate positive values (regions where α spins are dominant) and the dashed curves show negative ones (β spins predominate). Relative energies of α-Fe as a function of the lattice parameter (b) within a non-spin-polarized (dashed curve) and spin-polarized (full curve) description. Note the experimental lattice parameter (vertical dashed line).

Just as the α AOs/MOs of atomic/molecular oxygen were more spatially contracted than the corresponding β AOs/MOs (see Section 2.9), the α elec-

7) The fact that almost the entire α d block is occupied, including the Fe–Fe antibonding states at the top, also plays a large role in the difference in bonding strengths between α and β spins. So the α integrated COHP, which is smaller than β to start with, due to the spatial contraction of the α spin orbitals, is weakened even further by the population of antibonding states. For the β spin sublattice, these are lying above the Fermi level. Despite the weaker α interactions, the strengthening of the nearest-neighbor interactions in the β sublattice is sufficient to result in an integrated COHP value in ferromagnetic Fe which is higher than that in the nonmagnetic structure.

tron density of ferromagnetic Fe is larger near the nuclei, while the β electron density is larger in the regions further away from the nuclei, particularly around the nearest-neighbor contacts. Because the more diffuse β spin orbitals are mostly responsible for the Fe–Fe bonding, the magnetic structure *must* exhibit larger interatomic distances than the nonmagnetic structure, shown in Figure 3.21(b), simply because the optimization of the chemical bonding between these enlarged orbitals needs a wider distance to acquire optimum overlap. Despite having a larger internuclear distance (second chemical bonding myth), the total energy E is the only valid measure of the bond strength, and E is lower for the more strongly bonded magnetic state.

When the above analysis is repeated for the other two ferromagnetic transition metals (Co and Ni) assuming nonmagnetic ground states, the Fermi level is also found in antibonding levels. On spin-polarization, the exchange hole shifts the spin sublattices – α going down, becoming more contracted, β going up, becoming more diffuse – such that it removes the antibonding levels from the vicinity of the Fermi level, and the bond strengths increase by 4.0 and 0.5%. The theoretical moments are 1.60 and 0.62 Bohr magnetons, also closely matching the experimental ones [301]. It seems that the symmetry reduction of the wave function is similar to what happens during a structural Jahn–Teller distortion although only the *electronic* – not nuclear! – coordinates are involved in breaking the symmetry. Similar to the case of molecular triplet/singlet oxygen and counterintuitive only at first sight, we reiterate that the strengthening of the chemical bonding in these metals results from the *unpairing* of spins (first chemical bonding myth) which otherwise would be kept in antibonding crystal orbitals.

The results may be generalized to the entire series of the 3*d* transition metals, and this is presented in Figure 3.22, offering the corresponding DOS (a) and COHP (b) figures. For the early transition metals such as Ti, the Fermi level lies low in the COHP curve and thereby falls in the metal–metal bonding region. There is no tendency towards ferromagnetism for these metals. For Cr (a typical antiferromagnet, see below), the Fermi level in the COHP curve precisely separates the bonding and antibonding regions, whereas for the metals from "Mn" (hypothetical bcc-Mn) to Ni, the Fermi level lies in the region that is clearly responsible for metal–metal *antibonding*: while Fe, Co, and Ni are all ferromagnetic, bcc-Mn is predicted to be ferromagnetic as well, and a numerical calculation (with a lattice parameter of a = 2.88 Å) indicates a saturation moment of about one Bohr magneton.[8] Face-centered cubic (fcc) Fe does *not* show a similar Fe–Fe COHP because here the Fermi level is close to the nonbonding regime, similar to an antiferromagnet like chromium (see below) and, indeed, when fcc-Fe is embedded in a Cu matrix, it appears to

8) There is a high-temperature bcc-phase of Mn which is stable at 1200 °C but it has not been possible to quench this down to low temperature and measure its magnetic properties [302].

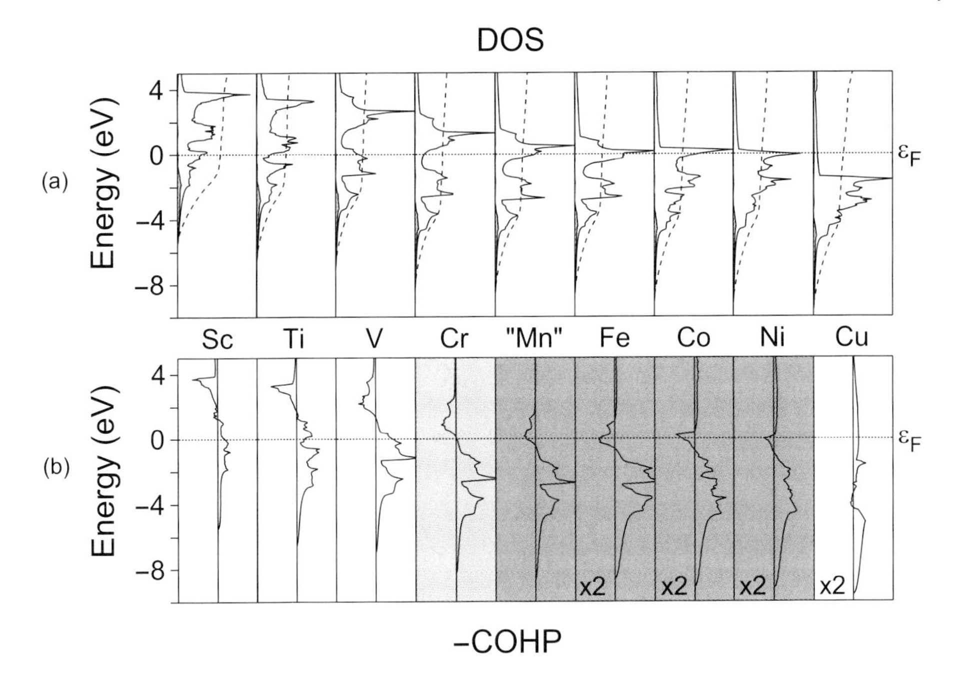

Fig. 3.22 Non-spin-polarized DOS (a), and M–M COHP curves (b) for the $3d$ transition metals. Mn was calculated using a bcc structure with a lattice parameter of 2.88 Å. The shaded area in each DOS curve indicates the minute contribution of the $4s$ orbitals because the DOS is almost entirely $3d$ in character. The M–M COHPs for Fe, Co, Ni, and Cu have been multiplied by a factor of two.

be antiferromagnetic at low temperatures [303]. Let us look more closely at itinerant antiferromagnetism:

By the time we get to Cr, the Fermi level in the COHP curve occurs at the crossover point between bonding and antibonding. Body-centered cubic Cr adopts a magnetic structure in which the neighboring magnetic moments are oriented antiparallel to each other.[9] The Cr band structure has been calculated quite early [304] and the Fermi surface – the constant-energy surface within reciprocal space separating occupied and nonoccupied wave vectors [305] – was evaluated [306]. (Commensurate) antiferromagnetic band-structure calculations [307] and further models expanding on the idea of Fermi surface nesting are well-known [308]. We will here focus on the chemical bonding of bcc-Cr, like we did for bcc-Fe. The calculated (TB-LMTO-ASA and GGA; the LDA is too crude for this delicate problem) band structure, DOS and COHP curves of *nonmagnetic* bcc-Cr is replotted in Figure 3.23, and it is based on the primitive (one-atom) unit cell [309]. Again, the DOS curve exhibits the typ-

9) The magnitudes of the spin moments vary in a sinusoidal manner. This so-called spin-density wave runs along one of the cubic axes, and it is transverse above 120 K but longitudinal below 120 K.

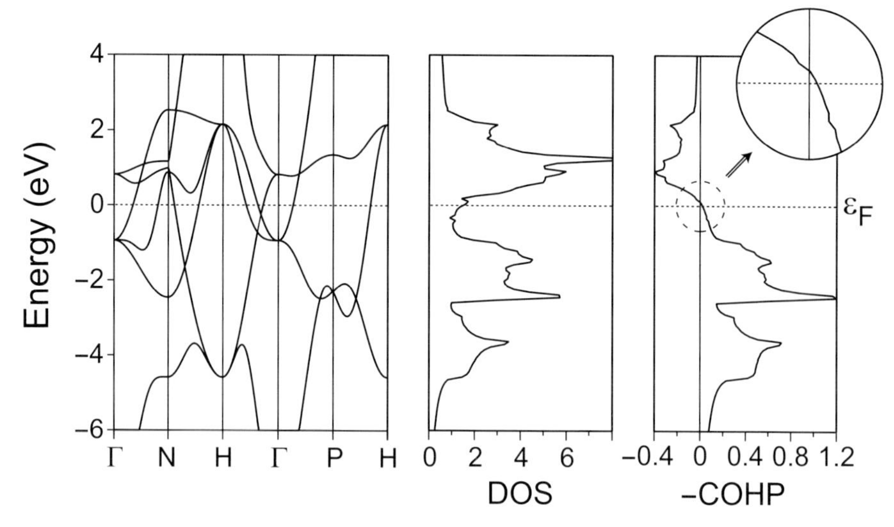

Fig. 3.23 Non-spin-polarized band structure, DOS and Cr–Cr COHP curves for nonmagnetic body-centered cubic Cr based on a primitive unit cell.

ical bcc "three-peaked" shape and, most importantly, the Fermi level lies almost exactly at the crossover from bonding to antibonding states in the COHP curve. At the Fermi level, the COHP curve exhibits a rather steep slope (insert of Figure 3.23, right).

When the band structure is re-calculated using a conventional (two-atom) unit cell, the twice as complicated band structure (not shown) exhibits a number of degeneracies and band crossings at the Fermi level. In contrast to an "ordinary" Peierls distortion involving changes in atomic positions (such as the one of elemental Te, Section 3.4), an electronic perturbation (spin-polarization) reshuffles the spins, and a spin-polarized antiferromagnetic calculation converges to a spin-magnetic moment of ca. 0.96 Bohr magnetons, alternating from the center to the corner of the cubic unit cell. Figure 3.24 shows this antiferromagnetic band structure, DOS, and COHP of bcc-Cr.

Besides broken degeneracies and avoided crossings close to the Fermi level, internal band gaps (grey shading) are visible along some lines. The shape of the DOS curve does not exhibit any drastic changes compared with the non-spin-polarized one but a few "twin peaks" reflect the broken degeneracies, and the lowered DOS around the Fermi level indicates the above-mentioned internal band gaps. In addition, the magnetic COHP curve resembles the non-spin-polarized one but its *slope* at the Fermi level has decreased (see insert in Figure 3.24). Also, the energetic changes for such an antiferromagnetic system are much smaller than for a ferromagnetic one (the reason for having used the GGA instead of the LDA), and this is easily understandable because rearranging the charge density from within the *nonbonding* states to form the

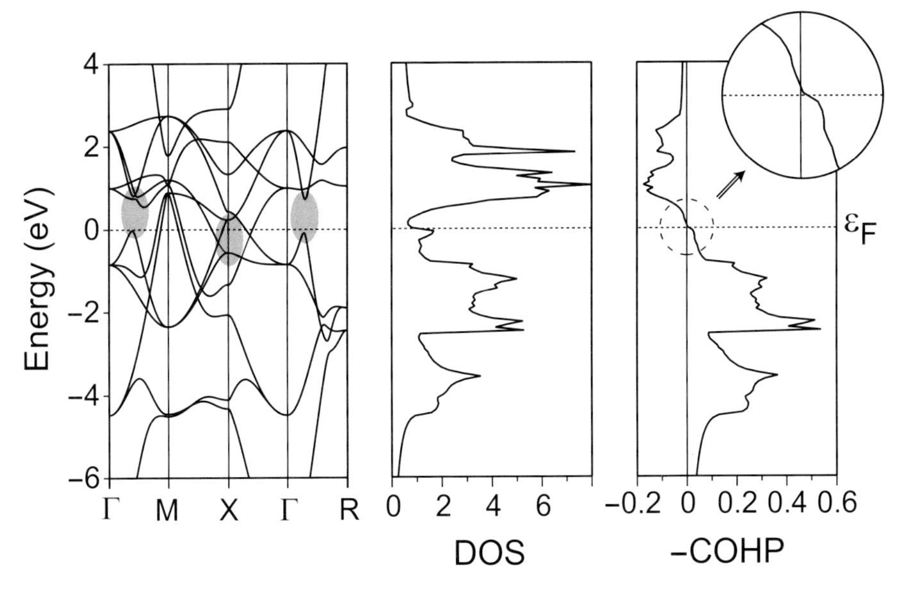

Fig. 3.24 The same as Figure 3.23, but for antiferromagnetic Cr described by a two-atom unit cell; here, the center and corner atoms of the bcc unit cell exhibit opposite spin directions and thus are no longer symmetry-equivalent.

antiferromagnetic electronic structure does not have a profound impact upon the total energetics [309].

Summarizing itinerant ferromagnetism (antiferromagnetism) of the 3*d* metals, the presence of M–M antibonding (nonbonding) states at the Fermi level in the nonmagnetic electronic structure of a transition metal, serves as a signpost indicating that the metal would prefer to be ferromagnetic (antiferromagnetic). When extending these results to the higher homologues of Cr, Mn, Fe, Co, and Ni from the 4*d* and 5*d* rows of the transition metals, it is reasonable to assume that these elements should have COHP curves similar in shape to those of the corresponding 3*d* elements. This is correct, yet neither of the 4*d*/5*d* elements becomes ferro- or antiferromagnetic. Numerical calculations on these systems making use of artificially introduced (locked) spin moments, immediately reveal the cause for the different behavior. The exchange splittings for the 4*d*, and especially the 5*d* elements, are much smaller. Therefore, the presence of a magnetic moment does not result in significant energetic changes, and the heavier metals remain without spin-polarization.[10] This difference in behavior between a 3*d* and a 4*d* (or 5*d*) metal may be understood in the same way as we have explained those between O_2 and S_2 or Se_2 (see Section 2.9).

10) This statement is correct for the bulk elements but not necessarily for very small particles. For example, clusters of rhodium atoms (Rh_n with $9 \leq n \leq 34$) do exhibit spin-polarization [310].

The first-row transition metals have the special $3d$ orbital set which is not shielded by any symmetry-related orbital closer to the atomic nucleus such that the $3d$ orbitals are fairly contracted. This is nicely reflected in the $3d/4d$ ζ orbital exponents given in Table 2.1. Small changes to the shieldings of the $3d$ set caused by spin-polarization give rise to comparably large changes in their energies and spatial extents. The heavier transition metals, on the other hand, have inner d functions which help to screen their valence d orbitals. Therefore, perturbations from spin-polarization do not have such a dramatic effect upon these well-shielded orbitals.

This is probably the right place to stress an obvious chemical analogy. The special chemical as well as physical behavior of the light main-group elements (B, C, N, O, F) can be traced back to the "missing" $1p$ orbital. Consequently, the $2s$ and $2p$ orbitals have similar spatial requirements, with well-known consequences for element–element multiple bonding and organic chemistry. The situation of the $3d$ elements seems qualitatively comparable. Because of the "missing" $2d$ orbital, spin-polarization leads to large exchange splittings and the occurrence of magnetic behavior, easily predicted from COHP curves. In this respect, the exceptional finding of spin-polarization (and itinerant ferro- and antiferromagnetism) may be looked upon as the transition metals' analogue of the special situation which the chemistry of carbon has within chemistry as a whole, but this message may be a bit difficult to take on board [311].

3.6 Itinerant Magnetism: Transition-metal Compounds

Within the last section, we have bravely re-interpreted ferromagnetism (and also antiferromagnetism) as a kind of chemical phenomenon, arising only in elemental systems with a high exchange splitting and a critical electron concentration – one which causes the Fermi level to fall in a region of antibonding (for ferromagnetism) or nonbonding (for antiferromagnetism) states, and the resulting instability results in an expected symmetry reduction. However, rather than altering the spatial coordinates of the atoms, the *electronic* symmetry is lowered by making the α and β spins inequivalent. This breakdown in the electronic symmetry leads to energy shifts, a repopulation of the spin sublattices and a redistribution of charge density. Before we investigate the symmetry changes in more detail (see Section 3.8), we will do some more chemistry to demonstrate that the same ideas which were developed in order to understand itinerant ferromagnetism and antiferromagnetism of the elements can also be applied, successfully, to intermetallic systems [301,312]. Our intention should be very clear, namely the wish to open the way for the synthesis of new magnetic alloys. For simplicity, we use the term "alloy" in a rather naive sense and do not strictly distinguish between intermetallic compounds and alloys. Only within intermetallic *compounds*, are the different atoms mixed on an atomic scale, whereas intermetallic *alloys* (the engineer's delight) can also be formed by two metals that do *not* form any intermetallic compound. The binary system Sn/Zn is a nice example because Sn/Zn can indeed be alloyed although there is not a single Sn/Zn compound (see Section 3.11).

In order to create new ferromagnetic and antiferromagnetic intermetallic materials based on the COHP reasoning introduced in the last section, the synthetic recipe for all these reads as follows. Try to adjust the Fermi level of the starting compound in order to position it within the antibonding states (ferromagnets) or nonbonding states (antiferromagnets) between the magnetically active atoms, i.e., those which have narrow band widths in their elemental forms [301, 309]. The synthetic tuning of the valence electron concentration may result from either electronic enrichment or depletion, and that is what solid-state chemists are rather good at.

We will start with three intermetallic compounds that have been known for a long time and then move over to brand-new materials. If we remind ourselves of the electronic structure of fcc-Fe mentioned in Section 3.5, the reader will recall that the Fermi level was positioned close to the nonbonding states, a fingerprint of antiferromagnetism. In order to make fcc-Fe a ferromagnet, one needs to keep the structure (almost) untouched but still raising the Fermi level such that it senses the antibonding states. One way of realizing this is given by the binary phase $FeNi_3$. It crystallizes in the [$AuCu_3$] structure [5], an fcc arrangement of nickel atoms where every fourth Ni has been replaced by Fe in an ordered fashion (see Figure 3.25(a)). The $FeNi_3$ lattice constant is less than

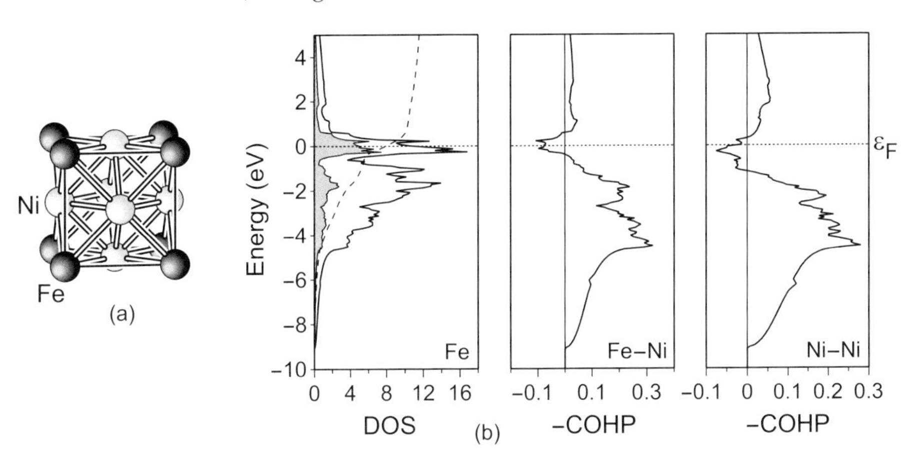

Fig. 3.25 The unit cell of $FeNi_3$ (a) and the DOS and Fe–Ni and Ni–Ni COHP curves (b) for nonmagnetic $FeNi_3$. The shaded region and dashed line in the DOS curve correspond to the projected DOS of the Fe atom and its integration, respectively.

one percent larger than that of elemental Ni, and its experimental magnetic moment is 1.18 Bohr magnetons [313]. If we ignore the fact that $FeNi_3$ is a structurally *ordered* phase, its valence-electron concentration (9.5 electrons per atom) allows us to simplify it as either a heavily electron-enriched fcc-Fe (by 1.5 electrons) or a slightly electron-poor fcc-Ni (by 0.5 electrons).

The electronic structure of $FeNi_3$ is easily calculated (TB-LMTO-ASA, LDA) for the nonmagnetic case, and we give the corresponding DOS and the Fe–Ni and Ni–Ni COHP analyses in Figure 3.25(b). One might expect $FeNi_3$ to be electronically similar to elemental Ni, and its DOS is indeed almost superimposable to the one of Ni, presented before in Figure 3.22.

In contrast to Ni, the Fermi level is located in a narrow DOS peak at the top of the 3*d* block, and the Fe states (see projected DOS in Figure 3.25) are spread almost evenly throughout the bands, *except* for the sharp peak at the Fermi level, which is almost half Fe. The electronic structure in the frontier bands is more iron-like, and since only one-fourth of the atoms are Fe, this is quite a large contribution. However, given that Ni is a little more electronegative than Fe (see the Pearson absolute electronegativities in Table 2.2) we should have expected the upper regions of the *d* block to be mostly Fe. Turning to chemical bonding, the Fe–Ni and Ni–Ni COHP curves are very similarly shaped, and the peak at the top of the *d* block is strongly *antibonding* for both Fe–Ni and Ni–Ni bonds; this should give rise to an instability towards ferromagnetism.

Indeed, a spin-polarized LDA calculation of the electronic structure results in an average spin moment of 1.18 unpaired electrons, in quantitative agreement with the experimental data. In the magnetic DOS and COHP curves (not shown for brevity), the α and β DOS curves are quite differently shaped [301]. Although the Fe states are still spread throughout the α block, they are responsible for the high local moment of Fe in $FeNi_3$ (2.88 Bohr magnetons), and the Ni moment is much smaller (0.62). The high moment on Fe, compared with that in metallic iron, is typical for such alloys between Fe and other late transition metals, simply because Fe is allowed to be more spacious in this intermetallic phase, supporting the α/β splitting of Fe. The formerly antibonding states at the Fermi level for Fe–Ni and Ni–Ni interactions have been strongly reduced and, consequently, have stabilized the overall structure on spin-polarization. A numerical integration reveals that the Fe–Ni interactions stay about the same whereas the Ni–Ni interactions strengthen by more than 15%.

Although the $FeNi_3$ example illustrates the validity of the attempted electron enrichment of Fe in order to make a ferromagnetic intermetallic, the alert reader has surely noticed that $FeNi_3$ stands for a binary compound made from two ferromagnetic metals. Thus, $FeNi_3$ may be looked upon as a too obvious trick. Let us therefore replace Ni by a nonmagnetic metal and still try to enrich fcc-Fe so that it becomes a ferromagnet. One obvious choice is given by the iron/palladium system where phases with average valence electron concentrations of 8.5 (Fe_3Pd), 9.0 (FePd), and 9.5 ($FePd_3$) electrons per metal atom are known. Fortunately, all of these phases adopt fcc-like structures. We will concentrate on FePd [314] in the [AuCu] structure type [5], shown in Figure 3.26(a). While FePd has the same electron count as cobalt (9.0), it contains a 3*d* (Fe) and a 4*d* atom (Pd), thereby representing – if we neglect the atomic ordering – some 3*d*/4*d* pseudo fcc atom lying between rhodium and cobalt.

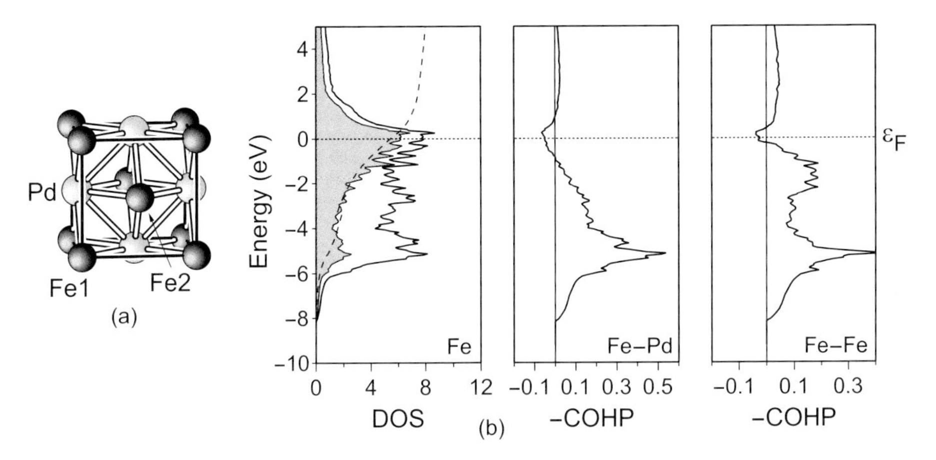

Fig. 3.26 The same as Figure 3.25, but for FePd in the [AuCu] structure.

The DOS and COHP curves for the nonmagnetic calculations of FePd are presented in Figure 3.26(b). There are clearly antibonding interactions of the type Fe–Pd and Fe–Fe (and also Pd–Pd, not shown) at the Fermi level. A prediction of ferromagnetism would thus be justified, and a spin-polarized LDA calculation yields magnetic moments of about 2.26 (Fe) and 0.17 (Pd) Bohr magnetons. The experimental data [284, 315] and more sophisticated calculations [316], however, suggest that the above LDA approach underestimates the sizes of the individual moments because in all the above-mentioned ferromagnetic Fe/Pd alloys, the moments scatter around 0.3 Bohr magnetons for Pd and 2.6 for Fe.

Using spin-polarization, the total energy of FePd is lowered by 0.32 eV, and this is paralleled by an increase in the overall bonding strengths of 1.1 % (calculated by integrating all COHPs in the unit cell). With the exception of the Fe–Pd bonds, the Fe–Fe and Pd–Pd bonds are strengthened because the antibonding states are shifted away from the Fermi level (not shown). Also, as expected from the calculations on the pure elements in Section 3.5, the exchange splitting of the Fe states is much larger than those of the Pd states. This large difference in exchange splitting, along with the Fe domination of the states at the Fermi level, leads to a much larger (by about an order of magnitude) magnetic moment on Fe.

We can also cover ferromagnetic compounds which do not contain any ferromagnetic transition metal (i.e., do not contain Fe, Co, or Ni). Among such phases, the so-called Heusler alloys X_2YZ (for example, Cu_2MnAl) are probably the most prominent [317, 318]. In the Heusler alloys the Y atoms form a face-centered cubic lattice with the X and Z atoms occupying all of the tetrahedral and octahedral voids respectively. MnSb, however, represents an even simpler case. It has a Curie temperature of 600 K [319] and crystallizes in the [NiAs] structure type (see Figure 3.27(a)). Here, Mn is octahedrally coordinated by Sb at 2.81 Å and Sb is found in a trigonal prism. One-dimensional Mn–Mn chains with a 2.88 Å bond distance [320] are easy to recognize. While the experimental bulk magnetic moment of $Mn_{1.09}Sb$ is 3.25 Bohr magnetons, polarized neutron diffraction experiments reveal a higher Mn-centered moment of 3.45 Bohr magnetons [321]. If this is true, each Sb atom carries an *oppositely* directed moment of about −0.20 Bohr magnetons and MnSb should rather be called a *ferrimagnetic* alloy, but this does not render the bonding analysis more difficult, though. The electronic structure (DOS and COHP) of non-magnetic MnSb is shown in Figure 3.27(b).

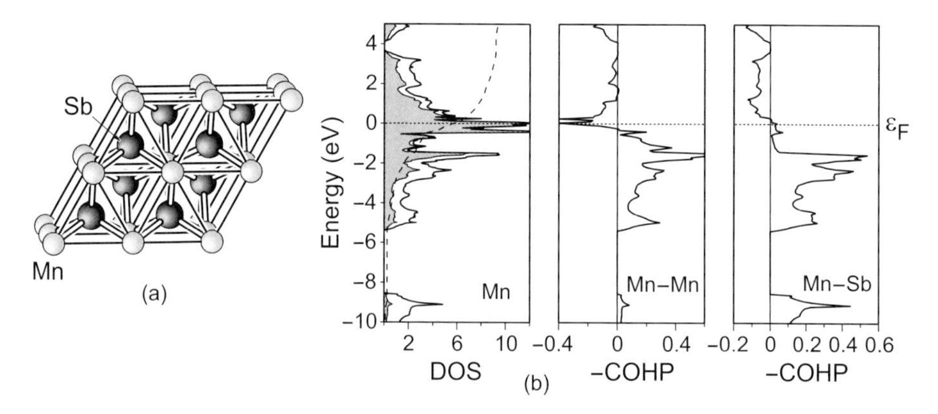

Fig. 3.27 The same as Figure 3.25, but for MnSb in the [NiAs] structure.

First, the region around the Fermi level is heavily dominated by Mn, and the states arising from the Sb 5*s* orbitals are partially visible at the bottom of the DOS curve. Because of a good electronic match between Mn 3*d* and Sb 5*p* orbitals and also strong Mn–Mn interactions, the octahedral 3*d* splitting is too wide to be visible. The absolute electronegativities of Mn and Sb (see Table 2.2) indicate that Sb should be slightly charged (−0.08 by Equation (2.101)). All occupied bands are Mn–Sb bonding but the Fermi level falls in a *strongly* Mn–Mn antibonding region, the latter pointing towards ferromagnetism; recall that elemental Mn is an antiferromagnet! In addition, the large Mn contribution at the top of the valence band indicates a large Mn moment.

A spin-polarized calculation for MnSb gives a magnetic moment of 3.24 Bohr magnetons, composed of 3.37 Bohr magnetons on Mn and an additional but *oppositely* directed −0.13 Bohr magnetons on Sb, in nice agreement with the experimental notion of a ferrimagnetic phase.[11] The small size of the Sb moment simply reflects the fact that Sb has no magnetically active orbitals; the Sb 4d orbitals are low-lying and completely occupied, and the valence 5s and 5p orbitals are well-shielded and do not have significant exchange splittings. In the magnetic DOS and COHP analyses, only the Mn contributions exhibit such splitting and one might think of MnSb as being composed of ferromagnetic Mn chains surrounded by an almost nonmagnetic sea of Sb atoms.[12]

The last three cases of ferromagnetic compounds ($FeNi_3$, FePd, MnSb) have served to exemplify itinerant magnetism from a chemical perspective. Admittedly, the weak spot in these arguments is that the magnetic properties of all intermetallic phases were known *before* theoretical "prediction". To proceed more systematically, a new family of intermetallic compounds is needed which allows a wide range of electronic filling and, at the same time, is robust enough *not* to change its crystal structure upon atomic substitution. Fortunately, such a class of compounds has been ingeniously brought into existence by the efforts of Jung's group [322], and these compounds are given by the quaternary family of intermetallics $A_2MRh_5B_2$, crystallizing with an ordered substitutional variant of the $[Ti_3Co_5B_2]$ structure type. Let us illustrate the structural peculiarities by discussing $Mg_2MnRh_5B_2$, an electron-poor representative crystallizing in the space group $P4/mbm$, like all others. Within the tetragonal unit cell of that nicely complex structure (see Figure 3.28(a)), there are trigonal, tetragonal, and pentagonal Rh prisms stacked on top of each other along the [001] direction. The triangular prisms are centered by B atoms, and the pentagonal prisms accommodate Mg atoms; the tetragonal prisms contain the magnetically important Mn atoms (Figure 3.28(b)). As a result, there are one-dimensional chains running along the c axis with Mn–Mn intrachain distances of about 2.9 Å, while the interchain distances are much larger (ca. 6.6 Å).

Experimentally, $Mg_2MnRh_5B_2$ is characterized by Curie–Weiss behavior above 160 K with a Weiss constant of −130 K. The latter information stands

11) The small differences between experimental and theoretical moments do not necessarily go back to theory problems but might also be explained by the fact that the experimental measurements were carried out on a nonstoichiometric (Mn-rich) compound.

12) Upon integration of the Mn–Sb and Mn–Mn bonds for the spin-polarized case, however, it is found that both kinds of bonds *weaken* upon spin-polarization although antibonding states are removed from the Fermi level. This is a clear sign that the antibonding character of the states around the Fermi level provide the motivation for ferromagnetism, although the "pay back" does not necessarily have to come from a strengthening of bonds. A closer analysis shows that the stabilization in MnSb results from atom-centered (on-site COHP terms, see Equation (2.58)) contributions.

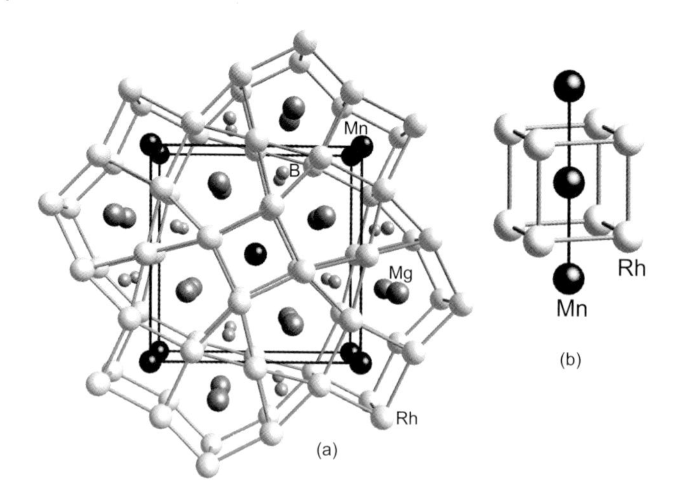

Fig. 3.28 Perspective view of the $Mg_2MnRh_5B_2$ crystal structure (a) with medium-grey Mg atoms, black Mn atoms, light-grey Rh atoms, and small grey B atoms. Tetragonal Rh prisms accommodate Mn atoms (b).

for considerable *antiferromagnetic* Mn–Mn exchange interactions, whereas the Curie constant corresponds to a paramagnetic moment of 3.2 Bohr magnetons per Mn atoms. It also needs to be mentioned that characteristic features for one-dimensional magnetic behavior are not obvious, contrary to the first impression that one gets by looking at the crystal structure.

Nonmagnetic LDA band-structure calculations on $Mg_2MnRh_5B_2$ using the TB-LMTO-ASA method fully corroborate these findings [323]. The Fermi level in the COHP analysis (Figure 3.29(a)) is clearly positioned in the region of *nonbonding* Mn–Mn interactions, the aforementioned fingerprint for antiferromagnetism. Thus, this complex phase imitates, when it comes to Mn–Mn bonding, the scenario known from bcc-Cr. While the true antiferromagnetic structure of $Mg_2MnRh_5B_2$ is still unknown, one may come up with a computational model by doubling the short tetragonal axis and switching over to spin-density-functional theory. Starting from a crude guess for the antiparallel moments on the neighboring Mn atoms and a zero net moment, the calculation converges to an antiferromagnetic structure with a self-consistent saturation moment of 3.23 Bohr magnetons per Mn atom; all other atoms are magnetically almost inactive.

The composition of a new *ferromagnet* should be easily predictable based on the valence electron count of $Mg_2MnRh_5B_2$. It contains 2×2 (Mg) + 1×7 (Mn) + 5×9 (Rh) + 2×3 (B) = 62 electrons. Given that a rigid-band behavior holds upon chemical substitution (this worked nicely for the entire $3d$ row), the experimentalist needs to push the Fermi level up to make it cut through the strongly antibonding states around +0.4 eV as given in the shaded region

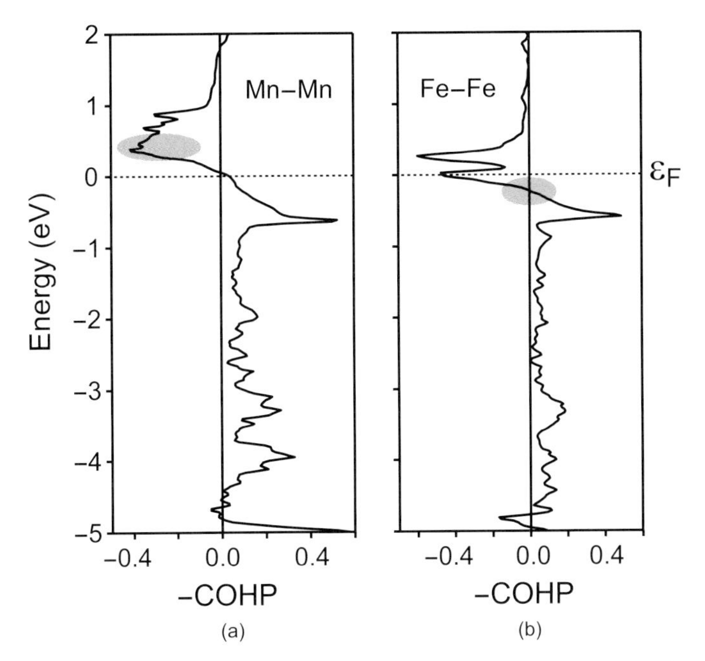

Fig. 3.29 COHP analysis of Mn–Mn bonding in non-spin-polarized $Mg_2MnRh_5B_2$ (a) and of Fe–Fe bonding in non-spin-polarized $Sc_2FeRh_5B_2$ (b). The shaded regions indicate charge enrichments/depletions by ca. three electrons with respect to the (zero) Fermi level.

of Figure 3.29(a). A numerical COHP integration reveals that this upward shift corresponds to *three* additional electrons, to be realized by replacing Mg against Sc and Mn against Fe, such that "$Sc_2FeRh_5B_2$" with 2×3 (Sc) + 1×8 (Fe) + 5×9 (Rh) + 2×3 (B) = 65 valence electrons would be an ideal synthetic goal.

$Sc_2FeRh_5B_2$ was indeed synthesized by a classic high-temperature route, and the susceptibility measurement yields that it is *ferromagnetic*, as predicted, below 450 K with a maximum value of 3.3 Bohr magnetons. The complicated magnetization curve does not follow the Curie–Weiss law. The shape of the corresponding Fe–Fe COHP analysis within a nonmagnetic electronic-structure calculation (Figure 3.29(b)) is in accordance with the ferromagnetic properties of $Sc_2FeRh_5B_2$ since there are strongly *antibonding* Fe–Fe interactions around the Fermi level. This is very similar to the Fe–Fe bonding in bcc-Fe. On running a spin-polarized band-structure calculation, these antibonding states are annihilated (not shown) and the total energy is lowered. When made fully self-consistent, one finds a theoretical saturation moment of 3.96 Bohr magnetons to which the Fe atom contributes the largest amount (2.93) as well as a small Rh contribution [323].

One may imagine other synthetic attempts based on the Fe–Fe COHP curve of ferromagnetic $Sc_2FeRh_5B_2$ (Figure 3.29(b)). A further electronic enrichment should populate even more antibonding states which would destabilize the structure. In fact, these syntheses mostly lead to competing perovskite-related phases despite the existence of the iridium-based compound $Sc_2CoIr_5B_2$. Recent synthetic attempts, yet to be published, prove that both $Ti_2FeRuRh_4B_2$ (66 valence electrons) and $Ti_2FeRh_5B_2$ (67 electrons) can also be achieved [324]. An electronic depletion by three electrons, however, would readjust the Fermi level back into a region of nonbonding metal–metal interactions (shaded region of Figure 3.29(b)), the fingerprint of antiferromagnetism. Thus, starting from $Sc_2FeRh_5B_2$ with 65 valence electrons, the subtraction of three electrons could be realized by another synthetic target "$Sc_2MnRu_2Rh_3B_2$" with 2×3 (Sc) + 1×7 (Mn) + 2×8 (Ru) + 3×9 (Rh) + 2×3 (B) = 62 valence electrons, which is very likely to be an antiferromagnet. Alternatively, one might also stay with the Fe atom, thereby lowering the valence electron count solely through substitutions in the $4d$ elemental substructure. Thus, yet another antiferromagnetic alternative would then read "$Sc_2FeRu_3Rh_2B_2$" corresponding to 2×3 (Sc) + 1×8 (Fe) + 3×8 (Ru) + 2×9 (Rh) + 2×3 (B) = 62 valence electrons.

Both $Sc_2MnRu_2Rh_3B_2$ and $Sc_2FeRu_3Rh_2B_2$ can be synthesized in almost quantitative yields. As predicted, the Mn-containing phase exhibits *antiferromagnetic* behavior with a Weiss constant around -300 K and a paramagnetic moment of 2.3 Bohr magnetons per Mn. Also, the magnetic properties of the isoelectronic Fe-containing phase characterize it as being an *antiferromagnetic* material, as predicted. Here, the Weiss constant lies at about -90 K while the paramagnetic moment is 4.0 Bohr magnetons per Fe. Clearly, the type of the magnetic behavior of both phases does *not* depend on the type of transition metal (Mn or Fe) but is solely a function of the electron filling, that is, an electron count which positions the Fermi level in nonbonding Mn–Mn or Fe–Fe states, thereby inducing antiferromagnetism [323].

It therefore seems that rational ways of making complex itinerant ferromagnets and antiferromagnets are in our hands, through a combination of chemical bonding theory for solids, moderately accurate density-functional band-structure calculations and, not forgetting, skillful synthetic techniques for intermetallic compounds. Independent research by the Miller group in the realm of GaM (M = Cr, Mn, Fe) intermetallics point in exactly the same direction [325]. It remains to be seen how far we can push this rigid-band approach in the design of itinerant magnets.

3.7
Atomic Dynamics in Fe:AlN Nanocomposites

The search for new magnetic materials as described in the last section may be looked upon as a typical example of the chemist's orthodox strategy. Namely, take your existing compound, analyze the interactions between the atoms, understand them using some theoretical framework – and then move on to rationally synthesize yet *another* compound with a new composition! Making new compounds is at the heart of chemistry, and chemists really love to fabricate them. And yet, there are other ways of arriving at new materials.

Within the realm of materials science, the search for the new is often much more focused on the combination of *existing* phases (which chemists may not find exciting at all) but the skill of the materials scientist manifests through the new properties resulting from the synergetic *blend* of the individual components within the composite.[13] Another, related, strategy (with a remarkable hype) is given by the attempt to optimize the properties of well-established, traditional materials by their particle size. The domains of nanochemistry and nanosciene have greatly profited from the advent of tunnelling and atomic-force microscopes which have allowed the study of these size effects. Let us therefore look at a magnetically interesting *composite* material from the nanoworld and see why dynamical processes, i.e., atomic movement, may become very important.

Because Fe is the archetypal ferromagnet and aluminum nitride, AlN, is a promising semiconductor, the combination of these two may result in new physical properties, especially if the two ingredients are combined at the nano scale. Indeed, there has been enormous interest in such "magnetic semiconductor" nanocomposites, and their synthesis is accomplished by the sequential deposition of metallic iron and aluminum nitride on glass substrates by radio frequency (rf) sputtering. The individual layer thicknesses of these nanocomposites, $\cdots$Fe:AlN:Fe:AlN:Fe$\cdots$, may vary between 6 and 500 nm (60–5000 Å) while periodicities can reach up to 35 repeat units. A puzzling effect appears upon characterizing these nanoscaled superstructures[14] because one finds a totally unexpected formation of iron nitrides at the Fe:AlN interface and a partial reduction of aluminum nitride to metallic aluminum, paralleled by a significant lattice expansion of bcc-Fe [326,327].

The chemical reaction found at the nano-interface – formation of Fe_xN from Fe by the presence of AlN – is a complete mystery, taking into account the

13) Engineers are conservative people, and they prefer to deal with things they know (at least in part) instead of having to adapt to something *really* new (I am, again, joking here).

14) The physical techniques for the structural characterization of such nanomaterials largely differ from conventional ones (such as X-ray diffraction). In the present example, indirect information is collected from X-ray absorption fine structure (XAFS) spectroscopy, conversion electron Mößbauer spectroscopy, and also X-ray photoelectron spectroscopy.

thermochemical data for bulk nitrides, because such a reaction is clearly *impossible* under equilibrium conditions. The Gibbs free formation energies ΔG_f indicate that AlN is thermodynamically stable (−287 kJ/mol) while Fe_2N is unstable (+12.6 kJ/mol) [328], so why should a stable phase decay in favor of an unstable one? Because we already sense that effects of nonequilibrium will play an important role, let us model the sputtering of an iron target with aluminum nitride using an atomistic computer simulation in order to observe, as a function of time and on a very short time scale, what is going on between Fe and AlN.

Instead of relying on an expensive quantum-chemical method such as the Car–Parrinello approach, the system's modelling may proceed using a classical strategy, but only if the chemical bonding is adequately described. Which kind of chemical bondings will be present here? We already know that, in the final nanocomposite, one is likely to observe solid AlN but also iron nitrides of varying compositions. For both, a significant charge transfer occurs between metal and nonmetal atoms. In addition, we need to properly model the presence of both metallic iron as well as metallic aluminum for which there is no interatomic charge transfer. Eventually, the formation of metallic iron–aluminum compounds may also result, for which one expects very small electron flow between Fe and Al.

One obvious idea – at least for a crystal chemist – would be to describe the metal–nonmetal interactions as existing in AlN (or any iron nitride) by the bond-valence concept (see Section 1.5). Recall that, largely independent of the amount of covalent or ionic bonding, all metal–nonmetal interactions are mapped into the same calculus. If one identifies the bond-valence sum v_i (as given by Equation (1.15)) to be identical with the atomic charge for any cation (metal atom) or anion (nonmetal atom), we have a recipe for the energetic description of the solid in purely electrostatic terms, augmented by a short-ranged exponential Pauli repulsion (with $\rho = 0.37$ Å) to prevent structural collapse and ensure equilibrium geometry (see Section 1.2). Note that, by generalizing the bond-valence concept for any given time step, all atomic charges then become *dynamic* variables!

Applying the time-dependent bond-valence concept for AlN yields excellent numerical results. Under standard conditions, AlN crystallizes with the wurtzite structure in the hexagonal system and, just as for cubic zinc-blende AlN, there is tetrahedral coordination for both aluminum and nitrogen. For a correct structural simulation, a bond-valence parameter of $r_0 = 1.786$ Å for the Al–N combination (close to the tabulated one from Table 1.4) ensures that the experimental Al–N distances are *exactly* reproduced. After having set up a supercell with 576 Al and N atoms, all atoms are randomly shifted by ca. 0.5 Å to partially destroy the structure. When they are given kinetic energies corresponding to room temperature, however, some thousand time steps of

0.01 fs are sufficient to re-achieve the perfectly crystalline structure of undistorted wurtzite AlN, as shown in Figure 3.30. The radial distribution func-

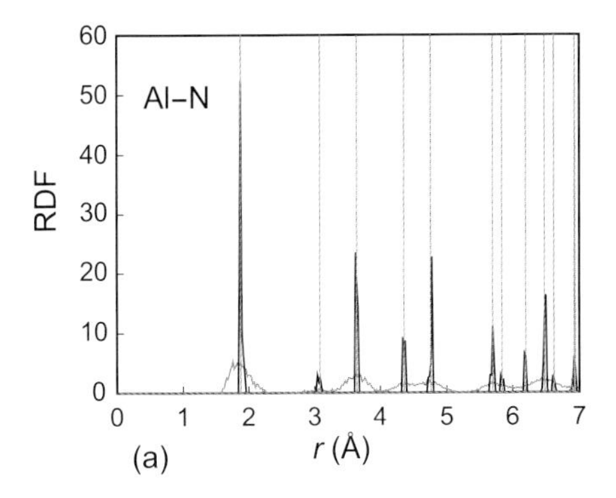

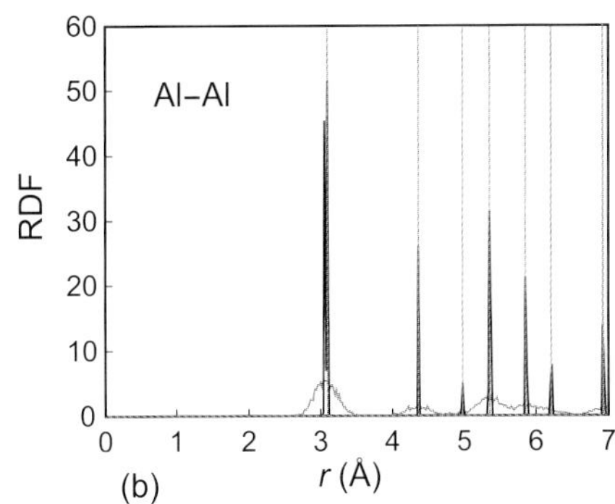

Fig. 3.30 Radial distribution functions (RDF) for the Al–N (a) and Al–Al (b) combinations in the simulation of crystalline AlN adopting the wurtzite structure. While the starting RDFs (grey) indicate a heavily distorted system, the RDFs after 190 000 time steps (full lines) show good radial agreement with the positions of an idealized crystalline AlN (thin vertical lines).

tions, which may be regarded as computationally derived histograms of the system's interatomic distances, reveal sharp peaks both for the Al–N (a) and Al–Al (b) combinations because all atoms are located very close to their equilibrium positions. We will stay with the RDF representation because it is so much easier to analyze than a direct view into the real-space structure.

When it comes to the metal–metal interactions (such as Fe–Fe and Al–Al), a whole group of empirical force fields are known from the literature. One of the most intuitive is the universal binding-energy–distance relationship [329,330] which simply reads

$$E = E_0 \left(1 + \frac{r_{ij} - d_0}{l}\right) \exp\left\{-\frac{r_{ij} - d_0}{l}\right\}, \qquad (3.2)$$

describing the course of the total energy E as a function of the interatomic distance r_{ij}. For the case of metallic solids, the zero energy E_0 may be easily parametrized from the experimental cohesive energy. The parameter d_0 is the interatomic distance at the minimum of the potential energy, and l is the tabulated [330] screening length of the electron gas, which effectively determines the range of the bonding interaction. Although such a rather simple force field always ends up in close-packed (fcc or hcp) structures, it may also be modified (for example by a superposition of two such force fields) to describe a body-centered cubic structure, if needed.

When simulating the structures of metallic aluminum (fcc: $E_0 = -156$ kJ/mol; $d_0 = 2.864$ Å; $l = 0.34$ Å) and of metallic iron (bcc: -198 kJ/mol; 2.482 Å; 0.27 Å), the two above force fields *exactly* reproduce the experimental lattice parameters and cohesive energies, because the latter two are the internal parameters of the simulations. Figure 3.31 illustrates the radial distribution functions for aluminum (a) and iron (b) both at the beginning and the end of the simulations. The simulations are based on supercells containing 256 (fcc) and

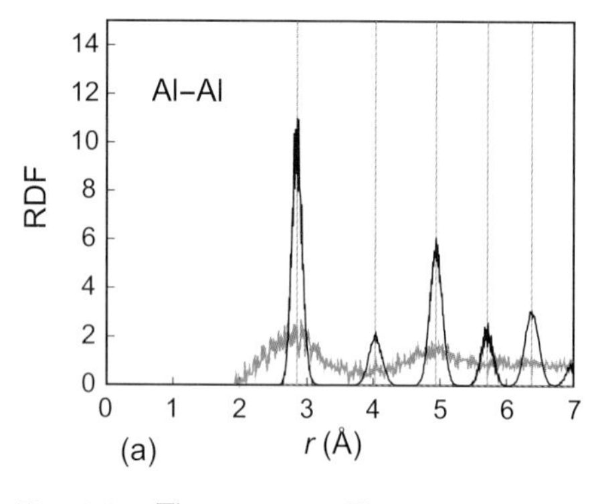

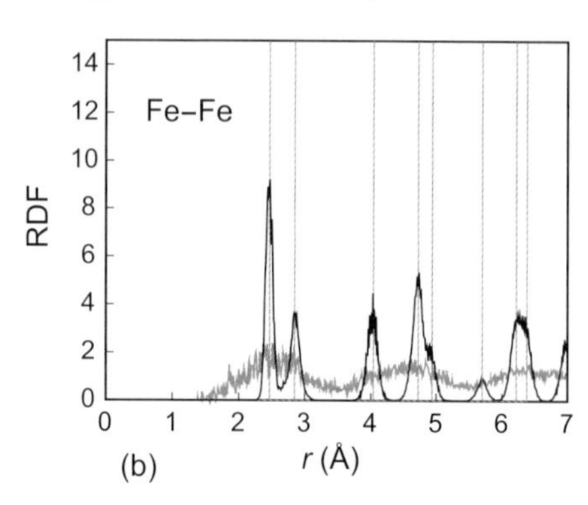

Fig. 3.31 The same as Figure 3.30, but for the Al–Al (a) and Fe–Fe (b) combinations in face-centered cubic aluminum and body-centered cubic iron. Compared with the starting RDFs (grey) of the molten systems, the RDFs after 200 000 time steps (full lines) are very close to the radial positions of the idealized crystalline structures (thin vertical lines).

250 (bcc) atoms, all of which were strongly distorted away from the equilibrium positions at the very beginning. About 200 000 time steps of 0.01 fs are needed to reach full structural relaxation. A numerical analysis of the energy–volume curves of the two systems further yields bulk moduli of 130 and 324 GPa for Al and Fe, respectively, to be compared with the experimental values of 72 and 168 GPa [331]. The discrepancy of a factor of about two is very acceptable taking into account the simplicity of the force field.[15]

Even the inclusion of another possible system such as FeAl is straightforward. Its force field may be derived by averaging over those of Fe and Al, and the influence of the small charge transfer from Al to Fe (because Al is less noble than Fe) will generate an internal (but screened) Madelung field whose size can be estimated from absolute electronegativities and hardnesses (Equation (2.2)). For FeAl, the charge transfer amounts to only 0.063 electrons, and the simulation of an fcc-type FeAl (not shown) yields a theoretical equilibrium volume which is only 3% smaller than the experimental one; its bulk modulus, however, is also overestimated [331].

Being able to handle all above systems numerically, the modelling of the Fe:AlN sputtering process may now be attempted by a molecular-dynamics approach [333]. To do so, a supercell of 768 bcc-Fe atoms is constructed from

15) It is a common feature of such simple bonding potentials that they cannot fully describe all features of the energy surface, especially in regions of anharmonicity. In fact, anharmonic effects suggest a rescaling of the bond stretching and bending forces in regions further away from the minimum energy [332].

6 vertical and 8 × 8 horizontal layers, depicted in Figure 3.32(a). This system is periodically repeated in the two horizontal directions, whereas a large amount of empty space is allowed above the Fe crystal. It is held at a temperature of 370 K and continuously bombarded with AlN molecules at a speed which matches the applied rf energy of about 2 eV. The individual time step is 0.2 fs and more than 370 000 time steps are collected such that the whole simulation (aixCCAD) covers a time frame of 74 ps. All above tested parametrizations are kept but the Coulomb interactions *within* the iron host are appropriately screened using an exponential decay parameter of 2.9 Å for interatomic distances larger than 2 Å. For Fe–N interactions, a bond-valence r_0 parameter of 1.769 Å is derived from the FeN structure, which is slightly different from the tabulated one (Table 1.4).

In Figure 3.32(b), the first few AlN molecules have hit the iron surface, and they break apart into Al and N atoms upon arrival. The nitrogen atoms immediately penetrate into the iron host, leaving the aluminum atoms on the surface. Although the fast nitrogen penetration is, in part, due to the very high kinetic energy of the rf-sputtered species, a similar reaction also results when the incoming AlN molecules are supplied with only five percent of their original kinetic energy. They shortly pause on the iron surface, fall apart, and again there is a migration of N atoms into the iron host. The N atoms penetrate at least into the second layer of the iron lattice and acquire an approximate sixfold coordination. Because of the resulting charge transfer, the *nitride ions* thus experience a much stronger chemical bonding inside the Fe bulk than within the AlN molecule or on top of the Fe surface. This behavior seems to be the clue to the unprecedented reaction. While it is thermodynamically clear that bulk FeN cannot compete with bulk AlN in terms of thermochemical stability, a single nitrogen atom leads to a larger lattice energy if embedded as a nitride ion, in three dimensions, compared with a weaker covalent bonding in a molecule or on a surface.

Because of a complex interplay of three different effects, namely the high energy of the incoming AlN, the slow energy dissipation to the surrounding heat bath, and the geometric distortion upon nitrogen penetration, some of the surface Fe atoms eventually rise to positions above the original surface, and the arrival of even more Al atoms finally leads to a heavily distorted Fe/Al "alloy" zone incorporating the above-mentioned N atoms. This is most easily seen from Figure 3.32(c) which shows the final atomic configuration. The Fe content of this 10 Å thick reaction zone slowly decreases while moving up vertically, and the zone structurally resembles a bcc lattice with statistically disordered Fe and Al atoms, coordinating the N atoms in their distorted octahedral vacancies.

Some N atoms penetrate the Fe lattice even more deeply because they are "pushed" by above-lying N atoms. Their majority is found at a depth of

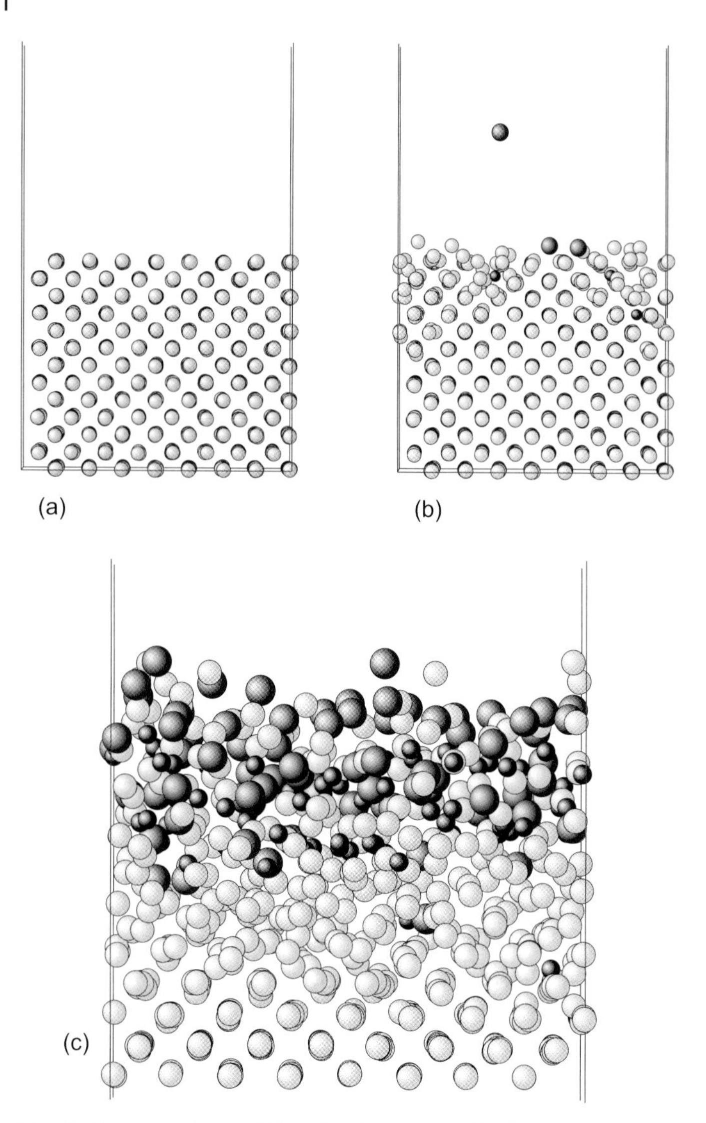

Fig. 3.32 Snapshots of the simulation of radio-frequency sputtering of an iron target with aluminum nitride. Fe atoms are depicted as medium grey spheres, Al as large dark spheres, and N as small black spheres. The initial geometry is shown in (a) and the impact of the first aluminum/nitrogen atoms is given in (b). A magnified view of the reaction zone after 370 000 time steps is depicted in (c).

roughly 2 Å from the surface but there is one particular atom (Figure 3.32(c)) which penetrates down to about 5 Å. For a more rigorous analysis, Figure 3.33(a) gives the RDFs for the Fe–N and Al–N combinations inside the aforementioned reaction slab. The first RDF maximum for Fe–N lies around 2.1 Å, whereas that for Al–N is a little larger, in accordance with the longer Al–N

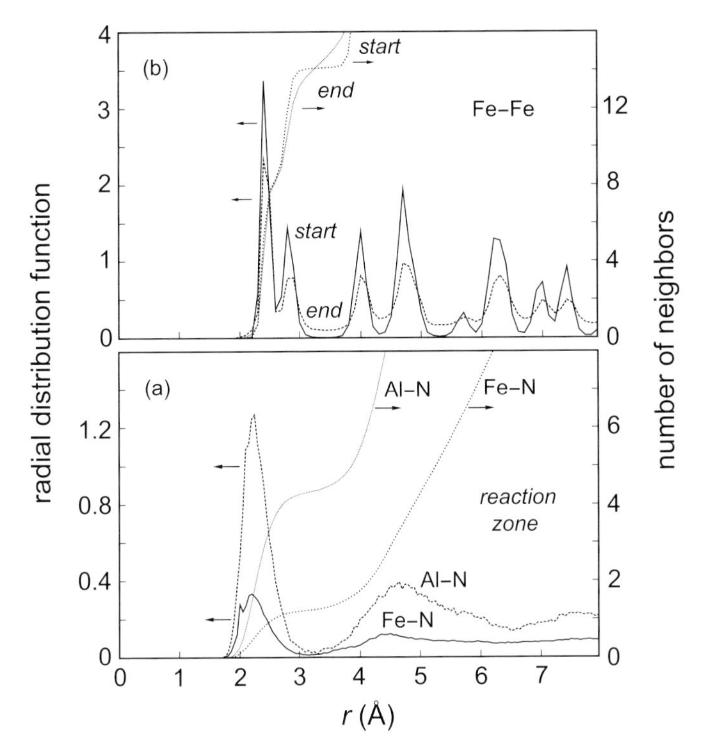

Fig. 3.33 Radial distribution functions and their integrals (number of coordinating atoms) as a function of the interatomic distance for the combinations Al–N and Fe–N (a) as well as Fe–Fe (b). The metal–nitrogen RDFs have been derived using only the geometrical data of the reaction zone.

bond. The first RDF minima at 3.3 Å are almost zero for both combinations, indicating no significant nitrogen movement between neighboring octahedral sites. The integration of the RDFs up to the first minima results in a nitrogen coordination number of 1.2 (with respect to Fe) and 4.4 (with respect to Al) such that the notion of an approximate octahedral coordination (5.6) for nitrogen–metal interactions is justified.

As noted before, the Fe surface distorts due to the impact of Al and N atoms, but the geometrical distortion is relatively short-ranged and decays after about 2.5 layers, leaving the lower-lying iron atoms almost untouched. This effect is also seen in Figure 3.32(c) and numerically evaluated in Figure 3.33(b). The RDF of Fe–Fe is perfectly ordered at the beginning of the simulation and shows very distinctive peaks, but it reveals both reduced peak heights and a significant size *between* the peaks at the end. Because the latter RDF has been evaluated over the entire system, the disordering of the layers close to the surface is underrepresented here. In addition, the insertion of N atoms leads to a small (2.5%) widening of the Fe crystal, best seen at the second RDF maximum which has increased by almost 0.1 Å if compared with the undistorted lattice.

Summarizing, dynamical effects may be theoretically observed in such nanocomposites, despite the fact that these phenomena are impossible to understand from thermodynamic criteria, assuming chemical equilibrium. This section merely serves to emphasize that, clearly, such kinetic processes can only be properly dealt with if time plays an *explicit* role in the modelling of the system, to be accomplished by molecular-dynamics simulations. The reaction phenomenon found here appears to arise both from the drastic synthetic conditions as well as from the preference of nitrogen to be three-dimensionally bonded to the iron host. The numerical analysis of such dynamical processes will certainly gain importance in the coming years.

3.8
Structural versus Electronic Distortions: MnAl

Let us now return to itinerant magnetism and time-independent problems. The perceptive reader will have noticed that the chemical phrasings for these magnetic phenomena of the elements and their compounds were based on concepts of electronic instabilities, and the appearance of the Fermi level in antibonding/nonbonding levels served as a "red flag" indicating these instabilities. Then, however, one might adopt a rather different point of view and ask whether or not alternative *structural* distortions to more stable but still nonmagnetic structures might also be possible. Indeed, such "conventional" distortions could also be expected to remove the antibonding M–M states of itinerant ferromagnets, for example. In elemental Te, there were no magnetically active orbitals (see Section 3.4) but for an itinerant magnet there must be *two* alternative ways to lower the overall symmetry, an electronic distortion (spin-polarization) *and* a purely structural one.

The question is indeed much more general and, once again, touches upon the importance of Jahn–Teller and Peierls distortions in structural chemistry [62, 73, 77] (see also Section 2.6.2). Both phenomena represent different structural mechanisms to lower the symmetry through electronic–vibrational coupling, and both may set in when there are partially occupied, degenerate molecular or crystal orbitals.[16] When the Jahn–Teller distortion has occurred, the system under study has lost *point-group* symmetry; the classic example being given by an octahedrally coordinated Cu(II) atom which distorts to an elongated octahedron because of Cu–X antibonding e_g orbitals. Even when

16) In reality, the theory is more complicated than this because both effects depend on many-body states (not orbitals), and the Jahn–Teller case needs a finer distinction between first order, second order and pseudo Jahn–Teller flavors [62].

Cu(II) is part of a crystal structure, only a local *point-group* symmetry element is lost, and the translational part is still intact. For a Peierls distortion, this is quite different because *translational* symmetry is lost upon distortion. The classic examples are given by the chain of hydrogen atoms (see Section 2.6.2), by polyacetylene which distorts to a conjugated chain of single and double bonds arising from partially filled and nonbonding frontier orbitals, and by elemental Te (Section 3.4). The role of all these electronic instabilities has recently been highlighted for the simple structures of the *nonmagnetic* metallic elements [334].

If this is so, a competition between "normal" Jahn–Teller and Peierls distortions on the one hand and distortions involving spin-polarization on the other, is plausible, and one may think of more general Jahn–Teller or Peierls scenarios. Within an *extended* Jahn–Teller (JT) case, a system may either distort along its nuclear coordinates (regular JT effect) or, if there is sufficient exchange splitting, distort along its electronic coordinates and spin-polarize. Recall that in going from nonmagnetic bcc-Fe to ferromagnetic bcc-Fe the translational symmetry is fully retained because the size of the unit cell is still the same (Jahn–Teller-like). Similarly, a system may have two choices within an *extended* Peierls case, namely either distorting along its nuclear coordinates (regular Peierls effect) or, if spin-polarization is possible, reshuffling its electronic coordinates. Let us remind ourselves that upon moving from nonmagnetic bcc-Cr to antiferromagnetic bcc-Cr, translational symmetry is lost because the unit cell doubles, resembling the Peierls case. In the above examples, both regular (nuclear coordinates) and "electronic" Jahn–Teller instabilities result from antibonding frontier levels whereas the regular and "electronic" Peierls distortions rest on nonbonding levels.

What happens if a system may choose between regular and "electronic" distortions? This is an exciting question, and it may be answered, at least for the Jahn–Teller case, by looking at manganese aluminide, MnAl. This intermetallic compound, known since 1948, is part of the extraordinarily rich Mn/Al system in which seventeen (!) phases have been described [335], and the ferromagnetic, ordered τ-MnAl phase [336] is used as a permanent magnetic material. It is a strong ferromagnet with a Curie temperature of 653 K, a manganese moment of 1.94 Bohr magnetons and, surprisingly, an enhanced magnetic anisotropy, that is, a coupling between magnetic and mechanical properties along certain directions. Figure 3.34(a) shows its crystal structure belonging to the [AuCu] structure type. The face-decorated cell can also be described using a primitive tetragonal unit cell denoted by dashed lines, and its primitive lattice constants are $\bar{a} = 2.77$ Å and $c = 3.54$ Å. This tetragonal cell itself can be regarded as being a distorted cubic cell ($\bar{a} \approx 3.01$ Å) of the [CsCl] type.

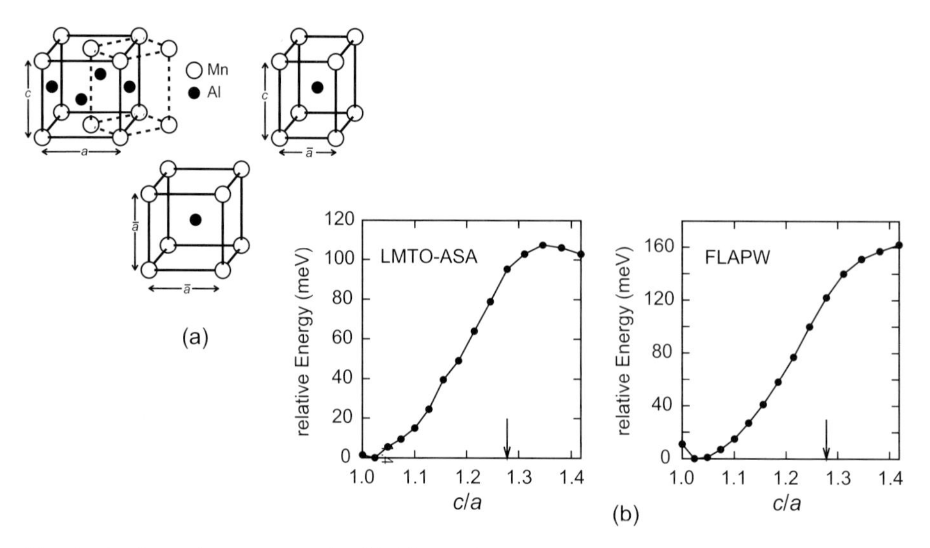

Fig. 3.34 Crystal structure of τ-MnAl (a), with the dashed lines denoting the reduced tetragonal cell, and also including an idealized simple cubic structure of MnAl at the bottom. The total energy of MnAl (b) according to nonmagnetic LMTO (LDA) and FLAPW (GGA) calculations as a function of the axial ratio c/a. The arrows indicate the experimentally-found ratio.

The synthesis of τ-MnAl is quite complicated, starting from a high-temperature phase and followed by a number of metallurgical steps, eventually reaching the ferromagnetic, *metastable* phase by a martensitic transformation [337]. It has been shown already that the magnetocrystalline anisotropy energy of τ-MnAl depends on the axial ratio c/a and that a tetragonal distortion from a cubic structure stabilizes the system [338]. Let us rephrase our initial question for the specific case of MnAl: Why does MnAl not adopt the cubic [CsCl] structure but instead prefer the tetragonal [AuCu] structure? Why is MnAl ferromagnetic? Are these two questions related to each other?

The analysis is based on a series of nonmagnetic (LDA) total energy calculations of MnAl using the TB-LMTO-ASA method for various c/a ratios but keeping a constant cell volume [339]. Since ASA calculations are known to show weaknesses upon structural deformation because of the underlying ASA shape approximations for the atomic potentials, the whole series *must* be repeated using the much more accurate FLAPW method (WIEN2k) and the GGA; both results have been plotted in Figure 3.34(b). Although there is a difference in numbers, both methods agree, qualitatively, in that nonmagnetic MnAl should crystallize in the cubic – not tetragonal! – system, adopting the [CsCl] structure, not the [AuCu] structure. For this cubic structure, there is a small charge transfer (ca. 0.19 electrons) from Al to Mn, in agreement with the course of the absolute electronegativities from Table 2.2.

The LMTO density-of-states of MnAl is shown in Figure 3.35(a), and it is composed of a wide *s*-like part in the lower half of the valence region, plus the contributions of the five Mn 3*d* bands in the proximity of the Fermi level. Because of the simple cubic Mn environment by Mn, the 3*d* splitting follows a t_{2g} above e_g mode, which is reversed compared with the one found in the elemental 3*d* ferromagnets. In MnAl, the lower-lying e_g-set orbitals (d_{z^2}, $d_{x^2-y^2}$) generate σ-type Mn–Mn interactions and thus show larger dispersion. The t_{2g} set (d_{xy}, d_{xz}, d_{yz}) interacts more weakly (π-like) and is positioned in the frontier band regime. At the Fermi level, the DOS is quite large, a sign of instability. We reiterate, however, that the nonmagnetic total energy calculations *exclude* the possibility of a structural deformation into the tetragonal system. To gain more insight into the various interactions, a COHP analysis of the Mn–Mn, Mn–Al, and Al–Al bonds is also given in Figure 3.35(b).

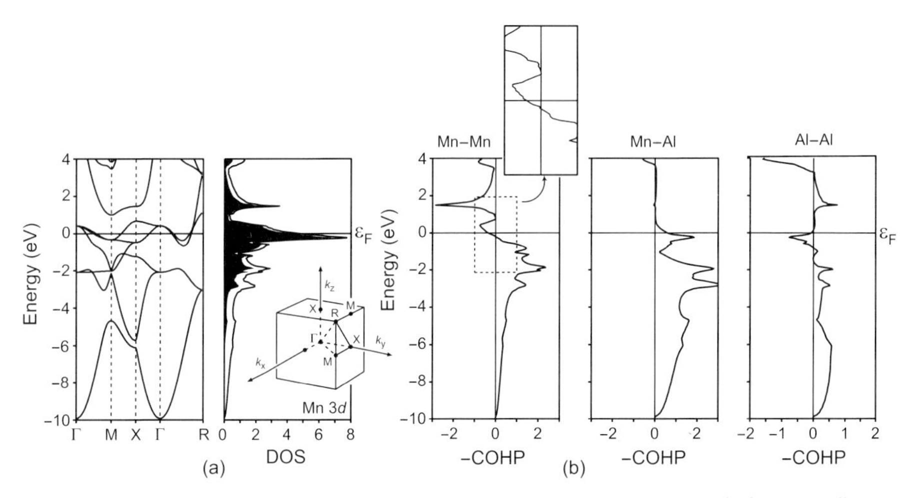

Fig. 3.35 LDA band structure and DOS (a) of nonmagnetic cubic MnAl according to TB-LMTO-ASA calculations with local projections of the Mn 3*d* contributions and COHP analyses (b) of the electronic structure.

Clearly, the strongest bonding interactions (eight Mn–Al bonds with an integrated COHP of −11.38 eV) are bonding throughout the entire valence band region, and they are almost perfectly adjusted with respect to the Fermi level. A further slight electronic enrichment would stabilize Mn–Al even more. Al–Al interactions (six bonds, −2.71 eV) are much weaker, and even some antibonding states have been filled already. For the six Mn–Mn bonds (−5.71 eV) it is interesting to note that, besides the low-lying bonding interactions, there are populated *antibonding* interactions located exactly at the Fermi level, and it is these destabilizing effects which point towards an underlying electronic instability. Thus, spontaneous spin-polarization towards a ferromagnetic structure is very much indicated for cubic MnAl, *not* a structural distortion (see above), which is impossible.

We therefore switch to the spin-polarized case but still stay with the primitive cubic structure. It turns out that the total energies are lowered by 0.243 eV (LMTO-ASA, LDA) and 0.393 eV (FLAPW, GGA), simply due to the different occupations of the α and β spin sublattices. The number of unpaired electrons is 1.94 (LMTO-ASA, LDA) and 1.85 (FLAPW, GGA), respectively, and the charge transfer between Al and Mn is almost unchanged. The new DOS plot on the basis of the LMTO electronic structure can be found in Figure 3.36(a).

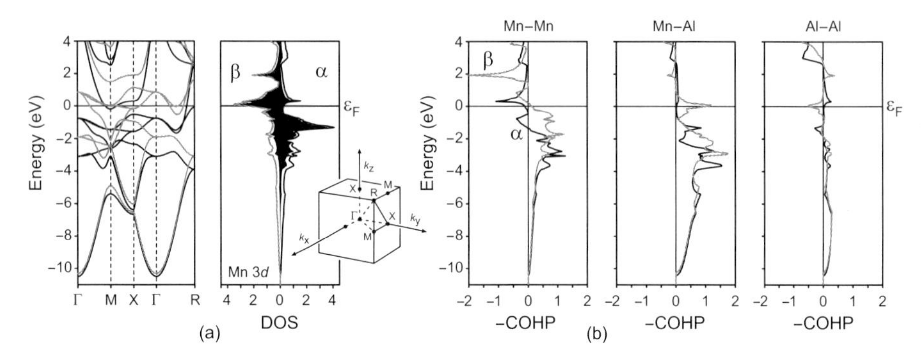

Fig. 3.36 The same as Figure 3.35, but now based on a spin-polarized calculation for ferromagnetic cubic MnAl. Majority/minority levels are given by black/grey lines.

As expected, the exchange splitting between the majority α (black line) and minority β (grey line) levels is small for the Mn 4*s* functions and much larger (1–2 eV) for the Mn 3*d* bands, such that most of the π bands (β spins) have been emptied because they lie above the Fermi level. The shapes of the spin-resolved DOS still resemble the nonmagnetic one. The COHP bonding analysis (Figure 3.36(b)) within cubic ferromagnetic MnAl also fulfills expectations. Al–Al bonding is almost untouched, and Mn–Al bonding changes only slightly due to the varied local electronic structure of the Mn atom. Mn–Mn bonding, however, is characterized by the *removal* of antibonding interactions at the Fermi level because of the repopulated spin sublattices.[17] Still, the ferromagnetic cubic structure of MnAl suffers from nonoptimized Al–Al bonding. At the Fermi level, there are visible antibonding levels only for this kind of interaction. Note, however, that a further spin-based electronic adjustment of the Al–Al bonds is impossible since the Al atom does not have a magnetically active orbital; *structural* changes must now come into play.

17) Curiously, a numerical integration indicates that only the covalent part of the Al–Al bonding has slightly increased upon spin-polarization; the covalent bond strengths of Mn–Mn and Mn–Al have both decreased a little. This situation therefore resembles the situation in FePd discussed in Section 3.6.

When the total energy calculations as a function of the c/a axis ratio are repeated for the spin-polarized *ferromagnetic* scenario, one arrives at the results plotted in Figure 3.37. Both density-functional methods agree in that a tetragonal MnAl structure with $c/a \approx 1.2$ (LMTO-ASA, LDA) and $c/a \approx 1.28$ (FLAPW, GGA) is preferred; the full-potential method is even quantitatively correct. There is also a very small decrease in charge transfer from Al to Mn which arrives at 0.18 for the experimental lattice parameters.

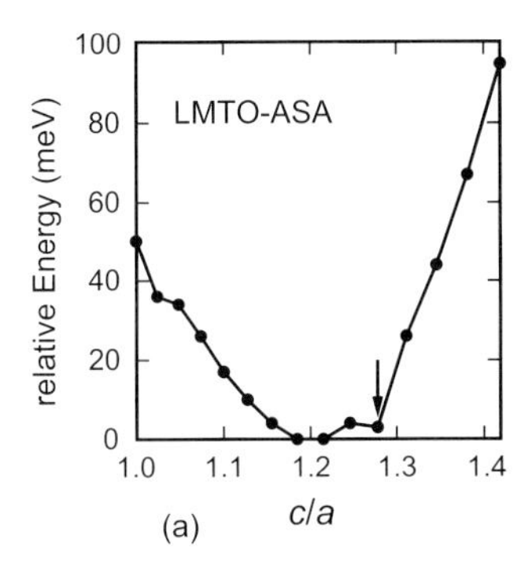

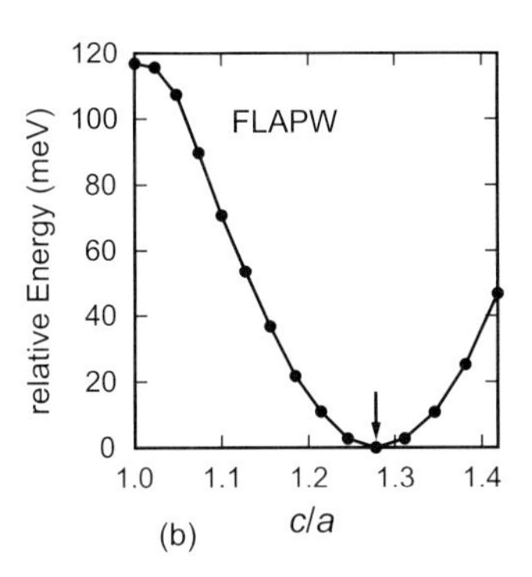

Fig. 3.37 The total energy of MnAl according to spin-polarized TB-LMTO-ASA (LDA, (a)) and FLAPW (GGA, (b)) calculations as a function of the axial ratio c/a. The arrows indicate the experimentally observed ratio.

The energetic gain in becoming tetragonal is an additional 0.047 eV (LMTO-ASA, LDA) and 0.117 eV (FLAPW, GGA), indicating strong magneto-elastic coupling. The magnetic moments converge to 2.32 (LMTO-ASA, LDA) and 2.23 Bohr magnetons (FLAPW, GGA), overestimating the experimental one. The corresponding DOS diagram (LMTO) is given in Figure 3.38(a), showing that the structural distortion has pushed the π-bonding part of the β spins even further above the Fermi level, namely to ca. +0.8 eV. These were originally touching the Fermi level and centered at around +0.3 eV (compare with Figure 3.36).

More insight is given by the COHPs (Figure 3.38(b)) of the tetragonal structure, to be compared with their cubic counterparts (Figure 3.36(b)). Due to the tetragonal distortion we now have shorter and longer bonds for the Mn–Mn, Mn–Al, and Al–Al combinations but we simplify them by a single, averaged COHP curve for each interaction. The integrated numerical data reveal that both Mn–Mn and Al–Al bonds have significantly strengthened, namely by ca. 7% (Mn–Mn) and 30% (Al–Al) when compared with the nonmagnetic cubic structure. The formerly antibonding Al–Al states at the Fermi level (Figure 3.36) have been totally removed by the symmetry reduction. Note that these were driving the cubic/tetragonal distortion.

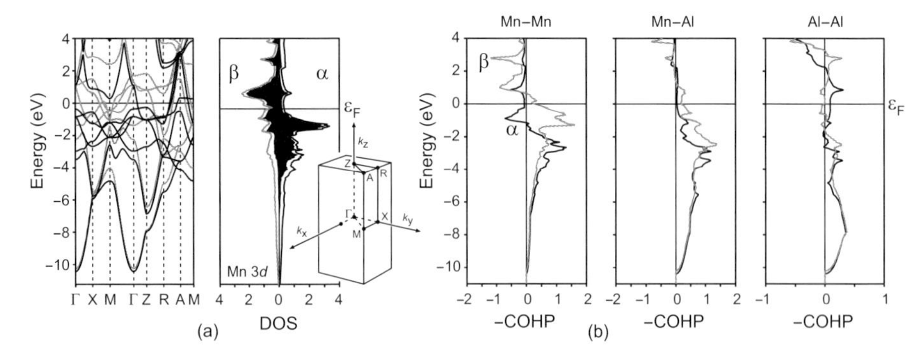

Fig. 3.38 As for Figure 3.36, but now based on a spin-polarized calculation for ferromagnetic tetragonal MnAl.

The covalent part of the Mn–Al bonds, however, has been weakened during the entire process, namely by ca. 6% when compared with the ferromagnetic cubic, and by ca. 7% when compared with the nonmagnetic cubic crystal structure. It is self-evident that we should interpret the strengthening of the homoatomic (Mn–Mn, Al–Al) bonding and the weakening of the heteroatomic (Mn–Al) bonding as an electronic signpost for the beginning *phase separation* into elemental Mn and Al. In fact, τ-MnAl is reported to be only metastable.

Returning to our opening questions, the tetragonal crystal structure of τ-MnAl and the phase's ferromagnetic properties may be understood to originate from an idealized structure (nonmagnetic cubic MnAl) of [CsCl] motif through a two-step symmetry-lowering process. While the nonmagnetic cubic

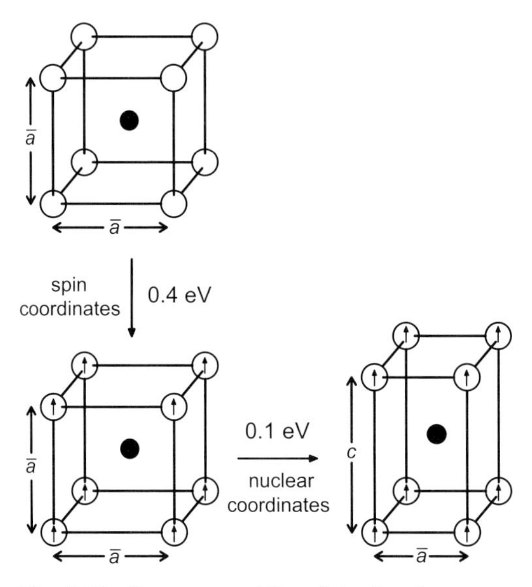

Fig. 3.39 Sequence of the distortion from nonmagnetic cubic MnAl to ferromagnetic tetragonal MnAl.

structure is *stable* against a structural deformation in terms of a normal Jahn–Teller distortion, antibonding Mn–Mn interactions at the Fermi level lead to spin-polarization and the onset of magnetism, i.e., a symmetry reduction taking place solely in the *electronic* degrees of freedom, and thus emptying antibonding Mn–Mn states and gaining ca. 0.4 eV. During this "electronic" distortion, the translational symmetry is retained. Nevertheless, the symmetry reduction is insufficient for a full optimization of the electronic structure, reflected by residual antibonding Al–Al states at the Fermi level, and they can only be removed by a subsequent, energetically smaller, structural deformation for the *nuclear* coordinates towards the tetragonal system. The energetic gain of this normal Jahn–Teller distortion is smaller (ca. 0.1 eV) but also retains the translational symmetry since the volume of the unit cell stays the same. The process as a whole is schematically sketched in Figure 3.39. Eventually, homonuclear bonding is strengthened and heteronuclear bonding is weakened.

3.9 Challenging Theory: Mercury Carbodiimide and Cyanamide

From the preceding sections, the reader might have received the impression that electronic-structure theory, especially when carried out using sophisticated parametrizations for exchange and correlation, is able to describe accurately the structure and properties of practically every material. This is a very appealing idea but before it settles too deeply, let us deal it a firm blow, at least with respect to density-functional theory. Although we have already alluded to failures of electronic-structure theory (and will re-visit them when discussing correlated oxides, see Section 3.10), it is time to meet another interesting DFT problem. This flaw is particularly nasty because it is not very easy to spot.

Extended materials containing infinite networks made from only carbon and nitrogen atoms are a hot topic within materials chemistry, and a large number of (not only theoretical) papers have been written on carbon nitride, C_3N_4, although bulk syntheses in combination with convincing structure determinations are missing till now [340, 341]. A potential precursor to C_3N_4 and related compounds might be given by the simple *cyanamide* molecule, H_2NCN, shown in Figure 3.40(a). H_2NCN is so fundamental that it can be found in virtually any textbook of inorganic and organic chemistry.

H_2NCN is an inexpensive, slightly acidic compound with an N–C single bond of 1.31 Å on the left side of the molecule and a C≡N triple bond of 1.15 Å on the right side. Another structural possibility for a compound containing two H atoms, two N atoms, and one C atom, however, is given by the

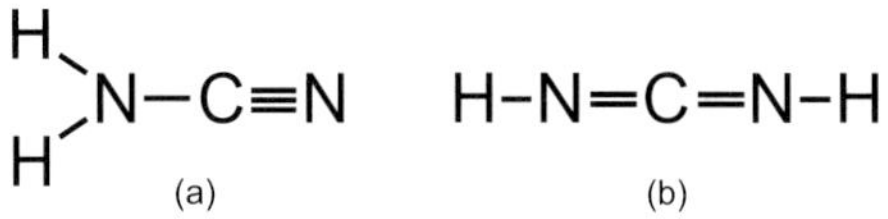

$$H_2N{-}C{\equiv}N \qquad H{-}N{=}C{=}N{-}H$$

(a) (b)

Fig. 3.40 The molecular structures of cyanamide (a) and carbodiimide (b).

composition HNCNH, called *carbodiimide*, and this molecule has also been included in Figure 3.40(b). This other *isomer* (that is how chemists would phrase it because HNCNH has the same composition as H_2NCN but exhibits a different structure) includes two N=C double bonds. H–N=C=N–H, however, is so unstable that it can only be observed under bizarre laboratory conditions and, in fact, it has never been isolated at ambient temperature because it immediately back-transforms ("isomerizes") to H_2NCN, the cyanamide. Accordingly, quantum-chemical calculations (CI-level of accuracy) show the carbodiimide molecule to be *less* stable than the cyanamide molecule by 33 kJ/mol [342].

The most important inorganic derivative of H_2NCN is given by the salt calcium cyanamide, CaNCN, an early fertilizer [343]. Quite simply, the NCN^{2-} anion replaces the O^{2-} anion in the crystal structure of CaO (see Section 3.1), which would correctly suggest a distorted sodium chloride structure type for CaNCN with six-fold Ca^{2+} coordination. Even this quasi-binary phase is rock-salt-like in terms of structure! But there are also many other metal derivatives of H_2NCN, and we briefly mention those containing alkaline (Li, Na) [344] or alkaline-earth metals (Mg, Ca, Sr, Ba) [345], main-group elements (In, Si, Pb) [346], transition metals (Ag) [347], and also rare-earth metals (Eu, Ce) [348]; the list grows all the time. The crystal structure of CaNCN is given in Figure 3.41(a).

Within CaNCN, the NCN^{2-} anion exists in the so-called *carbodiimide* form, that is, a linear $N{=}C{=}N^{2-}$ with two N=C double bonds. Thus, calcium "cyanamide" should rather be called a carbodiimide, but the name had been given to CaNCN many years before the crystal structure was clarified. We reiterate that *molecular* carbodiimide (H–N=C=N–H) is highly unstable. The same (false) naming applies to most other metal "cyanamides" of which most are, in fact, carbodiimides because they contain the $N{=}C{=}N^{2-}$ anion.

This finding is easily rationalized using Pearson's concept of hard and soft acids and bases (see Section 2.14). Whenever NCN^{2-} is coordinated by chemically hard cations (for example, Ca^{2+}, with η = 19.52 eV, see Table 2.3) there results a considerable amount of *ionic* bonding between the cation and a symmetrical $N{=}C{=}N^{2-}$ anion. Chemically softer cations such as Pb^{2+} (8.46 eV) or Ag^+ (6.7 eV) will bond more *covalently* to one end of the NCN^{2-} unit, inducing the less symmetrical ($N{-}C{\equiv}N^{2-}$) cyanamide shape; this species must be chemically softer. Figure 3.41(b) shows the PbNCN structure with a seven-

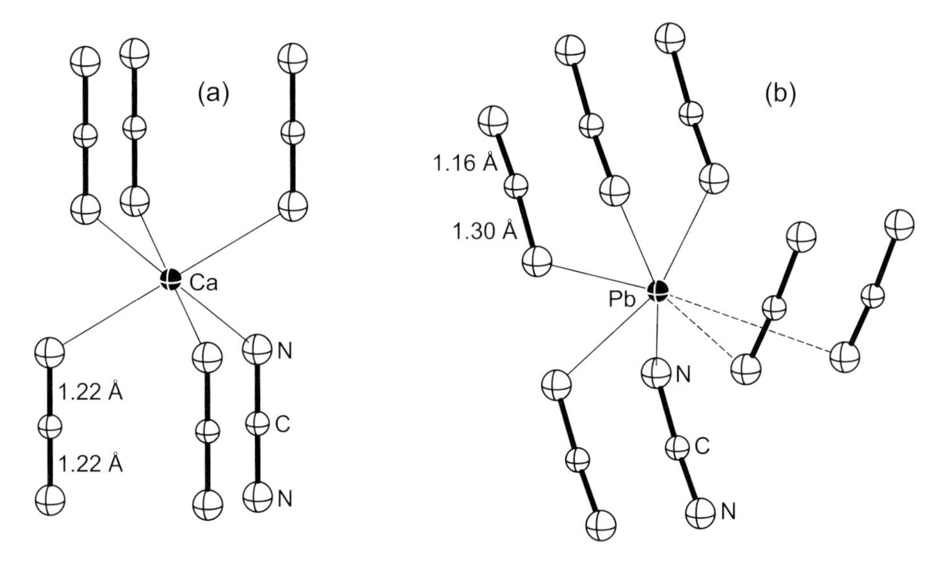

Fig. 3.41 Cuts from the crystal structures of CaNCN (a) and PbNCN (b).

fold (!) coordination of the softer Pb^{2+}; both PbNCN and also Ag_2NCN are cyanamides by composition *and* structure.

Eventually, the start of the millennium saw the synthesis of two new cyanamide derivatives of mercury, of which the first one, HgNCN(I), was made through double-deprotonation of H_2NCN in a strongly basic (pH $\approx$ 14) medium, followed by the coordination of the "naked" NCN^{2-} by Hg^{2+} [349], schematically following:

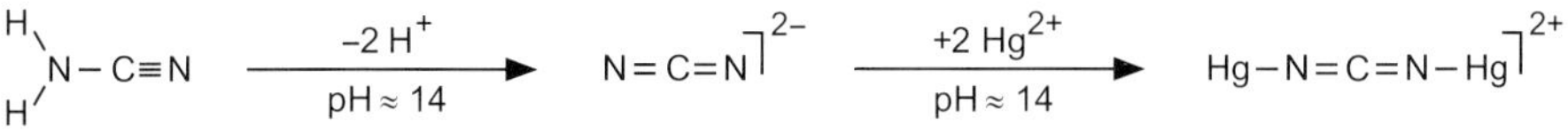

$$\mathrm{H_2N{-}C{\equiv}N} \xrightarrow[\mathrm{pH} \approx 14]{-2\,\mathrm{H^+}} [\mathrm{N{=}C{=}N}]^{2-} \xrightarrow[\mathrm{pH} \approx 14]{+2\,\mathrm{Hg^{2+}}} [\mathrm{Hg{-}N{=}C{=}N{-}Hg}]^{2+}$$

Scheme 3.1

The synthesis of the other derivative, HgNCN(II), utilized considerably softer acid–base conditions (pH $\approx$ 6) in the spirit of a step-wise replacement of hydrogen by mercury [350], shortened as:

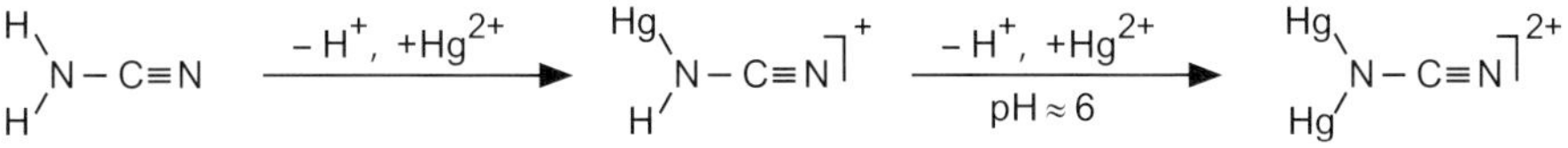

$$\mathrm{H_2N{-}C{\equiv}N} \xrightarrow{-\,\mathrm{H^+},\ +\mathrm{Hg^{2+}}} [\mathrm{(Hg)(H)N{-}C{\equiv}N}]^{+} \xrightarrow[\mathrm{pH} \approx 6]{-\,\mathrm{H^+},\ +\mathrm{Hg^{2+}}} [\mathrm{Hg_2N{-}C{\equiv}N}]^{2+}$$

Scheme 3.2

Following our nomenclature, HgNCN(I) stands for the mercury *carbodiimide* (Scheme 3.1) and HgNCN(II) represents the mercury *cyanamide* (Scheme 3.2). In fact, the two different compounds (or two different solid-state isomers, as some might want to call them) can be made on purpose, and their structural motifs are depicted in Figure 3.42.

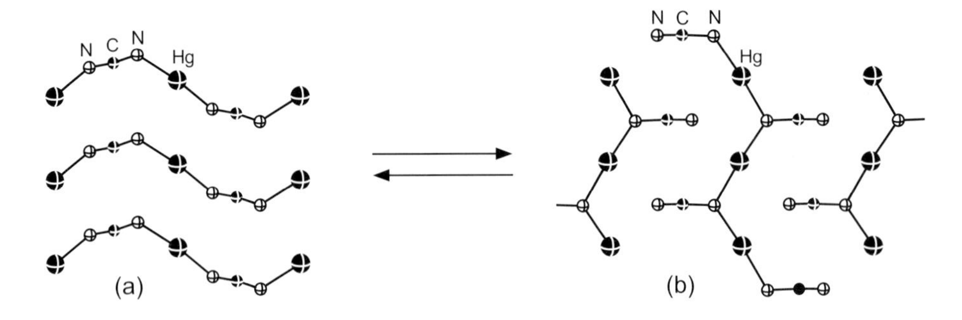

Fig. 3.42 Structural motifs from the crystal structures of mercury carbodiimide (a) and mercury cyanamide (b).

Mercury carbodiimide (Figure 3.42(a)) is characterized by a symmetrical $N{=}C{=}N^{2-}$ carbodiimide unit with two identical N=C distances of 1.22 Å. The two nitrogen atoms are coordinated by divalent Hg atoms on the left and right sides of the molecular anion at about 2.06–2.07 Å. Mercury cyanamide, in Figure 3.42(b), contains the asymmetrical $N{-}C{\equiv}N^{2-}$ unit with C–N distances of 1.35 and 1.12 Å, and only one N atom is covalently bonded to two Hg atoms at about 2.05–2.11 Å. Both compounds are three-dimensional Hg counterparts of the H_2NCN and HNCNH molecules at the beginning, and we are witnessing an unprecedented carbodiimide/cyanamide solid-state polymorphism! Both HgNCN phases can be easily distinguished from each other by means of infrared/Raman vibrational or crystallographic investigations (combined X-ray and neutron diffraction) but not by the naked eye because both are white insulators. Both, of course, yield identical chemical analyses.

The first question of any solid-state chemist thinking in terms of structure and energetics will definitely be: What is the *stable* polymorph, mercury carbodiimide, HgNCN(I), or mercury cyanamide, HgNCN(II)? Unfortunately, it is impossible to answer this question by means of differential thermal analysis because both polymorphs decompose, *prior* to interconversion,[18] at about 230 °C to yield a white polymer and mercury metal. Thus, theoretical reasoning and/or electronic-structure theory is needed. Let us attempt to argue using both classical and quantum-chemical means.

First, the crystallographic density of HgNCN(I) is 3.5% higher than that of HgNCN(II). Usually, higher stability accompanies better packing and smaller interatomic distances but this is just a crude rule (Pauling's bond-length–bond-

18) On the molecular level, such a transition leading to chemical equilibrium was reported long ago. When a silylcarbodiimide is synthesized and purified by distillation, storing it at room temperature leads to the spontaneous formation of a carbodiimide/cyanamide mixture [351].

strength or bond-valence recipe, see Section 1.5) to which many exceptions are known. *Second*, the number of mercury–nitrogen bonds per NCN^{2-} unit may be compared (see Figure 3.42) but they are identical (= two) for both polymorphs, i.e., no preference for either cyanamide or carbodiimide. *Third*, one might directly analyze the strength of the nitrogen–carbon bonds in the two polymorphs by the bond-valence concept (see again Section 1.5). Taking a bond distance of 1.22 Å (from the CaNCN structure) as the most appropriate choice for the N=C double bond, the bond orders in the $N{=}C{=}N^{2-}$ unit of HgNCN(I) are 1.96 and 2.02, summing up to 3.98; alternatively, the bond orders of the single and triple bonds in N–C≡N of HgNCN(II) then come to 1.40 and 2.62, summing up to 4.02. Thus, C–N bonding also seems to be perfectly adjusted in both anions, and no energetic ranking can be performed. However, a primitive energetic estimate based on the tabulated bond energies of N=C double bonds (616 kJ/mol) and those of N≡C (892 kJ/mol) and N–C bonds (305 kJ/mol) [63] shows that – *fourth* – two double bonds (and also the carbodiimide anion) should be more stable by 35 kJ/mol. We note that the ideas of single and triple nitrogen–carbon bonds are simplifying exaggerations. This estimate would also mean, however, that the unstable molecule H–N=C=N–H is energetically lower than the stable $H_2N{-}C{\equiv}N$, which is clearly incorrect. Tabulated bond energies do not help but actually mislead us.

Fifth, the calculation of Madelung energies also points in the direction of the carbodiimide. Using the formal oxidation states as ionic charges (i.e., Hg^{2+}, N^{3-}, and C^{4+}), the HgNCN(I) Coulomb part of the lattice energy (MAPLE from Section 1.2), and possibly also the (negative) total energy, is larger than that of HgNCN(II) by 17 kJ/mol but the ionic charges must be *very* far from reality. Eventually, quantum mechanics should provide the correct answer. Thus, *sixth*, density-functional total-energy calculations using plane-wave basis sets and ultra-soft pseudopotentials (VASP) on the basis of the LDA and also the GGA are performed, taking great care to converge them in terms of basis functions and $\mathbf{k}$ points.

Density-functional theory reproduces the geometries of HgNCN(I) and HgNCN(II) not too badly, and the theoretical lattice parameters do not deviate from the experimental ones by more than about 0.3 Å (root-mean-square value) using the LDA and about 0.4 Å using the GGA. For HgNCN(I), the bond distances/angles are reproduced within 0.01 (LDA) or 0.02 Å (GGA) and 2/3° (LDA/GGA). For HgNCN(II), the N≡C triple bond is a bit too long (1.19/1.20 Å by LDA/GGA) and the N–C single bond is a bit too short (1.27/1.28 Å). Eventually, HgNCN(I) is predicted to be *more stable* by about 7 kJ/mol than HgNCN(II) based on either the LDA or the GGA. Using the more sophisticated PAW approach instead of the ultra-soft pseudopotentials, this preference of HgNCN(I) increases to 13 kJ/mol; the *carbodiimide* HgNCN(I) should be more stable. The problem is therefore solved, is it not?

Very unfortunately, this is completely wrong. Calorimetric dissolution measurements of HgNCN(I) and HgNCN(II) in aqueous HCl and the application of the Hess law [259] reveal that mercury *cyanamide*, HgNCN(II), is more stable both in terms of enthalpy and Gibbs energy, with an enthalpy difference of about 2–3 kJ/mol [352], rebutting the density-functional theory assessments and all other classical reasonings. On searching for the origin of the DFT error, a hint is already given by tabulated correlation energy increments which demonstrate that $N{=}C{=}N^{2-}$ is more correlated than $N{-}C{\equiv}N^{2-}$ [353]; also, both are anions which makes things worse for DFT-based methods.[19] Because applying post-Hartree–Fock methods to extended systems is challenging (see Section 2.13), calculations on hypothetical *molecules* reflecting the same structural ingredients are a good alternative. These are performed by using established *ab initio* methods, namely the GAUSSIAN package with basis sets of the same type. On the DFT side, the GGA (Perdew–Wang 91 [164]) and also the Becke three-parameter hybrid functional (B3LYP [166]) is utilized whereas, on the post-Hartree–Fock side, second-order Møller–Plesset perturbation theory (MP2 [129]) and quadratic configuration interaction calculations are performed, including single and double substitutions (QCISD [184]).

When such model molecules of the general structural formulae $R{-}N{=}C{=}N{-}R$ and $R_2N{-}C{\equiv}N$ are handled by both approaches (DFT and post-HF) including full structure optimizations, significant differences merely show up in the energy differences (not geometries) between the carbodiimide and cyanamide isomers. This is depicted in Figure 3.43, reflecting a substantial discrepancy between the DFT and MP2/QCISD results, the latter lying quite close to each other. Density-functional theory prefers the carbodiimide over the cyanamide isomer, and the DFT problem is largely independent of the chemical nature of the ligand and *also* independent of the DFT parametrization, such that a system containing two double bonds is overestimated in stability by about 40 kJ/mol. There is a systematic error of the density-functional approach because DFT prefers a "homogenized" electron density, namely the two double bonds.

An alternative strategy for investigating the above trend focuses on molecules containing the structural units of mercury carbodiimide/cyanamide in the series of hypothetical oligomers $-[Hg{-}N{=}C{=}N{-}]_n$ and $-[Hg{-}N{-}C{\equiv}N]_n$ (see Figure 3.44(a)) in which the chains are terminated with amine groups.

Again, there is a systematic error evident from the data, and the discrepancy between DFT and post-Hartree–Fock approaches scales almost linearly with the chain length (b). Similar DFT failures have been reported for questions of catalysis [355] and for conjugated hydrocarbons [356]. It is all due to the fact that *no* DFT parametrization whatsoever treats different bond or-

19) None of the currently used local approximations (including gradient-dependent ones) for exchange and correlation fully cancels the unphysical electronic self-interaction, and this effect shows up more severely for anions than for cations [354].

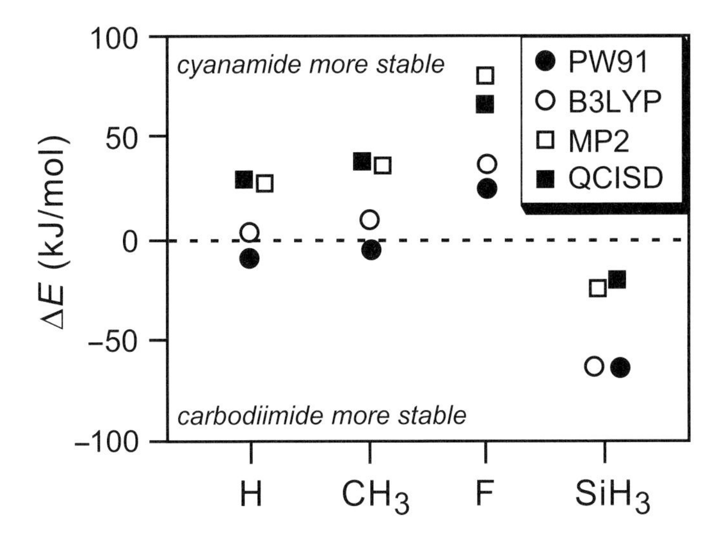

Fig. 3.43 The energy difference between the two molecules R–N=C=N–R and $R_2N–C{\equiv}N$ for various ligands, calculated by four different quantum-chemical methods. Marks below the dashed line indicate that the carbodiimide isomer is energetically preferred, while those above the dashed line favor the cyanamide unit.

ders with the same accuracy, and that is why the question of relative energetics of HgNCN(I) and HgNCN(II) *cannot* be answered by it. Because DFT gives almost the correct geometries, the error is likely to be overlooked, and these are certainly the worst computational errors that can occur. Thus, better exchange–correlation functionals are very much needed until a convincing

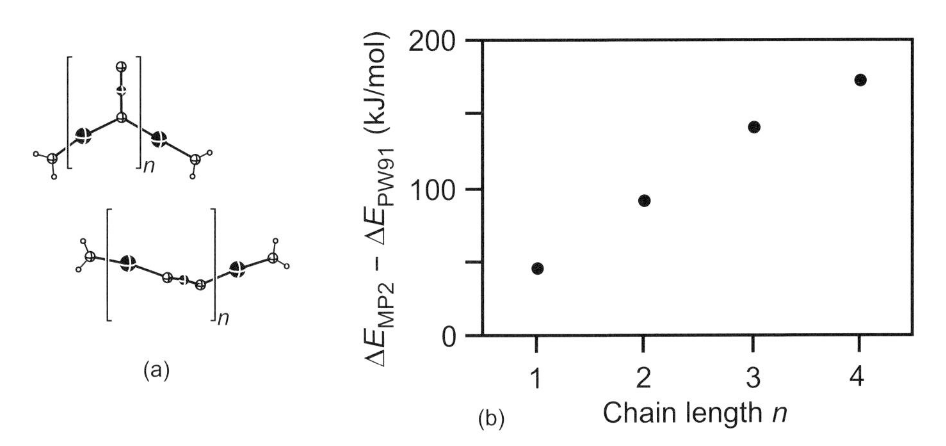

Fig. 3.44 Structures of mercury cyanamide and carbodiimide oligomers (a) resembling the –Hg–NCN– chains in the extended structures of HgNCN(I) and HgNCN(II). The energy difference between the PW91 functional and the MP2 method is plotted against the length of the oligomer chain (b).

quantitative theoretical assessment of the energetics of the two HgNCN polymorphs is possible by using DFT. Both phases, HgNCN(I) and HgNCN(II), might well serve as reference materials for that goal.[20]

Finally, let us not forget that there is a single, simple frame of understanding which *correctly*, albeit qualitatively, predicts the energetic order of mercury carbodiimide and mercury cyanamide, and it is based on chemical ideas, namely Pearson's concept of hard and soft acids and bases: soft prefers soft and hard prefers hard. The absolute softness of Hg^{2+} (7.7 eV, see Table 2.3) lies between those of Pb^{2+} and Ag^{+} such that a cyanamide anion and not a carbodiimide anion will be the preferred bonding partner for Hg^{2+}, just as for Pb^{2+} and Ag^{+}. I cannot refrain from noting that I find this quite remarkable.

3.10 Quasi-binary Oxynitrides: TaON and $CoO_{1-x}N_x$

We have started the applications part with a fundamental oxide (CaO) and continued with transition-metal nitrides and other N-based solids, and now is the right time to merge these two classes of solid-state compound. Metal *oxynitrides* may naively be considered as being simple combinations of oxides and nitrides but some of them exhibit fascinating chemical and physical properties which are not seen for oxides/nitrides alone, and therefore oxynitrides allow for new applications. In particular, the possibility of *designing* the material properties by cleverly varying the anionic (instead of cationic) constituents is both exciting and challenging for the synthetic chemist.

Thus, technological uses for oxynitrides have been identified in the fields of ionic conductivity, for various catalytic applications [357] and, in particular, in pigment production, thereby replacing traditional (and also toxic) cadmium-containing materials [358]. Simply stated, an oxynitride's color is controlled by changing the N:O ratio and, in order to maintain electroneutrality, three oxygens must be replaced by two nitrogens ($3\,O^{2-} \mathrel{\widehat{=}} 2\,N^{3-}$), thereby generating anionic vacancies. Alternatively, the increased anionic charge on nitrogen insertion may be compensated for by a cation with a higher oxidation number ($O^{2-} + M^{n+} \mathrel{\widehat{=}} N^{3-} + M^{(n+1)+}$) [359], and this approach gives access to a colored "Ta_3N_5" which is, in fact, $Ta_{3-x}Zr_xN_{5-x}O_x$. The color depends on the band gap of the solid, and the latter is a function of ionicity/covalency. Unfortunately, oxynitrides are not only more difficult to make, they are also much more difficult to structurally *characterize* because oxygen and nitrogen are al-

20) The false treatment of different bond orders is not the only possible source of systematic DFT errors. Another serious issue is the treatment of dispersion energies and forces, appearing in the calculation of sheet-like structures. Since the structures of HgNCN(I) and HgNCN(II) can be considered as one-dimensional packings of infinite chains, the interchain dispersion forces will also influence the relative energetics of the systems.

most indistinguishable from each other using X-ray diffraction because of the very similar scattering power of O and N. The question of the spatial distribution of the anions then calls for either neutron diffraction or, alternatively, electronic-structure theory.

Let us now see how density-functional theory performs for *the* classical oxynitride phase of the transition metals: tantalum oxynitride, TaON, with pentavalent Ta. The very first investigations appeared in the mid 20th century when two polymorphs were published by two competing groups [360, 361] almost coincidently. Both TaON polymorphs exhibit a 1:1 stoichiometrically precise O:N ratio. The first phase, β-TaON, is found as an intermediate in the ammonolysis of Ta_2O_5 but it can also be made by carefully oxidizing Ta_3N_5. β-TaON crystallizes in the monoclinic system, isostructural with baddeleyite (ZrO_2), and the pentavalent Ta experiences seven-fold coordination by three oxygens (1.99–2.15 Å) and four nitrogens (2.07–2.15 Å). A structural sketch is shown in Figure 3.45(a). The presence of a precise ordering of the O and N atoms over all positions was confirmed – a very rare experimental case – using neutron diffraction, later supported by theoretical calculations [362]. Only recently, however, it was discovered that β-TaON is yellow [363], and the green color often described in the context of β-TaON goes back to small contaminations with black niobium oxynitride, NbON. How fortunate that there are no accidental contaminations whenever theory is involved!

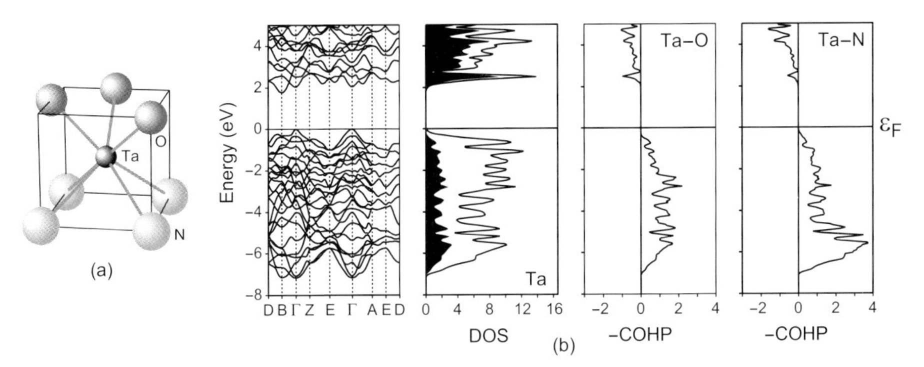

Fig. 3.45 The mono-capped TaO_3N_4 prism within the β-TaON structure (a) and the corresponding band structure, DOS, and Ta–N and Ta–O COHP analysis of β-TaON (b) based on TB-LMTO-ASA calculations and the LDA.

When performing density-functional calculations on β-TaON using plane waves and the GGA, optimizing the whole structure yields theoretical structural parameters (VASP) that are almost *indistinguishable* from experiment. The remaining differences between experimental and theoretical spatial parameters fall into the experimental uncertainty range, and the molar volume is overestimated (GGA) by only 1.5%. Also, the ordered anionic arrangement is significantly more stable than all unordered atomic arrangements. Luckily, even the experimental (direct) band gap of 2.4 eV is almost reproduced by the GGA (2.1 eV) although we stress that DFT should not be expected to give the correct result which is an excited-state property.

The chemical bonding analysis of β-TaON is based on a TB-LMTO-ASA calculation by means of the LDA because the latter is very close to the GGA; the results are plotted in Figure 3.45(b). Because of the low crystal symmetry, the band structure is too spaghetti-like for a detailed analysis but, in chemical terms, the stability of β-TaON clearly goes back to perfectly optimized Ta–O and Ta–N bonding up to the top of the valence band, and the COHPs also reveal the expected result that Ta–O bonding is more ionic than Ta–N bonding. First, the Ta–O dispersion is slightly more narrow and, second, the integrated COHP per Ta–O pair is considerably smaller than that of the Ta–N pair. The LDA band structure also reflects a smaller, indirect LDA band gap of 1.8 eV whereas the direct LDA band gap is 2.0 eV. Summarizing, β-TaON is an easy case for density-functional theory. This is not too surprising because the metal atom has no leftover valence electrons, but a noble-gas electron configuration, such that difficulties due to electron correlation on the metal atom should not arise.

The other polymorph, α-TaON, was synthesized by thermally decomposing NH_4TaCl_6 to give Ta_2N_3Cl which, on hydrolysis, results in red α-TaON, but the synthesis has been questioned [364]. The structure of α-TaON is hexagonal, and single-crystal X-ray data reveal space group $P6/mmm$ and the lattice parameters $a = 7.31$ Å and $c = 4.04$ Å. The Ta atoms are found on the Wyckoff positions $1a$ and $2c$, and O and N atoms reside at positions $1b$, $2d$, and $3f$. Because of the X-ray technique, the precise distribution of O and N atoms over the three latter sites (dubbed A1, A2, and A3) could not be given. Nonetheless, in this structure (see Figure 3.46(a)) the Ta2 atom is coordinated by five anions in a trigonal bipyramidal arrangement, with bond lengths of 2.02 and 2.11 Å, quite similar to those found in β-TaON. The other Ta1 atom exhibits a two-fold 2.02-Å-coordination by only two N/O atoms but the coordination sphere of Ta1 is probably saturated by six second-nearest N/O atoms at a rather large 3.66 Å distance.

Total-energy calculations (plane waves, pseudopotentials, GGA, VASP) allow the optimization of all lattice parameters and atomic positions. In addition, the energetically most favorable *ordered* arrangement of the N and O atoms over the three anionic sites is easily found. The calculated molar volume of α-TaON, however, turns out to be 20% smaller than the experimentally reported one, a real structural collapse! This DFT finding might have also been derived, although much more cheaply, from the volume increments of Ta^{5+}, O^{2-}, and N^{3-}, and the predicted 30 cm^3/mol is also 25% smaller than the reported volume for α-TaON (37.5 cm^3/mol). It is immediately clear that the

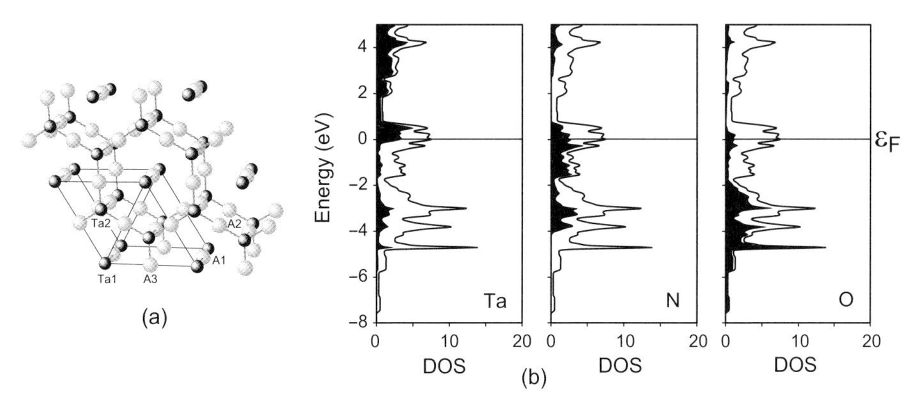

Fig. 3.46 Crystal structure proposal for α-TaON derived from single crystals (a) and the corresponding DOS of α-TaON (b), including Ta, N and O projections, on the basis of FLAPW calculations and the GGA.

difference between experiment and theory cannot be explained by theoretical deficiencies, in particular taking into account the superb DFT performance for β-TaON. Also, a theoretical density-of-states for the experimentally reported structure should yield a sizeable band gap characteristic for an electrical insulator but the theoretical DOS, calculated using the most accurate FLAPW-GGA method (WIEN2k), reveals a *metallic* material. The DOS is shown in Figure 3.46(b). Thus, the reported structure *cannot be correct* for the chemical composition TaON.

Obviously, theory does not support the existence of α-TaON. If it existed, it would lie higher in energy by 314 kJ/mol when compared with the well-established β-TaON. In fact, the successful literature search for the original X-ray data of α-TaON eventually supports the theoretical falsification of α-TaON existence [365]. We note that the falsity of this structure has not been spotted in the crystallographic databases for almost four decades, but electronic-structure calculations from first principles easily detect the problem. Independent structural checks, *without* having to re-do the synthesis, are now possible. Let us move on to another example.

From the perspectives of chemical bonding and also synthetic techniques, transition-metal oxynitrides *without* noble-gas electron configurations are even more interesting, and also more challenging to make. Thus, the remarkable "organometallic" strategy used to synthesize the first ever reported quasi-binary oxynitride of one of the ferromagnetic transition metals (Fe, Co, Ni) [366] proceeds through the thermolysis of molecular $Co(CO)_3(NO)$ and results in thin films of black, nanocrystalline, and electrically conducting oxynitrides of composition $CoO_{1-x}N_x$ with a broad phase width where the crystalline portion corresponds to $x \approx 0.5$. X-ray investigations of $CoO_{0.5}N_{0.5}$ suggest a cubic lattice (a = 4.415 Å) which is between those of the zinc-blende-like polymorphs of CoO (4.544 Å) and CoN (4.28 Å), and the Rietveld refinement of five broad reflections is in accordance with this lattice type and space group $F\bar{4}3m$.

Indeed, the two *binary* phases CoO and CoN were reported to exist both in the zinc-blende as well as in the rock-salt structure types [367] (see again Figure 1.1). The CoO ground state is an antiferromagnetic Mott insulator with [NaCl] structure [368] whereas cobalt nitride adopts the [ZnS] type [266] (see Section 3.2), and in the latter structure, CoN does not exhibit local magnetic moments.[21] Because of these variations for CoO and CoN, one might well ask whether the experimentally suggested [ZnS] structure of $CoO_{0.5}N_{0.5}$ is indeed the energetically most stable one, and other structural and physical properties also need to be clarified, in particular with respect to magnetism and thermal stability of these $CoO_{1-x}N_x$ materials.

The preference of the reported structure type is easily explained, and fast total-energy calculations using minimum unit cells containing only four Co, two O, and two N atoms, all with idealized high-symmetry positions, are accurate enough for this purpose. Whether these are based on all-electron or pseudopotential approaches is unimportant, and both LDA and GGA parametrizations perform quite satisfactorily. A corresponding overview of the course of the relative energies as a function of the cell volumes for the [ZnS]- and [NaCl]-type structures is given in Figure 3.47(a).

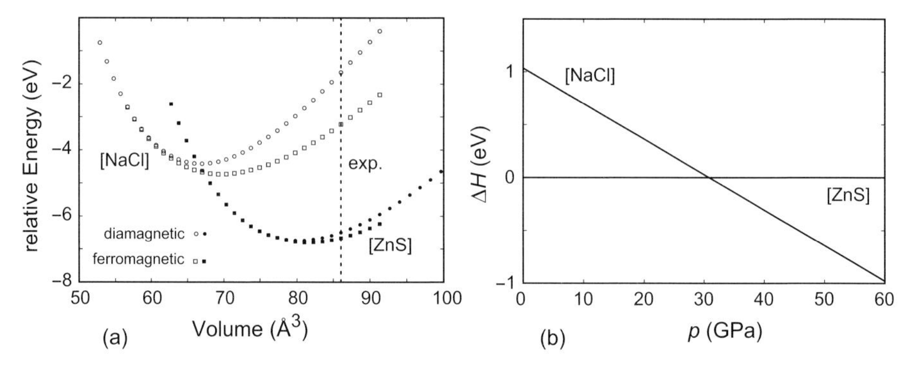

Fig. 3.47 Theoretical energy–volume diagram (a) of the rock-salt and zinc-blende type phases of $CoO_{0.5}N_{0.5}$ and the theoretical enthalpy–pressure diagram (b).

Theory unambiguously demonstrates that a four-coordinate Co (zinc-blende and also wurtzite type which is similar in energy but not shown in the figure) is clearly preferred to a six-coordinate Co (rock-salt type) by about 50 kJ/mol. Interestingly, theoretical volumes obtained using either the LDA or

21) Although CoN had been reported to be "paramagnetic" (no magnetic moment specified) initially [266], all DFT calculations indicate that CoN can only be either diamagnetic or Pauli paramagnetic. Indeed, pure Pauli paramagnetism (zero local moments) has been confirmed by independent NMR investigations for zinc-blende CoN [369].

GGA functional *both* underestimate the experimentally reported lattice volumes! Because numerical experience says that lattice parameters (volumes) are typically underestimated using the LDA but overestimated using the GGA, this may indicate that the investigated sample was probably richer in O content than the idealized composition $CoO_{0.5}N_{0.5}$. Recall that CoO is also slightly larger than CoN. In any case, zinc-blende $CoO_{0.5}N_{0.5}$ should be a metallic conductor due to a significant DOS at the Fermi level (not shown); this is in agreement with experiment. The minute preference for spin-polarization is a computational artifact due to nonoptimized atomic positions (see below).

For reasons of convenience, the energy–volume diagram is better converted into an enthalpy–pressure diagram, and the method of doing this has been explained in Section 2.18.2. Although the important thermodynamic variable is the Gibbs energy G, the entropic part can be safely neglected for the solid-state compounds. The resulting ΔH vs. p diagram is given in Figure 3.47(b), and it indicates that a pressure-induced phase transition from the zinc-blende to the rock-salt type is expected to occur at about 30 GPa.

In order to understand *why* the zinc-blende structure of $CoO_{0.5}N_{0.5}$ is preferred above the rock-salt structure, the electronic structures of both polymorphs should be investigated in more detail. Fortunately, we no longer have to perform the analysis, simply because $CoO_{0.5}N_{0.5}$ (with $9 + 0.5 \times 6 + 0.5 \times 5 = 14.5$ valence electrons) electronically very closely resembles CoN (14 valence electrons) and hypothetical NiN (15 electrons) such that the explanation for the structural change in the transition-metal *nitrides* (see Section 3.2) also holds for the cobalt oxynitride! Besides bonding Co–N and Co–O interactions over the entire valence-band region, high electron counts such as the one in $CoO_{0.5}N_{0.5}$ induce antibonding Co–Co interactions in the proximity of the Fermi level, and these unfavorable interactions become weaker in the zinc-blende type because it allows for wider Co–Co distances (3.06 Å) than in the rock-salt structure (2.90 Å). The rigid-band model is still valid.

The questions of absolute stability, phase width, magnetism, and anionic order of the $CoO_{1-x}N_x$ *range* of compounds need a larger computational effort. To fully scan the compositional range, a $2 \times 2 \times 2$ enlarged supercell (see Section 2.18) of the zinc-blende structure containing a total of 64 atoms is required. Of these, 32 are Co and 32 either N or O. The difference in internal energy ΔE of the ternary phase $CoO_{1-x}N_x$ in comparison with the binary phases of CoN and CoO is calculated according to

$$\Delta E = E_{CoO_{1-x}N_x} - ((1-x)E_{CoO} + xE_{CoN}), \qquad (3.3)$$

and it equals ΔH at zero pressure and temperature (see Section 2.18.2). We also emphasize the paramount importance of an accurate energetic characterization of the two binary boundary phases, CoO and CoN. Unfortunately, ordinary density-functional theory has reached its present limits. For the case of CoN, the Pauli paramagnetic [ZnS]-type ground-state phase can be satisfactorily described by a non-spin-polarized DFT calculation. For CoO, there are two polymorphs to consider: CoO may likewise adopt a (metastable) [ZnS]-type structure with an antiferromagnetically ordered ground state which can also be modeled by a spin-polarized GGA approach. Energetically even lower, however, lies the [NaCl]-type structure of CoO, an antiferromagnetically ordered and highly *correlated* Mott insulator. Here, standard DFT fails and does *not* give the correct energetic positioning (see Section 2.12). One may introduce an empirical Hubbard U correction for extra-intraatomic electron correlation (see Section 2.13) but the size of U is difficult to estimate. Another, more reliable way to circumvent the problem is simply *not* to calculate the energy of rock-salt CoO but to take advantage of the experimental, calorimetric result [370] showing that rock-salt CoO is more stable than the zinc-blende polymorph by approximately 0.38 ± 0.10 eV. Thus, both CoO boundary phases can be included in the calculation for the internal energy (or enthalpy) difference because essential pieces of information are provided by *experiment* in order to make the theory work.

When the set of oxynitrides $CoO_{1-x}N_x$ with varying x is calculated using either non-spin-polarized or spin-polarized approaches, including full relaxation of the cell and atomic site parameters, the energy differences between the nonmagnetic and magnetic calculations vanish because the atomic spin moments also vanish on the Co atoms. Upon full structural optimization for the given four-fold coordination, there is obviously no need to spin-polarize in order to lower the total energy through electronic-symmetry reduction, and structural optimization is already sufficient.[22] Although the structural principle of the [ZnS] type is still intact, not a single tetrahedron around any Co atom then remains regular, and one finds *spectra* of interatomic distances around 1.78–1.88 Å (Co–N) and 1.88–1.98 Å (Co–O), depending somewhat on the composition. The energetic result of the various anionic substitutions is presented in Figure 3.48. Negative/positive ΔE values indicate energetically stable/unstable ternary oxynitrides and, for zero pressure and temperature, these energy differences may be interpreted as differences in enthalpy, ΔH.

The upper curve refers to the two most stable binary phases (rock-salt CoO, zinc-blende CoN) and demonstrates that any compound belonging to the phase range $CoO_{1-x}N_x$ is enthalpically *unstable*. Thus, $CoO_{1-x}N_x$ should have never been observed or isolated due to the high stability of the competing rock-salt CoO! The lower curve of Figure 3.48, however, makes oxynitride formation immediately understandable since it stands for $CoO_{1-x}N_x$

22) If this qualitative argument is of general importance, one may predict that simple nonstoichiometric transition-metal oxynitrides are inferior candidates when it comes to the search for magnetic materials, simply because they possess more internal, structural degrees of freedom that will suppress spin-polarization.

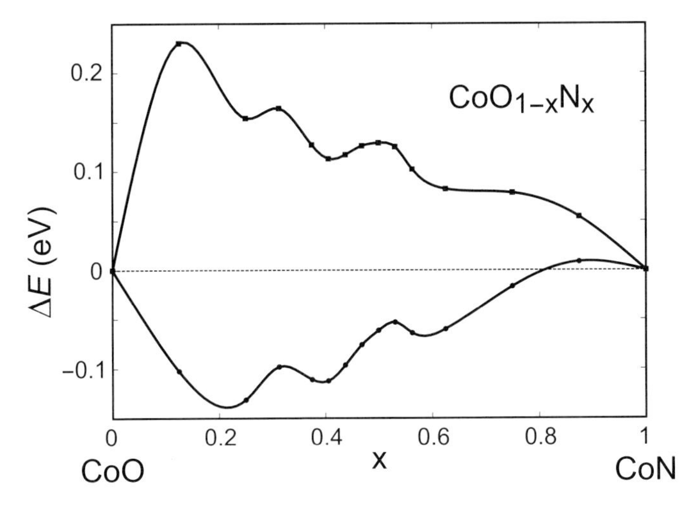

Fig. 3.48 Theoretical ΔE–x (or ΔH–x) diagram calculated for various $CoO_{1-x}N_x$ compositions. In the upper curve, the binary reference phases are taken as rock-salt CoO and zinc-blende CoN. In the lower curve, the zinc-blende-type phases of both CoO and CoN were used.

phase formation competing with *metastable* zinc-blende CoO and zinc-blende CoN. It further suggests that a broad range of metastable cobalt oxynitrides ($0.1 \leq x \leq 0.7$, that is, $CoO_{0.9}N_{0.1}$–$CoO_{0.3}N_{0.7}$) must be obtainable, with a maximum stabilization for the more oxygen-rich nonstoichiometric phases. The energetic values are admittedly small but they nevertheless seem reliable because no bond breaking/formation is involved on comparison with the binary zinc-blende phases. The energetic stabilization merely indicates the optimization of all atomic site parameters for Co, N, and O in the ternaries. The three local regions of stability in the lower curve, however, are probably due to the finite size of the computational model.

Summarizing, $CoO_{1-x}N_x$ oxynitride formation appears to be a kinetically controlled phenomenon [371], in accordance with the reported experimental observation of its thermal sensitivity [366]. The phases decompose into rock-salt CoO, metallic Co, and molecular nitrogen at higher temperatures. The broad phase width is also in accordance with theoretical calculations. Further entropy effects will additionally favor, but only very weakly, the formation of ternary phases due to substitutional entropy (O/N). In addition, there is no indication of anionic ordering in the calculations, and a random distribution of N and O is clearly preferred.

3.11 Into the Void: The Sn/Zn System

Being aware of the preceding theoretical successes, let us try to be extremely brave and support the experimentalists in a case where virtually nothing can be immediately used for electronic-structure theory, at least with respect to structural hints. For better or worse, the theorist's need to jump into a scientific problem is backed by enormous fundamental and economic reasons.

As any material scientist knows, corrosion is a *huge* problem for any advanced society. In Germany alone, corrosion costs of the order of € 50 billion each year, and the refinement, upgrading, and processing of metal surfaces is of absolute paramount importance. Since almost everything has a surface, there are lots of surfaces which have to be made more resistant against corrosion and abrasion, which need to become harder, more conducting, optically more appealing, magnetically more susceptible and so forth. At the beginning of the 21st century, corrosion protection needs also to be environmentally friendly, and one promising candidate for refining steel surfaces is given by the electrochemical deposition technique of Sn/Zn alloys. It turns out that this binary system owns a rich tradition in basic and applied research [372].

The complete thermodynamic assessment of the binary system Sn/Zn according to the CALPHAD technique – the calculation of phase diagrams from thermodynamic properties [373] – was performed only a few years ago [374] for standard conditions. These results are presented in Figure 3.49(a), based on thermodynamic data which use the standard element reference state. Here, only the liquid phase and the solubility of Sn in pure hexagonal Zn and also of Zn in pure body-centered tetragonal β-Sn (bct-Sn) was analyzed, and this is because there is *no experimental evidence* for *stable solid phases* in the system, either as an intermetallic compound or as intermetallic solutions with wider homogeneity ranges, even at high pressures [375]. Seemingly, Sn and Zn atoms cannot "get along" with each other on the atomic scale. The calculated phase diagram is in good agreement with all experimental reports [376], the eutectic point being at 199 °C and a liquid composition of 85.1 atomic percent tin. For completeness, we mention that a single *metastable* phase has been described in the Sn/Zn system, prepared by quenching experiments [377], but it already falls apart to form Sn and Zn at room temperature, as expected.

Given that the Gibbs energies G of all phases are known for all temperatures as a function of the composition, a Gibbs energy minimization yields the system's stable state. Neglecting pressure-dependent contributions, G of a solution phase is calculated by considering the contributions due to: a) the mechanical mixture of the phase constituents; b) the configurational entropy; and c) the excess Gibbs energy. If the liquid, Zn and β-Sn are all treated as substitutional solutions (that is, assuming a lattice with only one type of atomic

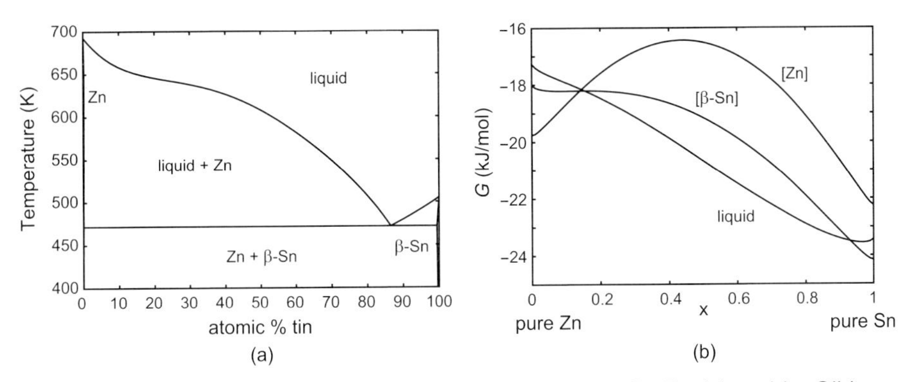

Fig. 3.49 Calculated phase diagram of the binary system Sn/Zn (a) and its Gibbs energy (b) at p = 1 bar and T = 180 °C for the liquid and the elemental structure types with random atomic occupation.

site on which Sn/Zn atoms may randomly mix) and if we known the energy of Sn in the Zn structure and also of Zn in the β-Sn structure, the calculation of minimal Gibbs energy may be performed, for example, for $p = 1$ bar and $T =$ 180 °C, a condition close to the eutectic point. The resulting diagram, largely based on experimental data, is shown in Figure 3.49(b).

First, there is only a *very* small mutual solubility of the elements and, second, one finds a strong tendency to de-mix even if only one crystal structure is considered, reflected by the "camel back" curves for each of the solid phases. In other words, even if there were no structural differences between Sn and Zn, they would immediately de-mix. The "camel back" is also observed for the liquid phase (!) such that Sn and Zn behave in a strongly xenophobic way to each other. In classical terms, Sn and Zn de-mix because of the very *positive* mixing enthalpies for each of the phases. These enthalpies ΔH (with reference to the respective pure solid elements) are shown in Figure 3.50(a), and all three curves ΔH vs. x were calculated from the above Gibbs energy data by taking the appropriate temperature derivatives.

Let us now find out whether these classical enthalpies may be reproduced by electronic-structure calculations (VASP) on Sn/Zn supercells using ultrasoft pseudopotentials, plane-wave basis sets and the GGA. We therefore have to theoretically determine the total energies of all crystal structure types under consideration (α-Sn, β-Sn, Zn) as a function of the composition Sn_xZn_{1-x} by a variation of the available atomic sites in terms of Sn and Zn occupation, just as for the preceding oxynitrides ($CoO_{1-x}N_x$). In the present case, supercells with a total of 16 atoms were generated, and nine different compositions per structure were numerically evaluated. Because this amounts to a significant computational task, the use of pseudopotentials is mandatory, and this also allows the rapid calculation of interatomic forces and stresses for structural

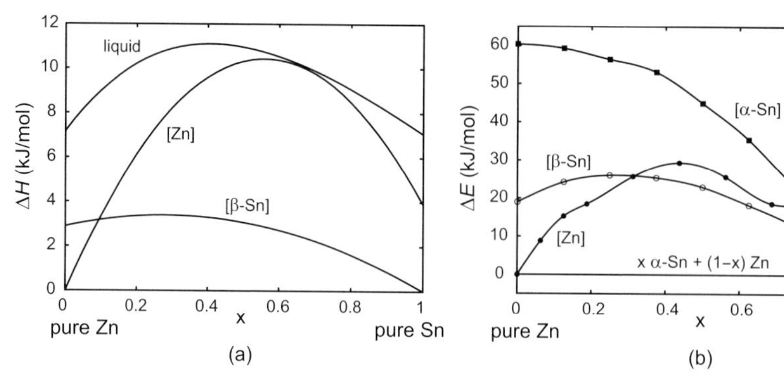

Fig. 3.50 Enthalpy diagram of the binary system Sn/Zn (a) in the liquid and in the elemental structure types with random occupation and Zn and β-Sn as reference states. Relative energies of Sn/Zn solid solutions adopting the elemental structures according to density-functional (GGA) total-energy calculations (b).

optimization. A plot of the relative energies is given in Figure 3.50(b), to be compared with the classical energies in (a).

Indeed, there is *qualitative* agreement between the classical enthalpy curves (a) and the quantum-mechanical energies (b). Whenever Sn is introduced into Zn, the energy increases and the system is more unfavorable. The same is true when Zn is going into the β-Sn structure, and it is even more unfavorable when Zn should adopt the diamond structure (α-Sn). In other words, all mixing energy values are clearly positive, reflecting that zinc and tin prefer not to form intermetallic solution phases with these structure types, largely independent of composition. Quantum-mechanically, Sn and Zn are xenophobic, too. Quantitatively, there are discrepancies. While it is difficult to say, reliably, where these really come from, probably theory (the exchange–correlation functional) is to blame, but this is just a cautious guess. Experience also shows that such comparisons between classical and quantum-mechanical results turn out much better if the systems under study *do* exhibit mixture on the atomic scale.

Let us continue to put more trust in the theory and use it for *narrowing design space* to help the experimentalist in trying to make new phases. Here, quantum chemistry is clearly needed because classical approaches do not allow a proper insight with respect to chemical bonding. We first assume the simple composition Sn_2Zn because it is not too far from the eutectic point. The next question applies to the total energies of Sn_2Zn if the latter hypothetical composition adopts a variety of known structure types. Thus, 30 different AB_2 structure types taken from some crystallographic tabulation [378] are numerically checked by total-energy calculations, and 23 of these structures are typically adopted by "ionic", namely [SiO_2], [$HgBr_2$], etc., and 7 by "intermetallic", namely [$InNi_2$], [$MgCu_2$], etc., compounds. Since the sizes of the Sn and

Zn atoms in these structures are unknown at the outset, full relaxation of all atomic positions and optimization of the cell geometries must be performed. When this is done, *none* of the investigated AB_2 structure types for Sn_2Zn is energetically lower than the pure elements – in accordance with experiment – because there is no such Sn_2Zn. Strangely, the "intermetallic" structure types are energetically most unlikely because of strong Sn–Zn antibonding, and the "ionic" structure types perform much better (that does not mean that the Zn $\rightarrow$ Sn charge transfer must be high because one would predict only 0.01 electrons transferred by using Pearson's estimate in Equation (2.101)). In particular, one structure type, [CaF_2], lies quite close in energy to the elements. After another quick LDA calculation with short-ranged basis functions (TB-LMTO-ASA), the chemical bonding of Sn_2Zn in this structure type may be observed in Figure 3.51.

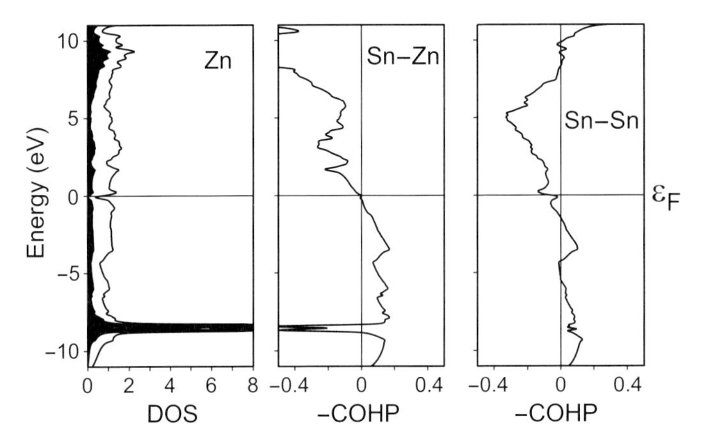

Fig. 3.51 Total density-of-states (DOS), partial DOS of Zn (in black), and COHP analysis for the Sn–Zn and Sn–Sn bonding in the calcium fluorite structure type of hypothetical Sn_2Zn.

For a [CaF_2]-like Sn_2Zn, the DOS is fairly low over the entire energy range, especially at the Fermi level, thereby reflecting its relative stability. It seems as if a band gap were almost opening up at the Fermi level. The chemical bonding (COHP curve) exhibits strongly attractive Sn–Zn interactions between -8 and -5 eV, arising mostly from Sn 5*s* and Zn 4*s* orbitals. Above that, the Sn–Zn bonding interactions are due to Sn 5*p* and Zn 4*p* orbitals. In addition, the Sn–Zn interactions are optimized over the valence band, saturating all bonding states, because ε_F stands for the crossover from bonding to antibonding. The COHP curve of the Sn–Sn combination, on the other side, reflects slightly weaker interactions because of the larger bond distances, and only the 5*p*–5*p* combination contributes to bonding. There is also a small antibonding, Sn–Sn-repelling region visible at the Fermi level. Finally, the Zn–Zn interactions (not shown) are quite weak due to the 4.67 Å wide Zn–Zn distance and thus prove irrelevant for the discussion of structural stability.

Even though Sn_2Zn is unstable, its bonding properties (the Sn–Sn part) show that the electronic structure needs only a little adjustment to make it stable. To remove the antibonding Sn–Sn interactions at ε_F, the atoms on the Sn sites should provide fewer valence electrons, and this can be accomplished by a Sn-neighboring element to the left in the periodic table. In order not to distort the lattice to strongly, it should be about the same size; also, a slightly lowered electronegativity would probably be advantageous because all "ionic" structures are preferred. One would therefore decide, as a *qualitative* chemical guess, to replace part of the tin substructure by indium atoms because In has one electron less than Sn, is larger by only 0.1 Å and is slightly more electropositive. If synthetic chemists tried to make a Sn_2Zn-derived phase, $Sn_{2-x}In_xZn$ (Sn_3InZn_2?, Sn_5InZn_3?) would be a reasonable starting point. This is not a great deal of information, agreed, but it is better than in desperation to dope the whole periodic table into the Sn_2Zn structure without *any* qualitative signpost. This is what we mean by saying that electronic-structure theory can *narrow* the design space for new compounds, which are yet to be synthesized [379].

It transpires that atomic substitution is the only possibility to enforce compound formation because alternative high-pressure syntheses of binary Sn/Zn phases are very likely to fail, which is easily shown from computer experiments. For example, let us again take the [CaF_2]-type of Sn_2Zn plus some other AB_2 structures and also, as a reference, the elemental structures. Upon quantum-chemically varying the volumes of all these systems, we theoretically study α-Sn, the high-pressure β phase of Sn, γ-Sn (another high-pressure allotrope), and the various Sn_2Zn structures. The final high-pressure diagram, given in Figure 3.52(a), was derived from calculating the energy E of each structure type as a function of the volume V.

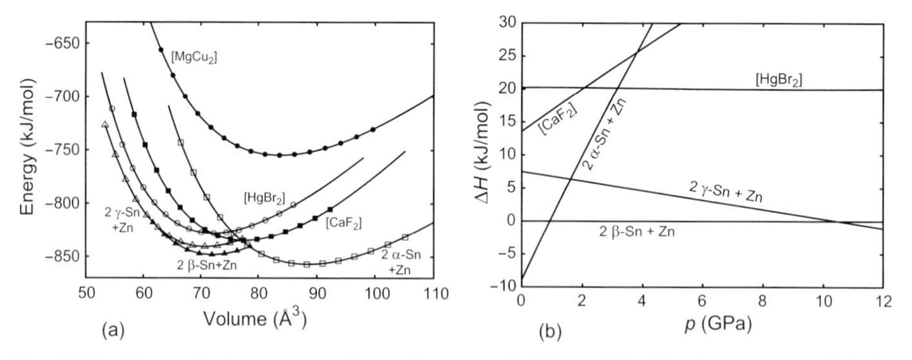

Fig. 3.52 Theoretical energy–volume diagram of different Sn_2Zn structures and their elemental constituents according to total-energy calculations (a) and the theoretical relative enthalpy–pressure diagram of Sn_2Zn for various structures (b).

The diagram states that a 2:1 mechanical mixture of α-Sn and Zn is lower in energy than any other option included in the figure. Although these structural alternatives are more densely packed, they are less stable. The transition pressures into the energetically higher-lying structures arrive as the gradients of common tangents between two curves but, for reasons of convenience, we also convert the data into an enthalpy–pressure diagram (see Section 2.18.2). It is presented in Figure 3.52(b).

A 2:1 mechanical α-Sn/Zn mixture is most stable at standard conditions but this changes with increasing pressure, favoring a Sn_2Zn phase of the [CaF_2]-type at approximately 4 GPa and of the [$HgBr_2$]-type at 3 GPa. Nonetheless, neither of the two are thermodynamically stable high-pressure Sn_2Zn compounds because of a phase change of tin! Theory predicts that α-Sn transforms into β-Sn at a pressure of about 0.9 GPa (experimental value: ca. 1 GPa [380]) and β-Sn transforms into γ-Sn at approx. 10.5 GPa (experiment: 10.0 ± 0.6 GPa [381]). Then, an established rule of crystal chemistry is fulfilled because the latter two phase transitions of Sn ($\alpha \rightarrow \beta \rightarrow \gamma$) are accompanied by an increase in the coordination numbers ($4 \rightarrow 4+2 \rightarrow 8$) and by an increase in interatomic distances (2.81 Å $\rightarrow$ 3.02/3.19 Å $\rightarrow$ 3.11 Å) and, trivially, by a decrease in volume.

The high accuracy of the predicted phase transitions of Sn supports the reliability of the quantum-mechanical method, despite the discrepancies between experimental enthalpies and theoretical energies upon mixing. The above example of theoretically *excluding* high-pressure binary phases is another sign for the theory being able to narrow the synthetic design space, and experimentalists are well advised *not* to waste time and effort in making such high-pressure phases. On the contrary, atomic substitution ($Sn_{2-x}In_xZn$) is one way to go, following theoretical pathways. Because syntheses of ternaries should be tried, the synthetic solid-state chemist should feel challenged but, at the same time, very much in control.

3.12 Predicting Oxynitrides: High-pressure Phases and VON

Despite being unable theoretically to detect new binary Sn/Zn intermetallic compounds in the preceding section, the astonishing numerical accuracy for the high-pressure phases of Sn by means of total-energy GGA calculations poses another question. Can one predict unknown high-pressure phases of other well-characterized *binary* compounds? For example, are there high-pressure polymorphs of, say, tantalum oxynitride, TaON? The theory of its ground-state baddeleyite structure, β-TaON, was covered in Section 3.10, and we recall that the Ta atom is perfectly bonded to seven O/N neighbors. Since the chemistry of TaON and NbON is very similar, what can also be said about possible high-pressure phases of NbON? What happens if TaON and NbON are pressurized?

These questions are easily answered by theoretically studying the energies of a large number of ABC structures for TaON and also NbON in which the metal has a *larger* coordination number than seven. We reiterate that high-pressure phases usually accompany larger coordination numbers. Then, the structure optimizations by total-energy calculations are carried out as noted before (pseudopotentials, plane waves, GGA, VASP), and the resulting energy–volume as well as enthalpy–pressure diagrams are shown in Figure 3.53.

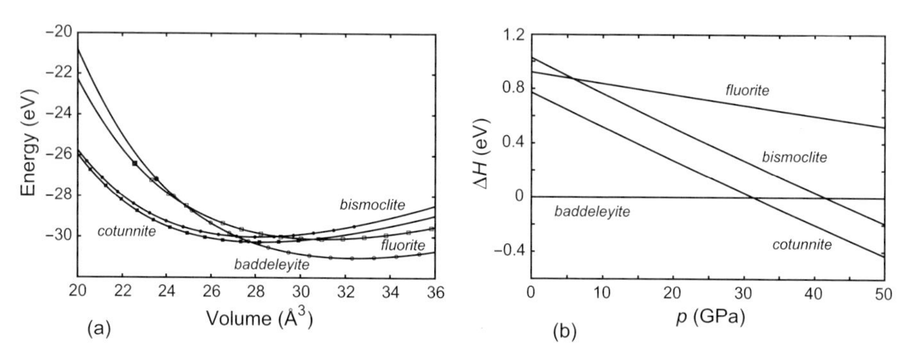

Fig. 3.53 Theoretical energy–volume (a) and enthalpy–pressure (b) diagrams for various structures of TaON on the basis of pseudopotential GGA calculations.

Under standard conditions (zero pressure), TaON in the baddeleyite structure has the lowest energy, as expected; the findings of Section 3.10 are therefore further supported. At higher energies, however, there are at least three other structures types (bismoclite, cotunnite, fluorite [5]) which acquire *smaller* molar volumes and are therefore accessible by high-pressure experiments. Indeed, the enthalpy–pressure diagram (Figure 3.53(b)) reflects that the high-pressure structures become enthalpically favored with increasing p, and a cotunnite-like TaON is predicted as the stable polymorph for pressures above 31 GPa [382]. This structure is characterized by a nine-fold coordination of the tantalum atom, and four oxygen and five nitrogen atoms form a trigonal prism, capped on all rectangular faces, such that there are Ta–O/N distances between 2.04 and 2.33 Å. The local Ta environment is given in Figure 3.54(a).

Figure 3.54(b) also offers the density-of-states and a closer look at the bonding situation. Cotunnite-like TaON proves to be a semiconductor with a very small bandgap of only 0.15 eV (the latter has almost vanished in the DOS figure but only because of the artificial smoothing of the curve). The chemical bonding looks fairly normal because all Ta–O and Ta–N interactions are strong and perfectly optimized up to the Fermi level. We also emphasize the Ta–Ta

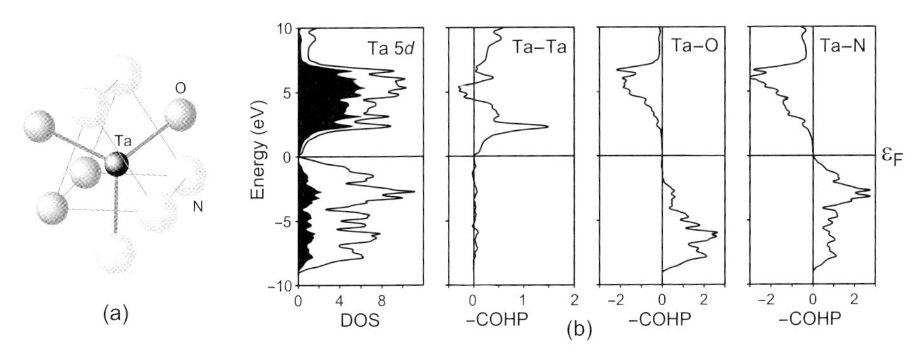

Fig. 3.54 The TaO_4N_5 structural fragment from the predicted cotunnite phase of TaON (a) and the corresponding DOS and also COHP analyses for the Ta–Ta/O/N bonds (b) by means of LMTO theory and the GGA.

interactions which are essentially nonbonding over the entire valence region, such that they do not take part in terms of a structural preference.

For NbON, the data (not shown for reasons of brevity) predict a very similar high-pressure behavior. This is not unexpected because both Ta^{5+} and Nb^{5+} have *identical* Shannon ionic radii (seven-fold coordination) of 0.69 Å, as given in Table 1.1. Beyond 27 GPa, NbON also adopts the cotunnite structure – as expected – but the density-of-states is nonzero at the Fermi level, such that cotunnite-like NbON will have metallic properties [382]. Similar cases of metalization for insulating (or semiconducting) materials under pressure have already been reported [383].

What about vanadium oxynitride, VON? Surprisingly, this daltonide phase has *not* been reported so far, although there has been serious research. The solid-state chemical literature contains a couple of nonstoichiometric vanadium oxynitrides $VO_{1-x}N_x$ [384] which all adopt the rock-salt type, and the electronic and magnetic properties of these have been determined [385]. For example, the nonstoichiometric phase $VO_{0.5}N_{0.5}$ is electrically conducting (with a weak temperature dependency) and paramagnetic with a local moment of about 0.25 μ_B per vanadium atom.[23] Once again, these experimental properties are easily checked on the basis of rock-salt-type supercell calculations. Given a face-centered cubic cell of vanadium atoms, the octahedral vacancies are alternately filled with oxygen and nitrogen atoms, such that the composition becomes the above $VO_{0.5}N_{0.5}$. In terms of structural properties, density-functional theory performs quite well. When based on the LDA, there is only a small 1.3% underestimation of the cubic lattice parameter (4.131 Å), and the GGA results in a slight overestimation of 0.8%. The spin-polarized GGA calculation also reproduces the experimental moment satisfactorily (ca. 0.3 instead of 0.25 μ_B) but the moment vanishes within the LDA. Such phe-

23) The small moment seems strange when thinking about the large effective moments of V^{2+} and V^{3+} (2.8 and 3.8 μ_B) but one needs to keep in mind the metallic nature of this phase.

nomena are well known, and the LDA often turns out less magnetically susceptible than the GGA.

In order to predict a *stoichiometrically precise* VON and the conditions of its synthesis, a moment of reflection is needed. With respect to binary oxides, only V_2O_5 (of which there are two polymorphs) contains vanadium in its highest (+5) oxidation state [386]. In addition, electronic-structure theory has been already used to predict the high-pressure phase V_3N_5 [387], also with pentavalent vanadium. A good estimate for the molar volume of a stoichiometric VON should therefore be based on the arithmetic average of those of V_2O_5 (54 cm^3/mol) and high-pressure V_3N_5 (44 cm^3/mol), namely 19.6 cm^3/mol. The empirical value derived from the tabulated volume increments (see Table 1.2) is a much larger 33 cm^3/mol but the N^{3-} increment is not very certain. The computational search for a structurally stable VON is then carried out on the basis of 20 known ABC structure types with smaller coordination numbers [378]. Since the effective ionic radii of V^{5+} for tetrahedral/octahedral coordinations are 0.355 and 0.54 Å (thereby significantly smaller than for Nb^{5+}/Ta^{5+} but between those of As^{5+} and Sb^{5+}), a lower coordination number is also expected for VON. Once again, atomic positions and lattice parameters are optimized to achieve lower energies for all structural alternatives.

Indeed, the quantum-chemical structure optimizations confirm the preference of V^{5+} for lower coordination numbers. Under the assumption of zero pressure, the lowest energy is found for the case of α-cristobalite (four-fold coordination, molar volume = 41.1 cm^3/mol), followed by quartz (four-fold, 33.2 cm^3/mol), followed by anatase (six-fold, 19.8 cm^3/mol), followed by baddeleyite (seven-fold, 16.6 cm^3/mol) and several others. In order to arrive at stable structures, however, a minimum thermochemical requirement is that the total energy of VON is lower than those of its educts. Here we rely on the plausible reaction

$$V_2O_5 + 3VN + N_2 \rightleftharpoons 5VON$$

and note that *none* of the above structures of VON fulfills this criterion by being exothermic. Under standard conditions, VON is energetically unstable and seemingly impossible to make. This must be the reason why VON has not been reported so far! We therefore continue with another set of calculations in order to see how the theoretical polymorphs respond when pressure is applied. Nothing spectacular results during these investigations except for the baddeleyite structure of VON. At extreme theoretical pressures, one finds two essentially unstable atomic configurations which do *not* have well-defined equilibrium volumes. One of these, very surprisingly, contains N_2 dimers with an N–N bond length of 1.34 Å. This value is between those of O=O (1.21 Å) and F–F (1.42 Å) but much larger than in the uncharged N≡N molecule (1.10 Å). Clearly, this numerical experiment indicates the presence of a pernitride N_2^{2-} species and an *internal redox competition* yielding trivalent vanadium according to $(V^{3+})_2(O^{2-})_2(N_2^{2-})$ compared to the simple $(V^{5+})(O^{2-})(N^{3-})$. At the same time, the vanadium coordination number increases to eight, and the larger V^{3+} cation is kept inside a doubly capped distorted trigonal prism (see Figure 3.55). In fact, similar pernitride anions have been experimentally reported for the cases of SrN_2 [388] and BaN_2 [389], also made through high-pressure syntheses!

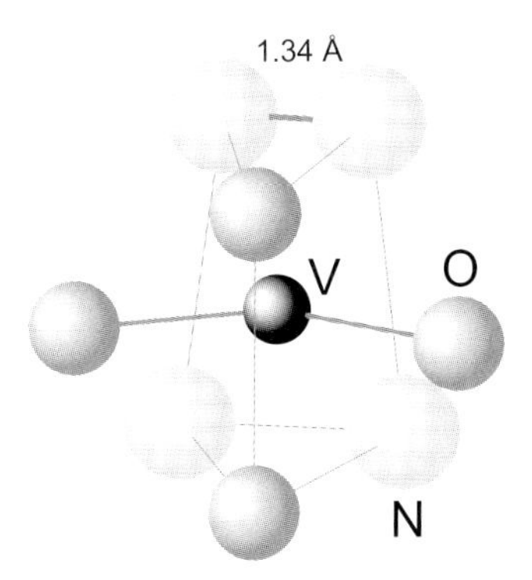

Fig. 3.55 Structural cut from a theoretically predicted, but essentially unstable (no equilibrium volume), high-pressure polymorph of VON with nitrogen dimers and trivalent vanadium; also see text.

With respect to the energetically low-lying polymorphs of VON, it is also important to note that those exhibiting four-fold coordinations *increase* in energy when pressure is applied. According to Le Chatelier's principle, high pressure makes them less likely. To nonetheless answer the question of a successful VON synthesis by using high pressure, all numerical information must be condensed into a theoretical enthalpy–pressure diagram where we have bravely replaced Gibbs energy differences ΔG by enthalpy differences ΔH and thereby neglected all entropic contributions arising from $\Delta G = \Delta H - T\Delta S$. This is rather problematic because it truly restricts the results to absolute zero temperature. Note that, at room temperature, the entropy of gaseous nitrogen stabilizes the educt side and strongly disfavors solid VON. The ΔH vs. p diagram is given in Figure 3.56(a).

At about 12 GPa, a pressure-induced synthesis of a baddeleyite-type VON is predicted. In accordance with the pressure–homologue rule, VON will adapt the ground-state structure of TaON. Other structural alternatives are energetically worse, including a distorted baddeleyite type with only six-fold V^{5+} coordination because it results as being significantly stiffer when pressurized. A final bonding analysis, given in Figure 3.56(b), immediately shows *why* higher pressure is needed to make VON. Besides the strong V–O/N bonding interactions (not shown) very much resembling those in the isotypical TaON (see

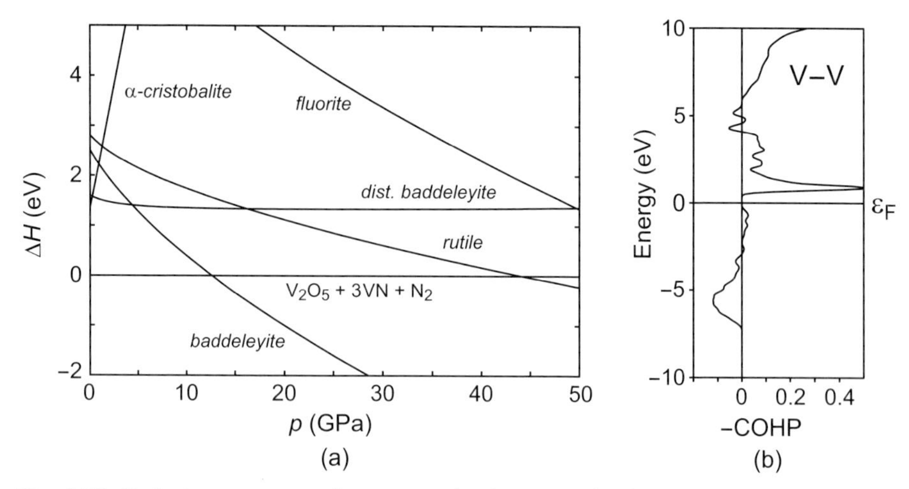

Fig. 3.56 Enthalpy–pressure diagram at absolute zero for the formation of VON (a) and V–V COHP bonding analysis of baddeleyite-type VON (b). The COHP includes the seven nearest V–V interactions between 3.07 and 3.30 Å.

Figure 3.45(b) in Section 3.10), there is an *extra* repulsion from antibonding V–V interactions due to the relatively close encounter of the vanadium atoms. In addition, baddeleyite-type VON is predicted as a semiconducting material with a band gap around 0.6 eV.

Summarizing, the pseudopotential GGA calculations allow the prediction of unknown phases for the oxynitrides of V, Nb, and Ta, and they also allow the provision of quantitative data for the required synthetic conditions in terms of pressure. Similar predictions would have been impossible to make using purely traditional approaches.

3.13
Predicting Magnetic Cyanamides and Carbodiimides

Continuing our attempts to utilize electronic-structure theory in order to support the synthetic solid-state chemist, let us reconsider cyanamide and carbodiimide chemistry which was already the focus of Section 3.9. The reader will recall that there has been a lot of research with respect to NCN^{2-} chemistry, and the recently characterized materials comprise either alkaline, alkaline-earth, rare-earth or main-group elements. Also, a few of the later (d^{10}) transition metals may give NCN^{2-}-based phases such as HgNCN. Negatively expressed, there is *no* corresponding compound incorporating a true 3*d* transition metal with some leftover electrons on the cation. "MNCN" compounds with M being either Mn, Fe, Co, etc., have never been reported so far, and there is no structural proposal to hand for hypothetical compounds. This is indeed a strange situation.

Obviously, it is unclear whether such quasi-binary compounds can be made although their potential magnetic properties would make their synthesis a quite attractive goal. Thus, we consider it important first to propose reasonable crystal structures for hypothetical quasi-binary cyanamides and carbodiimides incorporating the aforementioned 3*d* metals. Second, we will also try to distinguish the factors favoring lucrative MNCN structures. The third goal is theoretically to calculate whether the hypothetical MNCN materials are thermodynamically stable phases and, if so, to use this information to guide the synthetic chemist in choosing the optimum synthetic conditions.

As has been alluded to already, total-energy calculations must be carried out for the hypothetical MNCN compounds based on the correct crystal structures. Since the latter crystal structures are likewise unknown, this problem can only be solved by *generating* them through geometry optimization (that is, by electronic-structure theory) of some structural hypotheses which seem reasonable from a chemical perspective and, at the same time, which may be expected to cover large regions of phase space (structural diversity). To do so, all those crystallographic structures are taken into account that have been determined for the most common cyanamides and carbodiimides.[24] Second, other structures containing *similar* triatomic QCN anions are considered, namely the cyanate (OCN^-), thiocyanate (SCN^-), and selenocyanate ($SeCN^-$) anions.[25] Third, structural starting points are derived from the crystal structures of transition-metal *chalcogenides* MQ_x, with $x = 1$ or 2, where chalcogens are replaced by NCN^{2-} units.[26] Thus, our approach comes close to a *combinatorial* search for new compounds by theoretical means.

Technically, the DFT-based structure optimizations (VASP) are performed by means of plane waves with a high kinetic energy cutoff (700 eV) and the projector-augmented wave (PAW) approach. Because *all* initial structures are somewhat incorrect, a high accuracy is needed for describing the ion–electron interactions. All atomic coordinates and cell parameters are optimized but, to save computing time, the initially given space-group symmetry is kept throughout. Spin-polarization must be allowed because of the large exchange splittings of the magnetic 3*d* metals.

24) Namely those of alkaline (Li_2NCN and K_2NCN) and monovalent nonalkaline (Ag_2NCN) elements, alkaline-earth elements (MgNCN, SrNCN), d^{10} transition metals (ZnNCN, HgNCN(I), HgNCN(II)), and a main-group element (PbNCN).

25) Structural candidates are derived from monoclinic/orthorhombic AgOCN for the cyanates group, KSCN, CsSCN, hexagonal/orthorhombic/rhombohedral CuSCN, and orthorhombic/monoclinic AgSCN for the thiocyanates group, and KSeCN as the only selenocyanate.

26) These are tetragonal/orthorhombic FeS, CoS, NiS, CuS, Cr_5S_6, and TiS_2. The reason for proposing such structural models is that the equally charged cyanamide group NCN^{2-} exhibits a volume increment which is very close to the one of the sulfide S^{2-} anion: the values are 28 and 29 cm^3/mol, respectively [390].

In order to check the quality of the LDA against the GGA for this very problem, geometry optimizations are first performed for two existing compounds, the rather ionic carbodiimide CaNCN and the more covalent cyanamide PbNCN (see again Figure 3.41). It transpires that both DFT approaches yield acceptable lattice parameters, with a mean 1.9% underestimation (LDA) and mean 0.3% overestimation (GGA). The average error of the internal geometries (C–N bond lengths) are 0.2% (LDA) and a larger 1.2% (GGA) albeit that both exhibit the known deficiency. DFT tends to homogenize electron densities, such that the C–N bond lengths of PbNCN are not reproduced very well, making it look almost like a carbodiimide – the notorious problem mentioned in Section 3.9. Because the GGA results appear slightly worse than those of the LDA (its superior *structural* performance may be due to a fortunate error cancellation), the LDA functional is the better choice for the geometry optimizations of the MNCN models.

In terms of *energetic* performances of the LDA and GGA, another test serves for the archetypal (and industrially important) compound CaNCN. For this highly stable material, reliable thermodynamic data of its standard formation enthalpy ΔH_f are available, corresponding to the formal reaction

$$\mathrm{Ca} + \mathrm{C} + \mathrm{N_2} \rightleftharpoons \mathrm{CaNCN},$$

namely a strongly exothermic −350.6 kJ/mol [378]. When the total electronic energies of all educt (Ca, C, N_2) phases are subtracted from that of the product (CaNCN), a theoretical reaction energy at absolute zero (to be compared with the above formation enthalpy assuming negligible enthalpy corrections between 0 and 298 K, a reasonable approximation) of −424.7 kJ/mol is found for the LDA, whereas the GGA value is −359.9 kJ/mol. Thus, there is the LDA-typical *overbinding* which largely overestimates the CaNCN stability by more than 21%. On the other hand, the accuracy of the GGA must be considered to be excellent because the error is less than 3%.[27] Thus, the GGA is the right choice for *absolute* energetic questions.

We spare the reader from the sight of all 140 total-energy and structure data derived from 28 structural alternatives of the five hypothetical MNCN phases with M = Mn, Fe, Co, Ni, and Cu. Instead, Figure 3.57 demonstrates that the optimization procedure produces configurations in which the metal atom is coordinated by six, five, and four N neighbors There is also (not shown) another energetically high-lying structure with twelve nearest N neighbors.

For the whole family of compounds, the computed energy differences between all structural alternatives decrease from MnNCN to CuNCN. Thus, structural preferences in terms of differing coordinations are *less* pronounced while filling up the *d* block. In addition, most of the lowest-lying structures re-

27) Considering the serious DFT problem encountered in Section 3.9, the minute GGA error indicates another fortunate error cancellation.

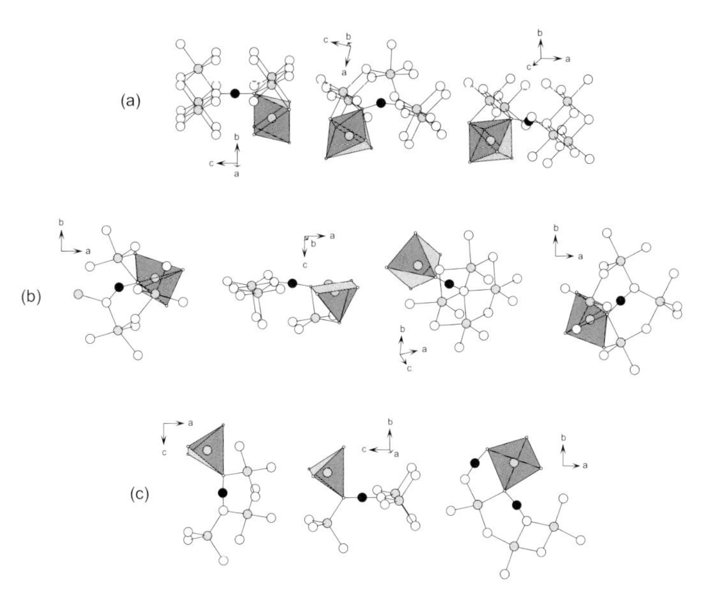

Fig. 3.57 Cuts from the computationally derived MNCN structures with M = Mn, Fe, Co, Ni, and Cu. For the Mn case, six-fold coordination (a) is found for structures derived from the initial [MgNCN] type, orthorhombic [SrNCN], [HgNCN(II)], five-fold coordination (b) from [HgNCN(I)], [PbNCN], [K_2NCN], [Ag_2NCN], and four-fold coordination (c) from [ZnNCN], [Li_2NCN], [Ag_2NCN]. C, N and M atoms are given in black, white, and grey circles, respectively.

fer to a (distorted) *octahedral* coordination for the transition metal although its preference is quite unexpected. For comparison, we note that the early binary MN nitrides (with M being an early 3*d* transition metal) crystallize in the rocksalt type, and the later ones (i.e., with M = Fe or one of the following metals) crystallize in the zinc-blende type with four-fold coordination [267]; this was the subject of Section 3.2. The finding is especially remarkable considering the fact that the NCN^{2-} anion is *larger* than the N^{3-} anion, which would favor, in a purely geometrical interpretation, smaller coordination numbers (here, the classical rule of thumb is also qualitatively incorrect). Figure 3.58 illustrates the preference of octahedral coordination for the case of MnNCN; the noninteger effective coordination numbers have been quantified by following the recipe of Brunner and Schwarzenbach [391].

Besides the preferred six-fold coordination, the figure also shows that a low molar volume, i.e., a high density, is indicative of a higher stability. This finding is a general one for the whole series of compounds and most structures are found to be *very dense* by the DFT calculations. Although the volume increment of the NCN^{2-} unit is about 28 cm^3/mol [390] and the presently known MNCN phases exhibit molar volumes equal to or higher than 30 cm^3/mol, most theoretical values presented here are far smaller, reflecting very compact

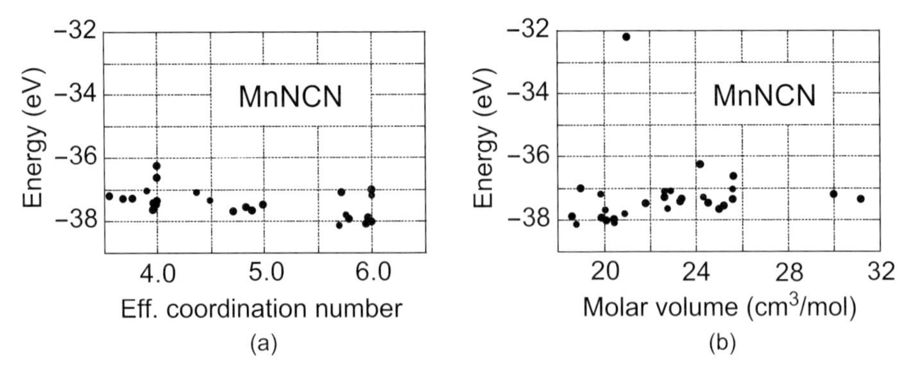

Fig. 3.58 Course of the total energies per formula unit of hypothetical MnNCN as a function of the effective coordination number (a) and the molar volume (b).

structures. For the lowest-energy candidates, these molar volumes decrease from CuNCN (about 23 cm^3/mol) to MnNCN (about 19 cm^3/mol), the second unexpected finding.

With regard to the cyanamide/carbodiimide shape difference in this group of MNCN compounds, four of the five most stable structures are *carbodiimides* because they exhibit two C=N bonds in the NCN^{2-} unit with about the same length (1.23 Å). The fifth one exhibits two different C–N bond lengths (1.20 and 1.27 Å), such that this structural type is close to a *cyanamide*. We have already emphasized that the discrimination between carbodiimide and cyanamide isomers *only* in terms of DFT is inappropriate because DFT overestimates the stability of the carbodiimide form (see also Section 3.9). On the contrary, the empirical correlation derived from the Pearson cation hardnesses [201] is predictive. Hard cations bond ionically to NCN^{2-} and yield carbodiimides, but soft cations bond covalently and result in cyanamides. Although the five transition metal cations considered here must be considered soft (η = 9.0, 7.2, 8.2, 8.5, and 8.3 eV for Mn^{2+}, Fe^{2+}, Co^{2+}, Ni^{2+}, and Cu^{2+}; see also Table 2.3) which would favor the cyanamide shape, mostly *carbodiimides* result from the geometry optimizations. If this must be taken seriously and is not a DFT artifact, it would be the first breakdown of the aforementioned NCN^{2-} shape rule, and this (third) unexpected finding is presumably related to the spin-polarized electronic configurations of the transition-metal ions. None of the known cyanamides/carbodiimides, in contrast, deviates from diamagnetic behavior if we neglect the (rather ionic) $4f$ compounds.

Indeed, spin-polarization plays a decisive role in lowering the total energies, and the energetically preferred structures also relate to *high-spin* electronic configurations. For the exemplary case of MnNCN, one finds five unpaired electrons (t_{2g}^3, e_g^2) and an antiferromagnetic coupling between the Mn^{2+} centers. For the less favorable structural models, ferromagnetically coupled low-spin configurations are also found. Whenever the coordination environment deviates substantially from a regular one, intermediate spin moments are also detected.

Given the computationally generated structural information for all phases, we may now answer the question of the synthetic pathways towards them. It is connected with the energetic positioning of MNCN in comparison with the elements or other competing compounds in terms of enthalpy and also entropy. For the known MN, MC and M_2C phases, the structures are taken from the crystallographic databases, whereas those of hypothetical nitrides/carbides – possibly in existence but not experimentally described so far – are also optimized, starting with those of FeN/CoN [266], VC [392], and Fe_2C/Co_2C [393]. In short, structural optimizations in combination with total-energy calculations are carried out for the reactants of the following four chemical reactions:

1: $M + C + N_2 \rightleftharpoons MNCN$
2: $MC + N_2 \rightleftharpoons MNCN$
3: $MN + \frac{1}{2}(CN)_2 \rightleftharpoons MNCN$
4: $\frac{1}{2}M_2C + \frac{1}{4}(CN)_2 + \frac{3}{4}N_2 \rightleftharpoons MNCN$

Theoretical reaction energies, identical with experimental reaction enthalpies ΔH_R at zero temperature (and close to them at room temperature), appear as the difference between the total electronic energies of the products and those of the educts. We also mention that all educts involve *gaseous* species ($(CN)_2$, N_2), having important implications for the Gibbs reaction energies ΔG_R of cyanamides/carbodiimides. High-temperature routes must be *avoided* because there is an entropic destabilization of MNCN formation due to the $T\Delta S$ stabilization of the gaseous educts according to the Gibbs–Helmholtz equation (see Section 2.18.2).

For (very) low temperatures, let us neglect the entropic term and concentrate on the enthalpies only, shown in Table 3.3, based on the GGA. Some of them (reaction # 4) are found to be positive, such that obtaining cyanamides/carbodiimides starting from M_2C carbides is enthalpically disfavored. On the contrary, all calculated reaction enthalpies for reactions # 2 and 3 are negative. Thus, simple 1:1 carbides and nitrides would lead to exothermic MNCN phases but a couple of such binary transition-metal compounds are still hypothetical. Most importantly, however, the direct route (reaction # 1) starting with the elements, evidences all reaction (or formation) enthalpies as *endothermic* such that, very unfortunately from a synthetic point of view, the whole set of MNCN compounds is predicted as *enthalpically unstable* relative to the elements because of the positive ΔH_f entries. Reconsidering the above-mentioned entropic stabilization of the gaseous educts, the hypothetical phases MNCN are even more *thermodynamically unstable*, in terms of ΔG_f.

Tab. 3.3 Reaction enthalpies ΔH_R (kJ/mol) calculated for hypothetical transition-metal MNCN compounds on the basis of the GGA at absolute zero temperature.

Reaction	Educts	M = Mn	Fe	Co	Ni	Cu
# 1	$M + C + N_2$	44.8	77.3	101.5	123.8	144.8
# 2	$MC + N_2$	−11.5	−30.5	−43.9	−86.4	−182.6
# 3	$MN + \frac{1}{2}(CN)_2$	−21.8	−19.8	−36.0	−89.4	−173.2
# 4	$\frac{1}{2}M_2C + \frac{1}{4}(CN)_2 + \frac{3}{4}N_2$	−11.8	5.0	32.2	31.7	−51.6

The course of the MNCN formation enthalpies with reference to the elements (along reaction # 1) is visualized in Figure 3.59 for all five phases. For comparison, we show the trends from both the GGA and also the LDA calculations.

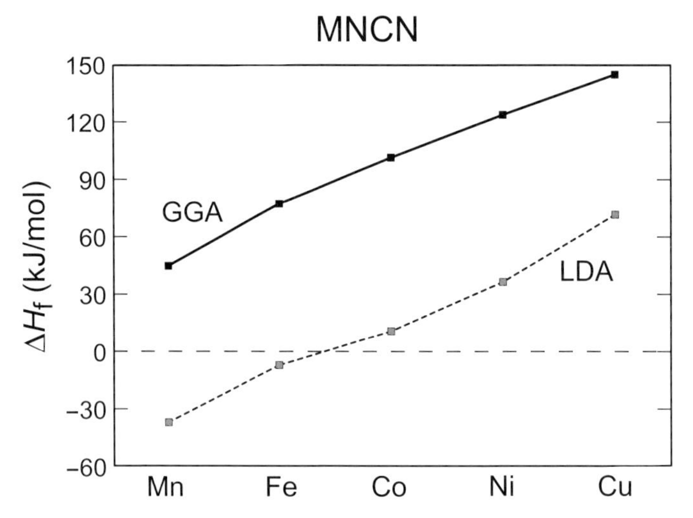

Fig. 3.59 Evolution of the formation enthalpies of MNCN compounds relative to the elements as a function of the transition metal M using the GGA (full line) and the LDA (broken line) approach.

It is obvious that the course of the GGA and that of the (energetically less accurate) LDA functional is about the same, and we also witness the expected LDA-typical overestimation of cyanamide/carbodiimide stability of the order of 70 kJ, just as for the preceding test case of CaNCN. Nonetheless, both functionals reveal an increasing destabilization of MNCN formation on adding more electrons to the $3d$ transition metal. This evolution of instability is a consequence of the continuous filling of antibonding (t_{2g}) and strongly antibonding (e_g) d levels for the octahedral M^{2+} coordination. Thus, the larger the number of d electrons, the less stable the MNCN phase.[28] The finding that all MNCN compounds are enthalpically unstable relative to the elements, is the fourth theoretical surprise, but also it explains *why* no such phase has been reported so far.

On the other hand, a positive GGA formation enthalpy between ca. +45 and +145 kJ/mol certainly does *not* exclude compound formation because there are likewise unstable solid phases which have been known for decades, e.g., AgSCN (ΔH_f = +88 kJ/mol), CuCN (+95 kJ/mol), and $Hg(SCN)_2$ (+201 kJ/mol) [378]. Obviously, MNCN is not too unstable to be accessible, at least not for the persistent synthetic chemist. Nonetheless, the synthesis of MNCN may not proceed from the elements but requires alternative strategies. For example, a formal exchange reaction (metathesis) such as

$$MX_2 + ANCN \rightleftharpoons MNCN + AX_2,$$

in which MX_2 is a transition-metal halide, ANCN symbolizes a, not exceedingly stable, cyanamide/carbodiimide, and AX_2 is an exceptionally stable halide – hopefully highly volatile such that it can be removed from the equilibrium – appears to be a reasonable starting point. Low-temperature routes are needed so that MNCN will not be entropically destabilized. Also, an electron-poor representative such as MnNCN as opposed to CuNCN is the best candidate to start with [394].

Indeed, the above reaction scheme involving ZnNCN as the carbodiimide precursor and $MnCl_2$ as the metal halide, leads to the formation of solid *green* MnNCN at only 600 °C where the freshly formed $ZnCl_2$ is volatile enough to let the chemical equilibrium fully shift to the right side [395]. As predicted, MnNCN contains octahedrally coordinated Mn^{2+} because the experimental structure is isotypic with CaNCN, one of the lowest-energy structures predicted theoretically. The NCN^{2-} unit adapts the carbodiimide shape. As also predicted from the GGA calculations, there are five unpaired electrons on Mn^{2+}, and an antiferromagnetic order predominates at low temperatures.

Even the very endothermic compound CuNCN can be synthesized and structurally characterized, but the greater instability excludes a route at elevated temperatures. Instead, CuNCN must be prepared through room-temperature oxidation of a Cu(I) precursor under aqueous conditions. As expected, the Cu(II) atom in solid *black* CuNCN exhibits a typical first-order Jahn–Teller effect [396].

28) The high-spin electronic configurations are $t_{2g}^3e_g^2$, $t_{2g}^4e_g^2$, $t_{2g}^5e_g^2$, $t_{2g}^6e_g^2$, and $t_{2g}^6e_g^3$ for the Mn, Fe, Co, Ni and Cu compounds.

3.14 Predicting Ternary Magnetic Nitrides

In this final section dealing with the practical applications of electronic-structure theory, we are going to combine several ideas and techniques that were developed in the preceding ones. While Section 3.2 has dealt with our understanding of the crystal structures of the transition-metal nitrides, the topic of Sections 3.5 and 3.6 was a chemical theory of itinerant magnetism. We now wish to *understand* itinerant magnetic nitrides using the above framework, and we also want to *predict* their existence using the quantitative methods already introduced on investigating oxynitrides (Sections 3.10 and 3.12), the Sn/Zn system (Sections 3.11), and magnetic cyanamides/carbodiimides (Section 3.13).

For such an attempt, the binary nitride phase Fe_4N is the perfect starting compound. It has been the subject of intense experimental [397] and also theoretical [398] studies for several decades. The interest in Fe_4N is easily explained not only because of its potential application as a high-density recording material (see also Section 3.2), but Fe_4N is also characterized by a large saturation magnetization (208 Am^2/kg), less than 4% smaller than that of metallic iron (α-Fe). Also, Fe_4N exhibits a low coercivity (460 A/m) and a chemical inertness that even exceeds metallic iron. The perovskite-like crystal structure of Fe_4N is given in Figure 3.60.

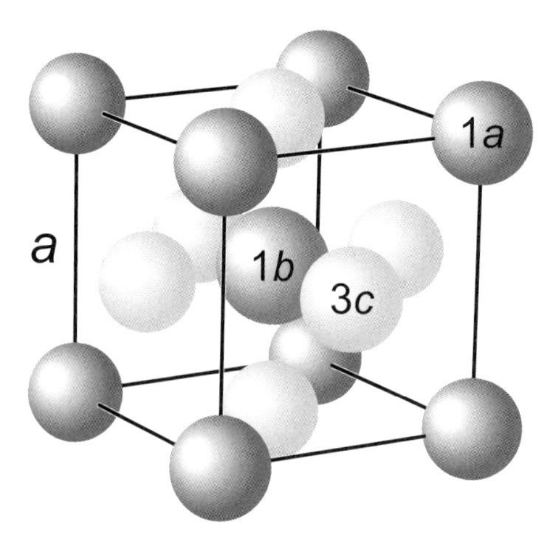

Fig. 3.60 The crystal structure of Fe_4N (a = 3.795 Å) in space group $Pm\bar{3}m$. The large nitrogen atom occupies the very center (Wyckoff position $1b$), and the dark/light grey iron atoms are found at the corners ($1a$) and face centers ($3c$).

There have been many attempts to improve the magnetic properties of Fe_4N through atomic substitution, and some nonstoichiometric compounds of general formula $M_xFe_{4-x}N$ (x < 1) with M = Mn, Ru, Os, Co, Ir, Ag, Zn were reported. In addition, a few stoichiometrically precise phases MFe_3N with M = Ni, Pd, Pt, Au, and In are known [399]. The structural idea of all these ternaries can be summarized as follows. Within the perovskite-like structure, only Mn occupies both 1*a* and 3*c* lattice sites upon substitution, but all other metals exclusively replace the corner twelve-coordinate Fe atom.

The attempted replacement of Fe by its higher homologues Ru/Os, and also by Ir, has been somewhat neglected by the experimentalist [399] because only berthollide compounds $M_xFe_{4-x}N$ with $x \leq 0.20$ have been prepared. On the theoretical side [400], a large number of studies has been published analyzing the role played by M guest atoms (Ni, Pd, Pt, Mn, Ag, Au, Sn, Zn, In, Cu, Co, Cr, Ti, Ru) in hypothetical MFe_3N phases. Thus, the presently available theoretical (and also experimental) knowledge of how to substitute Fe in Fe_4N by either Os, Rh, or Ir is unsatisfactory. In order to fill this important gap, let us therefore examine the complete set of daltonide compounds MFe_3N of the iron- and platinum-group metals (Fe/Ru/Os, Co/Rh/Ir, Ni/Pd/Pt) by means of total-energy density-functional calculations in order to determine stable atomic configurations, absolute stabilities, and also physical properties.

For this family of compounds, plane-wave basis sets and ultra-soft pseudopotentials are a good compromise in terms of accuracy and speed. The GGA functional is needed to determine optimized structural models (VASP). Potential high-pressure phases are accessible through compressing/expanding the unit cells, i.e., by scaling the lattice parameters below and above the minimum geometries. Statistically *disordered* MFe_3N compounds are investigated by setting up a $3 \times 2 \times 2$ enlarged supercell containing 48 sites for metal atoms and 12 nitrogen atoms; the 12 disordered M atoms are randomly placed in the 48 positions. Fortunately, the energy differences between differing configurations turn out to be quite small.

At the outset, the atomic ordering of the iron- and platinum-group metals M over the 1*a* and 3*c* sites are analyzed. To do so, the total energies of statistically *disordered* compounds designated $^{1a}(M_{\frac{1}{4}}Fe_{\frac{3}{4}})^{3c}(M_{\frac{3}{4}}Fe_{2\frac{1}{4}})^{1b}N$ – the Wyckoff superscripts indicate the positions in the random phase – are compared with two *ordered* configurations. The M atom may either exclusively replace the corner Fe atom, $^{1a}M^{3c}(Fe_3)^{1b}N$, or M may substitute one of the three facial Fe atoms, $^{1a}Fe^{3c}(MFe_2)^{1b}N$. The numerical results are given in Table 3.4 as relative energies with respect to the random occupation.

In the case of $CoFe_3N$, the energy differences between the three structural alternatives are so tiny that the small Co atom (very similar to Mn) does not prefer the 1*a* or 3*c* positions. An increase in atomic size, however, clearly disfavors the facial 3*c* position (see left energy column) for the heavier atoms Rh, Ir, Pd, and Pt but the corner 1*a* position is the best one for these, as shown by the strong energetic gains (see center energy column). This simple interpreta-

Tab. 3.4 Metallic radii r_M (Å), total energy differences ΔE (kJ/mol) between different atomic orderings relative to a random occupation, and the corresponding lattice parameters a (Å) of MFe_3N in the perovskite-like structure type.

distribution		$^{1a}Fe^{3c}(MFe_2)^{1b}N$		$^{1a}M^{3c}(Fe_3)^{1b}N$		$^{1a}(M_{\frac{1}{4}}Fe_{\frac{3}{4}})^{3c}(M_{\frac{3}{4}}Fe_{2\frac{1}{4}})^{1b}N$	
M	r_M	ΔE	a	ΔE	a	ΔE	a
Co	1.25	−2	3.79	−2	3.79	0	3.79
Ni	1.25	4	3.80	−16	3.80	0	3.80
Ru	1.34	−2	3.89	−11	3.85	0	3.88
Os	1.35	4	3.90	−7	3.82	0	3.88
Rh	1.34	10	3.91	−71	3.87	0	3.90
Ir	1.36	16	3.92	−79	3.85	0	3.91
Pd	1.38	33	3.94	−111	3.88	0	3.93
Pt	1.38	40	3.97	−149	3.88	0	3.95

tion is in accordance with independent suggestions [401] relating the atomic ordering with two factors, namely the relative affinity of M and Fe for nitrogen and, more importantly, the differences in the atomic sizes ($1a$ position: 1.42 Å; $3c$ site: 1.28 Å); note that the $1a$ position does not have neighboring N atoms. Small deviations from this qualitative rule of thumb seem to be related to differences in chemical affinities to nitrogen. Quantitatively, the known compounds $PdFe_3N$ and $PtFe_3N$ support the present calculations [399] because both follow the $^{1a}M^{3c}(Fe_3)^{1b}N$ scheme, with experimental lattice parameters of 3.866 and 3.857 Å, quite close to the theoretical ones but manifesting the GGA-typical small overestimation. For the remaining study, we stay with the ordered distribution, where the large M atom occupies the $1a$ corner site.

In a second step, the *absolute* stabilities of the MFe_3N family of compounds are studied by theoretically determining the total electronic energies of all participating phases. In the present case, the educts must also include the binary nitride FeN adopting the zinc-blende type (see Section 3.2). Theory clearly indicates that FeN is enthalpically more stable than the elements, in accordance with independent calorimetric measurements [402] because the phase designated $FeN_{0.91}$ is the most exothermic iron nitride ($\Delta H_f \approx -50$ kJ/mol). Given the simple chemical reaction

$$M + 2Fe + FeN \rightleftharpoons MFe_3N,$$

the absolute energetic stabilities of MFe_3N are the energy differences between products and educts, and they are shown in Figure 3.61.

To the theorist's delight, the reaction energies for the already synthesized compounds $NiFe_3N$, $PdFe_3N$, and $PtFe_3N$ turn out to be *negative* and thus confirm the GGA result to be correct (at least qualitatively), simply because all three have been reported as enthalpically stable phases. The fully ordered

Fe	Co	Ni
-2	7	-8
Ru	Rh	Pd
64	-23	-47
Os	Ir	Pt
108	2	-74

Fig. 3.61 Theoretical reaction energies ΔE (kJ/mol) of MFe_3N where M is a group VIII transition metal. Experimentally reported phases have been emphasized in grey.

compounds MFe_3N with M = Ru, Os, Co, and Ir are all predicted as being unstable, and these have only been reported as nonstoichiometric phases. Surprisingly, the phase $RhFe_3N$, yet to be made, appears as an exothermic and ordered compound. Also, all MFe_3N compounds (even the unstable ones) are predicted to be ferromagnetic. For a closer analysis, let us focus on the energetically almost stable $IrFe_3N$ and the energetically stable $RhFe_3N$.

$IrFe_3N$ is predicted to have a lattice parameter of $a = 3.85$ Å, corresponding to a 0.9% widening in comparison with Fe_4N, and its ferromagnetic ground state exhibits a saturation moment of 9.0 μ_B per formula unit (Ir: 0.92 μ_B, Fe: 2.68 μ_B). For Fe_4N, local Fe moments of 3.01 and 2.44 μ_B are found for the $1a$ and $3c$ positions. The increase in the $3c$ Fe moment for $IrFe_3N$ is a consequence of the very wide Fe–Fe distances but the total moment of $IrFe_3N$ is smaller since a former Fe atom on the $1c$ position has been replaced by Ir. Because $IrFe_3N$ is only slightly endothermic (i.e., by +2 kJ/mol), is there possibly a stable high-pressure phase? Thus, the total energy calculations are repeated but using cell volumes which vary around their equilibrium volumes. We also have to include the stable intermetallic compound $FeIr_3$ but no binary nitride as none is known and theoretical investigations predict them to be unstable. The resulting energy–volume diagram is presented in Figure 3.62(a). For a more convenient visualization of the transition pressures, the latter diagram is converted into an enthalpy–pressure diagram (see Sections 3.10, 3.11, 3.13), given in Figure 3.62(b).

At zero pressure, a mechanical mixture of Ir/Fe plus the binary phase FeN is the thermodynamically stable configuration. At higher pressures above 9 GPa, the intermetallic compound $FeIr_3$ becomes enthalpically favorable and leaves Ir/Fe/FeN behind. At even higher pressures beyond 37 GPa, $IrFe_3N$ eventually proves to be the thermodynamically stable phase. Because the energy–volume diagram of $IrFe_3N$ exhibits a *larger* volume than those of the others, the existence of a transition pressure towards $IrFe_3N$ requires a higher

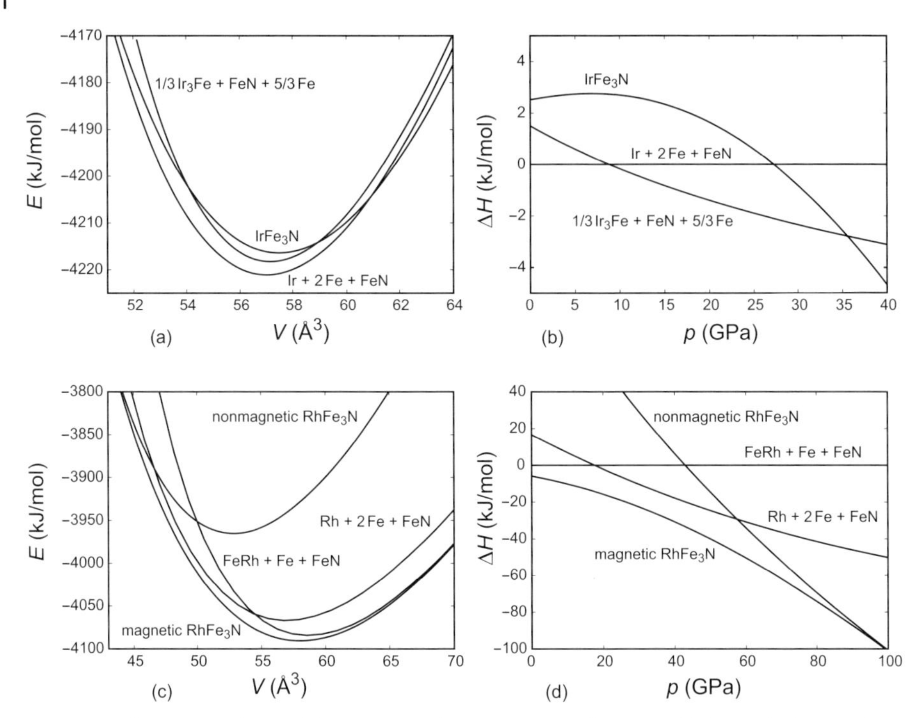

Fig. 3.62 Energy–volume (a) and enthalpy–pressure diagram (b) for $IrFe_3N$ and its competing phases. (c) and (d) as (a) and (b) but for $RhFe_3N$.

compressibility for this compound. Indeed, the calculated bulk modulus B_0 of $IrFe_3N$ is only 204 GPa, smaller than the corresponding values of Fe (235 GPa), FeN (264 GPa), and Ir (326 GPa). Thus, $IrFe_3N$ should be accessible for pressures higher than 37 GPa.

The energy minimization of $RhFe_3N$ yields $a = 3.87$ Å which is 1.4% larger than the theoretical Fe_4N value. The $RhFe_3N$ saturation magnetization is predicted as 9.2 μ_B per formula unit (Rh: 0.96 μ_B, Fe: 2.76 μ_B) and the very large Fe moment mirrors the widened lattice parameter. Although $RhFe_3N$ is more stable (by -23 kJ/mol) than its competitors FeN/Fe/Rh, the full picture requires energy–volume and enthalpy–pressure diagrams which are also shown in Figure 3.62(c) and (d). In these, another competing phase is also included, namely the known ferromagnetic intermetallic compound FeRh; no binary Rh nitride seems to be stable. Let us also include a hypothetically nonmagnetic (i.e., without any spin moments) $RhFe_3N$ in the figures. At zero pressure, FeRh is slightly lower in enthalpy than Fe and Rh, but magnetic $RhFe_3N$ is still the most stable phase (by -7 kJ/mol) in this chemical system. Due to the larger equilibrium volume of FeRh, its enthalpy increases faster on applying pressure than for the mechanical mixture Fe/Rh, such that the composition Fe/Rh/FeN is favored over FeRh/Fe/FeN already beyond 18 GPa. Nonetheless, the ferromagnetic, perovskite-like $RhFe_3N$ is the stable phase over the entire pressure range. An important stabilizing factor of $RhFe_3N$ is the onset of ferromagnetism (spin polarization) since the hypothetically nonmagnetic phase is less stable by +122 kJ/mol! For very small volumes below 45 Å^3, ferromagnetic and nonmagnetic $RhFe_3N$ energetically coincide, such that the ferromagnetism can be mechanically quenched by huge pressures.

The scaling of the magnetic moments by lattice expansion or compression, either by outer (pressure) or inner (atomic substitution) forces, is a typical phenomenon observed in itinerant ferromagnetic compounds. The stronger the exchange splitting, the more effective is the shielding of the minority (β) spins by the majority (α) spins [301] (see also Section 2.9). Because the more diffuse β electrons are mainly responsible for the chemical bonding, the interatomic distances will *increase* for a better overlap [68] (see also Section 3.6). Conversely, the continuous loss of magnetization on applying pressure, results from the suppressed exchange splitting by constraining the more diffuse β spins and the more localized α spins to the same spatial region. This is also the reason why the magnetic moments of the 3*c* Fe atoms decrease in the order $RhFe_3N > IrFe_3N > Fe_4N$, mirroring the shortening of the Fe–Fe distances. A better insight into the chemical bonding is presented in Figure 3.63 from DOS and COHP analyses, the latter based on electronic-structure calculations using the LMTO method.

Within the non-spin-polarized calculation, a high DOS is found at the Fermi level, mostly due to Fe 3*d* contributions, a clear sign of electronic instability. Its source is visible from the Fe–Fe COHP curves due to very strong antibonding interactions in the highest occupied bands. In contrast, the Fe–Rh combination only reflects bonding interactions, as does the Fe–N combination. The nonoptimized Fe–Fe bonding is indicative for a symmetry lowering, and the onset of spin-polarization leads to drastic changes. The DOS at the Fermi level shrinks significantly and the α/β spins move to lower/higher energies; nonetheless, the material stays metallic. While the bonding interactions for the Fe–N and also Fe–Rh combinations are not very affected (the integrated COHP values stay almost the same), the, formerly antibonding, Fe–Fe interactions at the Fermi level have been removed, thereby improving Fe–Fe bonding and stabilizing the ferromagnetic ground state of $RhFe_3N$, just as for the 3*d* metals and their intermetallic compounds.

Summarizing, first-principles theory is able quantitatively to describe the structural stabilities of new perovskite-like MFe_3N compounds, in particular the preference of the larger M atoms to go to the corner position, and ferromagnetic ground states result from an optimized chemical bonding due to the removal of formerly antibonding Fe–Fe states at the Fermi level. Due to the numerical accuracy of the pseudopotential GGA calculations, two unknown phases may be predicted as worthwhile synthetic goals. At ambient

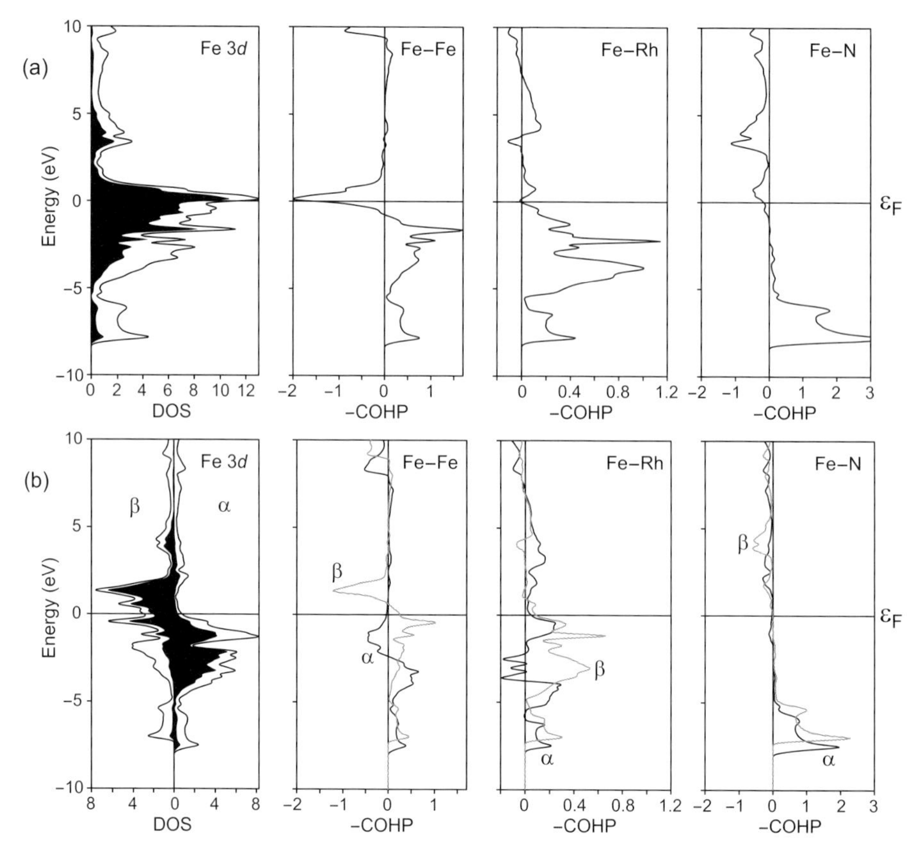

Fig. 3.63 Total and local (Fe 3*d*) densities-of-states (DOS) and COHP analysis for $RhFe_3N$ on the basis of non-spin-polarized (a) and spin-polarized (b) GGA calculations. For the magnetic case, majority/minority spin channels are given by black/grey lines.

pressure, ferromagnetic $RhFe_3N$ is an exothermic compound while ferromagnetic $IrFe_3N$ is accessible at pressures above 37 GPa [403]. Shortly after having published this prediction, $RhFe_3N$ was made under ordinary laboratory conditions [404], namely by the ammonolysis (NH_3/H_2) of a tight mixture of Rh and Fe_2O_3; the lattice parameter of $RhFe_3N$ is $a = 3.8292(2)$ Å at room temperature, and the material looks like a *semihard* ferromagnet. The high-pressure synthesis of $IrFe_3N$ still needs to be accomplished.

4
Epilogue

All theoretical chemistry is really physics,
and all theoretical chemists know it.
RICHARD P. FEYNMAN

All many-body physics is really chemistry,
and all theorists know it.
ANONYMOUS

If the reader has actually made it up to this point, he or she will have the impression that the whole universe of solid-state materials, i.e., insulators, semiconductors, metals, and intermetallic compounds can nowadays be studied by electronic-structure theory, and predictive conclusions are really in our own hands. Indeed, the numerical limitations of most classical approaches – in particular, the ionic model of everything – have been overcome. While the computational methods of today include very different quantum-chemical methods, their varying levels of accuracy and speed are due to differences in their atomic potentials and the choice of the basis sets that are involved. The latter may either be totally delocalized (plane waves) or localized (atomic-like), adapted to the valence electrons only (pseudopotentials) or to all the electrons. In order to *understand* structures and compositions of solid-state materials, the results of electronic-structure theory are typically investigated in terms of some quantum-*chemical* analysis.

Without any doubt, there is good reason to believe in a bright future for theory and computation within solid-state (materials) chemistry, even more so if a couple of great challenges are to be addressed in the coming years. It is fairly clear that structure, energetics, and physical properties of many compounds can be modeled, independently checked, and also chemically understood using the available tools. Likewise, theory's great power has been demonstrated with respect to (numerically reliable) phase prediction for relatively simple phases such as elements and binary compounds. It remains to

Computational Chemistry of Solid State Materials. Richard Dronskowski

ISBN: 3-527-31410-5

be seen whether similar successes can be repeated for *very* complex materials or whether astronomical complexities in terms of structure and composition will render computational approaches difficult or impossible, even if "combinatorial" (parallel) computer technologies will be explored, simply due to the "combinatorial explosion". Complexity *is* a problem but only if understanding is lacking. How does one theoretically or synthetically investigate a ten-component system [405, 406]? A similar problem exists whenever structural information is lacking; *amorphous* materials – even if the phase under study is only a binary or ternary one – will be an important issue for tomorrow's computational approaches.

It must also be stated that a decisively important numerical technique – density-functional theory – has proved to be extremely useful in the past, but some of its underlying operational assumptions are based on very simple ideas. Fortunately, the majority of electronically "inhomogeneous" systems seem to be quite tolerant of being described using homogeneous approaches. Serious DFT problems, however, are known, and it will be exciting to witness how these weaknesses can be overcome instead of being patched, for example, by bringing *molecular* quantum-chemical techniques for exchange and correlation into the solid state. Excited-state questions – the notorious band-gap problem – are another issue, to be solved by truly correlated methods. Other similar efforts are targeted at temperature-dependent electronic-structure theories. It does not need a great deal of imagination to understand that time-dependent phenomena will become increasingly important over the coming years, because one would wish to observe chemical reactions theoretically on the atomic scale. Remember that alternative *experimental* methods are not available for the solid state, let alone in real time. Think about the course of chemical reactions going on inside a heterogeneous catalyst, in detail, for any time scale, at any temperature. The reactivity of solid-state materials, theoretically (and also experimentally) neglected for a long time, is indeed an exciting topic [407, 408].

Finally, language and culture barriers between solid-state experimentalists and theorists must be overcome, one reason for having written this pocket guide. There has been progress, but a lot more needs to be accomplished. How does one *imagine* the composition of a room-temperature superconductor? Such a prestigious goal can probably not be reached solely from solving Schrödinger's equation but it will require a chemical language which captures – somewhat rephrased – the physical idea, and the first cautious steps have been taken [409].

It may seem strange, but there must have been times when crystallographic techniques were considered as not being part of solid-state and materials chemistry, and even contrary to its spirit. Fortunately, these obstinate prejudices have disappeared completely and in fact, computational techniques are now also becoming essential tools of the synthetic experimentalist.

Bibliography

1 Unfortunately, there are not many beginner's books on solid-state chemistry; a classic title is: A. R. West, *Solid State Chemistry and its Applications*, Wiley, New York 1984; *Basic Solid State Chemistry*, Wiley, New York 1997. Another excellent book, especially targeted at students, goes back to: L. Smart, E. Moore, *Solid State Chemistry: An Introduction*, 2nd ed., Chapman & Hall, London 1995. In addition, there are modern monographs: A. K. Cheetham, P. Day, *Solid State Chemistry: Techniques*, Clarendon Press, Oxford 1987, and A. K. Cheetham, P. Day, *Solid State Chemistry: Compounds*, Clarendon Press, Oxford 1992; C. N. R. Rao, J. Gopalakrishnan, *New Directions in Solid State Chemistry*, 2nd ed., Cambridge University Press 1997. A more materials-oriented perspective is taken by: R. J. D. Tilley, *Understanding Solids: The Science of Materials*, Wiley, New York 2004.

2 A classic text into Zintl chemistry is by: H. Schäfer, B. Eisenmann, W. Müller, *Angew. Chem. Int. Ed. Engl.* **1973**, *12*, 694; a more modern description is given by: S. M. Kauzlarich (Ed.), *Chemistry, Structure and Bonding in Zintl Phases and Ions*, Wiley-VCH, Weinheim, New York 1996.

3 The world's first book on quantum chemistry was written by: H. Hellmann, *Einführung in die Quantenchemie*, Franz Deuticke, Leipzig, Wien 1937, an abridged version of *Kvantovaya Khimiya*, ONTI, Moscow, Leningrad 1937. The scientific and private life of Hans Hellmann and his tragic fate has been described by: W. H. E. Schwarz, D. Andrae, S. R. Arnold, J. Heidberg, H. Hellmann Jr., J. Hinze, A. Karachalios, M. A. Kovner, P. C. Schmidt, L. Zülicke, *Bunsenmagazin* **1999**, *1*, 10; *ibid.* **1999**, *2*, 60. See also: K. Jug, W. Ertmer, J. Heidberg, M. Heinemann, W. H. E. Schwarz, *Chem. unserer Zeit* **2004**, *37*, 412.

4 A pleasant introduction into the quantum chemistry of solids (and surfaces), largely without mathematical apparatus, is due to: R. Hoffmann, *Solids and Surfaces: A Chemist's View of Bonding in Extended Structures*, VCH, Weinheim, New York 1988.

5 A classic and comprehensive (albeit slightly dated) overview of structural chemistry is by: A. F. Wells, *Structural Inorganic Chemistry*, 5th ed., Oxford University Press 1984. A concise and modern book, often used in the classroom, has been written by: U. Müller, *Inorganic Structural Chemistry*, Wiley, New York 1993.

6 These data (as well as many others) are contained in Pauling's classic monograph on the structures of molecules and solids and also early quantum chemistry: L. Pauling, *The Nature of the Chemical Bond*, Cornell University Press, Ithaca, New York 1960.

7 A. Simon, *Angew. Chem. Int. Ed. Engl.* **1983**, *22*, 95.

8 This is covered in greater detail in the following nice little book, now out of print, on structural (defect) chemistry:

N. N. Greenwood, *Ionic Crystals, Lattice Defects and Nonstoichiometry*, Butterworth & Co., London 1968.

9 The scientific career of Victor Moritz Goldschmidt and his private life have been depicted by: G. B. Kauffman, *Chem. Educator* **1997**, *2(5)*, 1.

10 V. M. Goldschmidt, T. Barth, G. Lunde, W. Zachariasen, *Skr. Nor. Vidensk.-Akad. Oslo I* **1926**, *2*; V. M. Goldschmidt, *Trans. Faraday Soc.* **1929**, *25*, 253.

11 J. A. Wasastjerna, *Z. Phys. Chem.* **1922**, *101*, 193; J. A. Wasastjerna, *Soc. Sci. Fennica Comm. Phys.-Nat.* **1923**, *1(38)*, 1.

12 L. Pauling, *Proc. Roy. Soc. London A* **1927**, *114*, 181.

13 J. C. Slater, *Phys. Rev.* **1930**, *36*, 57.

14 R. D. Shannon, *Acta Cryst. A* **1976**, *32*, 751.

15 J. K. Burdett, S. L. Price, G. D. Price, *Solid State Commun.* **1981**, *40*, 923.

16 E. Mooser, W. B. Pearson, *Acta Crystallogr.* **1959**, *12*, 1015.

17 A. Zunger, in *Structure and Bonding in Crystals* (A. Navrotsky, M. O'Keeffe, Eds.), Academic Press, New York 1981; A. N. Bloch, G. C. Schatteman, in *Structure and Bonding in Crystals* (A. Navrotsky, M. O'Keeffe, Eds.), Academic Press, New York 1981.

18 D. G. Pettifor, *Solid State Phys.* **1987**, *40*, 43; D. G. Pettifor, *Mat. Sci. Technol.* **1988**, *4*, 675; D. G. Pettifor, M. Aoki, *Phil. Trans. Roy. Soc. London A* **1991**, *334*, 439.

19 M. L. Huggins, J. E. Mayer, *J. Chem. Phys.* **1933**, *1*, 643.

20 P. P. Ewald, *Ann. Phys.* **1921**, *64*, 253.

21 F. Bertaut, *J. Phys. Radium* **1952**, *13*, 499.

22 D. E. Williams, *Acta Cryst. A* **1971**, *27*, 452.

23 H. D. B. Jenkins, K. F. Pratt, *Chem. Phys. Lett.* **1979**, *62*, 416.

24 B. P. van Eijck, J. Kroon, *J. Phys. Chem. B* **1997**, *101*, 1096.

25 A. F. Kapustinskii, *Q. Rev.* **1956**, *10*, 283.

26 R. Hoppe, *Angew. Chem. Int. Ed. Engl.* **1966**, *5*, 95; *ibid.* **1970**, *9*, 25.

27 G. Cordier, P. Höhn, R. Kniep, A. Rabenau, *Z. Anorg. Allg. Chem.* **1990**, *591*, 58.

28 O. Reckeweg, F. J. DiSalvo, *Z. Anorg. Allg. Chem.* **2001**, *627*, 371.

29 F. Liebau, *Structural Chemistry of Silicates*, Springer, Berlin 1985.

30 H. Yamane, F. J. DiSalvo, *J. Alloys Compd.* **1996**, *240*, 33; H. Huppertz, W. Schnick, *Chem. Eur. J.* **1997**, *3*, 249.

31 H. Huppertz, B. v. d. Eltz, *J. Am. Chem. Soc.* **2002**, *124*, 9376; H. Huppertz, *Z. Naturforsch. B* **2003**, *58*, 278; H. Emme, H. Huppertz, *Chem. Eur. J.* **2003**, *9*, 3623.

32 R. Stranger, I. E. Grey, I. C. Madsen, P. W. Smith, *J. Solid State Chem.* **1987**, *69*, 162; M. Saßmannshausen, H. D. Lutz, *Z. Anorg. Allg. Chem.* **2001**, *627*, 1071.

33 W. Biltz, *Raumchemie der festen Stoffe*, Verlag von Leopold Voss, Leipzig 1934.

34 W. Bronger, *Z. Anorg. Allg. Chem.* **1996**, *622*, 9.

35 A. Simon, R. Dronskowski, B. Krebs, B. Hettich, *Angew. Chem. Int. Ed. Engl.* **1987**, *26*, 139.

36 D. Fischer, M. Jansen, *Angew. Chem. Int. Ed.* **2002**, *41*, 3746.

37 W. Bronger, R. Kniep, M. Kohout, *Z. Anorg. Allg. Chem.* **2004**, *630*, 117.

38 A. Byström, K.-A. Wilhelmi, *Acta Chem. Scand.* **1951**, *5*, 1003; W. H. Zachariasen, *Acta Crystallogr.* **1954**, *7*, 795.

39 I. D. Brown, *The Chemical Bond in Inorganic Chemistry – The Bond Valence Model*, IUCr monographs on Crystallography 12, Oxford University Press 2002.

40 G. Donnay, R. Allmann, *Am. Mineral.* **1970**, *55*, 1003; I. D. Brown, K. K. Wu, *Acta Cryst. B* **1976**, *32*, 1957; I. D. Brown, D. Altermatt, *ibid.* **1985**, *41*, 244.

41 R. Dronskowski, *Angew. Chem. Int. Ed. Engl.* **1995**, *34*, 1126; M. Scholten, P. Kölle, R. Dronskowski, *J. Solid State Chem.* **2003**, *174*, 349.

42 D. W. J. Cruickshank, *Acta Crystallogr.* **1956**, *9*, 754; V. Schomaker, K. N. Trueblood, *Acta Cryst. B* **1968**, *24*, 63.

43 M. Trömel, *Acta Cryst. B* **1983**, *39*, 664.

44 N. E. Brese, M. O'Keeffe, *Acta Cryst. B* **1991**, *47*, 192.

45 S. Adams, *Acta Cryst. B* **2001**, *57*, 278.

46 There are many good books on group theory, and I would recommend just two, very different in style and content: F. A. Cotton, *Chemical Applications of Group Theory*, Wiley, New York 1990; M. Tinkham, *Group Theory and Quantum Mechanics*, McGraw Hill, New York 1964.

47 *International Tables for Crystallography, Vol. A: Space-group symmetry*, 5th ed. (Th. Hahn, Ed.), Kluwer, Dordrecht 2002.
48 H. Bärnighausen, *Match* **1980**, *9*, 139.
49 U. Müller, *Z. Anorg. Allg. Chem.* **2004**, *630*, 1519.
50 *International Tables for Crystallography, Vol. A1: Symmetry relations between space groups* (H. Wondratschek, U. Müller, Eds.), Kluwer, Dordrecht 2004.
51 There are many excellent textbooks on quantum chemistry, and some of the most popular ones are due to: I. N. Levine, *Quantum Chemistry*, 5th ed., Prentice Hall, Englewood Cliffs, New Jersey 2000; A. Szabo, N. S. Ostlund, *Modern Quantum Chemistry: Introduction to Advanced Electronic Structure Theory*, McGraw-Hill, New York 1989; W. Kutzelnigg, *Einführung in die Theoretische Chemie*, Wiley-VCH, Weinheim, New York 1993.
52 J. K. Burdett, *Chemical Bonding in Solids*, Oxford University Press 1995.
53 Among the many excellent textbooks on solid-state physics I would recommend: N. W. Ashcroft, N. D. Mermin, *Solid State Physics*, Holt, Rinehart & Winston, New York 1976; H. Ibach, H. Lüth, *Solid-State Physics*, 3rd ed., Springer, New York, Heidelberg 2003.
54 R. Hoffmann, *J. Chem. Phys.* **1963**, *39*, 1397; *ibid.* **1964**, *40*, 2745; *ibid.* **1964**, *40*, 2474.
55 J. K. Burdett, *Molecular Shapes*, Wiley, New York 1980; T. A. Albright, J. K. Burdett, M.-H. Whangbo, *Orbital Interactions in Chemistry*, Wiley, New York 1985.
56 E. Schrödinger, *Ann. Phys.* **1926**, *79*, 361.
57 H. Primas, U. Müller-Herold, *Elementare Quantenchemie*, Teubner, Stuttgart 1984.
58 F. Jensen, *Introduction to Computational Chemistry*, Wiley, New York 1999; M. Springborg, *Methods of Electronic-Structure Calculations – from Molecules to Solids*, Wiley, New York 2000; C. J. Cramer, *Essentials of Computational Chemistry*, repr., Wiley, New York 2004.
59 S. F. Boys, *Proc. Roy. Soc. London A* **1950**, *200*, 542.
60 E. R. Davidson, D. Feller, *Chem. Rev.* **1986**, *86*, 681.
61 W. J. Hehre, L. Radom, P. v. R. Schleyer, J. A. Pople, *Ab Initio Molecular Orbital Theory*, Wiley, New York 1986.
62 Many more interesting facets of chemical bonding are covered in Burdett's last book which is also a very entertaining read: J. K. Burdett, *Chemical Bonds – A Dialog*, Wiley, New York 1997.
63 A. F. Holleman, E. Wiberg, *Lehrbuch der Anorganischen Chemie*, 101st ed., Walter de Gruyter, Berlin, New York 1995.
64 M. Boggio-Pasqua, A. I. Voronin, Ph. Halvick, J.-C. Rayez, *J. Molec. Struct. (Theochem)* **2000**, *531*, 159.
65 M. Kaupp, B. Metz, H. Stoll, *Angew. Chem. Int. Ed.* **2000**, *39*, 4607.
66 A. Zerr, G. Miehe, G. Serghiou, M. Schwarz, E. Kroke, R. Riedel, H. Fueß, P. Kroll, R. Boehler, *Nature* **1999**, *400*, 340.
67 K. Ruedenberg, *Rev. Mod. Phys.* **1962**, *34*, 326.
68 R. Dronskowski, *Adv. Solid State Phys.* **2002**, *42*, 433; R. Dronskowski, *Int. J. Quant. Chem.* **2004**, *96*, 89.
69 F. Bloch, *Z. Phys.* **1928**, *52*, 555.
70 L. Fritsche, personal communication, Karlsruhe 2005.
71 C. J. Bradley, A. P. Cracknell, *The Mathematical Theory of Symmetry in Solids*, Clarendon Press, Oxford 1972.
72 M. Meyer, S. Glaus, G. Calzaferri, *J. Chem. Educ.* **2003**, *80*, 1221.
73 R. E. Peierls, *Quantum Theory of Solids*, Oxford University Press 1955.
74 R. Hoffmann, C. Janiak, C. Kollmar, *Macromolecules* **1991**, *24*, 3725.
75 K. Krogmann, H.-D. Hausen, *Z. Anorg. Allg. Chem.* **1968**, *358*, 67; K. Krogmann, *Angew. Chem. Int. Ed. Engl.* **1969**, *8*, 35.
76 S. Roth, D. Carroll, *One-Dimensional Metals*, 2nd ed., Wiley-VCH, Weinheim, New York 2004.
77 H. A. Jahn, E. Teller, *Proc. Roy. Soc. London A* **1937**, *161*, 220.
78 J. K. Burdett, *Inorg. Chem.* **1981**, *20*, 1959.
79 W. H. E. Schwarz, P. Valtazanos, K. Ruedenberg, *Theor. Chim. Acta* **1985**, *68*, 471; W. H. E. Schwarz, K. Ruedenberg, L. Mensching, L. L. Miller, R. Jacobson,

P. Valtazanos, W. von Niessen, *Angew. Chem. Int. Ed. Engl.* **1989**, *28*, 597.
80 D. J. Chadi, M. L. Cohen, *Phys. Rev. B* **1973**, *8*, 5747.
81 H. J. Monkhorst, J. D. Pack, *Phys. Rev. B* **1976**, *13*, 5188.
82 P. E. Blöchl, O. Jepsen, O. K. Andersen, *Phys. Rev. B* **1994**, *49*, 16223.
83 R. S. Mulliken, *J. Chem. Phys.* **1935**, *3*, 564; *ibid.* **1955**, *23*, 1833; *ibid.* **1962**, *36*, 3428.
84 P.-O. Löwdin, *J. Chem. Phys.* **1950**, *18*, 365.
85 E. R. Davidson, *J. Chem. Phys.* **1967**, *46*, 3320; K. R. Roby, *Molec. Phys.* **1974**, *27*, 81.
86 K. Jug, *Theor. Chim. Acta* **1973**, *31*, 63; *ibid.* **1975**, *39*, 301.
87 R. Heinzmann, R. Ahlrichs, *Theor. Chim. Acta* **1976**, *42*, 33; C. Ehrhardt, R. Ahlrichs, *ibid.* **1985**, *68*, 231.
88 J. Meister, W. H. E. Schwarz, *J. Phys. Chem.* **1994**, *98*, 8245.
89 R. F. W. Bader, P. J. MacDougall, C. D. H. Lau, *J. Am. Chem. Soc.* **1984**, *106*, 1594; R. F. W. Bader, R. J. Gillespie, P. J. MacDougall, *ibid.* **1988**, *110*, 7329; R. F. W. Bader, P. L. A. Popelier, T. A. Keith, *Angew. Chem. Int. Ed. Engl.* **1994**, *33*, 620.
90 T. Hughbanks, R. Hoffmann, *J. Am. Chem. Soc.* **1983**, *105*, 3528.
91 R. Dronskowski, P. E. Blöchl, *J. Phys. Chem.* **1993**, *97*, 8617.
92 G. G. Hall, *Proc. Roy. Soc. London A* **1952**, *213*, 113.
93 P. P. Manning, *Proc. Roy. Soc. London A* **1955**, *230*, 424.
94 C. A. Coulson, *Proc. Roy. Soc. London A* **1939**, *169*, 413.
95 W. V. Glassey, R. Hoffmann, *J. Chem. Phys.* **2000**, *113*, 1698.
96 D. Dürr, *Bohmsche Mechanik als Grundlage der Quantenmechanik*, Springer, Berlin 2001.
97 F. Hund, *Z. Phys.* **1925**, *33*, 345; *ibid.* **1925**, *34*, 296.
98 W. Pauli, *Z. Phys.* **1925**, *31*, 765.
99 R. J. Boyd, *Nature* **1984**, *310*, 480.
100 K. Hongo, R. Maezono, Y. Kawazoe, H. Yasuhara, M. D. Towler, R. J. Needs, *J. Chem. Phys.* **2004**, *121*, 7144.
101 G. Burns, *J. Chem. Phys.* **1964**, *41*, 1521.
102 W. Kutzelnigg, *Angew. Chem. Int. Ed. Engl.* **1984**, *23*, 272.
103 A. D. Becke, K. E. Edgecombe, *J. Chem. Phys.* **1990**, *92*, 5397.
104 W. L. Luken, J. C. Culberson, *Theor. Chim. Acta* **1984**, *66*, 279.
105 A. Savin, O. Jepsen, J. Flad, O. K. Andersen, H. Preuss, H. G. von Schnering, *Angew. Chem. Int. Ed. Engl.* **1992**, *31*, 187.
106 M. Kohout, A. Savin, *Int. J. Quant. Chem.* **1996**, *60*, 875.
107 A. Savin, A. D. Becke, J. Flad, R. Nesper, H. Preuss, H. G. von Schnering, *Angew. Chem. Int. Ed. Engl.* **1991**, *30*, 409; A. Savin, R. Nesper, S. Wengert, T. F. Fässler, *Angew. Chem. Int. Ed.* **1997**, *36*, 1808.
108 M. Kohout, F. R. Wagner, Y. Grin, *Theor. Chim. Acta* **2002**, *108*, 150.
109 T. F. Fässler, A. Savin, *Chem. unserer Zeit* **1997**, *31*, 110.
110 M. Kohout, A. Savin, *J. Comput. Chem.* **1997**, *18*, 1431.
111 J. K. Burdett, T. A. McCormick, *J. Phys. Chem. A* **1998**, *102*, 6366.
112 M. Wolfsberg, L. Helmholz, *J. Chem. Phys.* **1952**, *20*, 837.
113 W.-H. Whangbo, R. Hoffmann, *J. Chem. Phys.* **1978**, *68*, 5498; J. H. Ammeter, H.-B. Buergi, J. C. Thibeault, R. Hoffmann, *J. Am. Chem. Soc.* **1978**, *100*, 3686.
114 G. Calzaferri, L. Forss, I. Kamber, *J. Phys. Chem.* **1989**, *93*, 5366.
115 A. Vela, J. L. Gazquez, *J. Phys. Chem.* **1988**, *92*, 5688.
116 N. J. Fitzpatrick, G. H. Murphy, *Inorg. Chim. Acta* **1984**, *87*, 41; *ibid.* **1986**, *111*, 139.
117 A. Herman, *Modelling Simul. Mater. Sci. Eng.* **2004**, *12*, 21.
118 For excellent overviews, read: P. Pyykkö, *Adv. Quantum Chem.* **1978**, *11*, 353; P. Pyykkö, *Chem. Rev.* **1988**, *88*, 563; *ibid.* **1997**, *97*, 597. The relativistic chemistry of gold has also been reviewed: P. Pyykkö, *Angew. Chem. Int. Ed.* **2004**, *43*, 4412.
119 D. R. Hartree, *Proc. Cambr. Phil. Soc.* **1929**, *24*, 89; *ibid.* **1929**, *24*, 111.
120 V. Fock, *Z. Phys.* **1930**, *61*, 126; *ibid.* **1930**, *62*, 795.
121 J. C. Slater, *Phys. Rev.* **1930**, *35*, 210.

122 C. C. J. Roothaan, *Rev. Mod. Phys.* **1951**, *23*, 69.

123 M. Scholz, H.-J. Köhler, *Quantenchemische Näherungsverfahren und ihre Anwendungen in der Organischen Chemie: Quantenchemie – Ein Lehrgang, Band 3* (W. Haberditzl, M. Scholz, L. Zülicke, Eds.), Hüthig, Heidelberg 1981.

124 R. S. Mulliken, *Phys. Rev.* **1928**, *32*, 186.

125 W. Heitler, F. London, *Z. Phys.* **1927**, *44*, 455.

126 These problems are covered, in much greater depth, in: P. Fulde, *Electron Correlations in Molecules and Solids*, 2nd ed., Springer, Berlin, Heidelberg 1993.

127 R. Hoffmann, S. Shaik, P. C. Hiberty, *Acc. Chem. Res.* **2003**, *36*, 750.

128 J. A. Pople, R. K. Nesbet, *J. Chem. Phys.* **1954**, *22*, 571.

129 C. Møller, M. S. Plesset, *Phys. Rev.* **1934**, *46*, 618.

130 P. Coffey, K. Jug, *J. Chem. Educ.* **1974**, *51*, 252.

131 J. C. Inkson, *Many-Body Theory of Solids – An Introduction*, 2nd print, Plenum Press, New York, London 1986.

132 C. Pisani, R. Dovesi, *Int. J. Quant. Chem.* **1980**, *17*, 501; V. R. Saunders, *Faraday Symp. Chem. Soc.* **1984**, *19*, 79; C. Pisani, R. Dovesi, C. Roetti, *Lect. Notes Chem.*, Vol. 48, Springer, Heidelberg 1988.

133 C. Pisani, *Lect. Notes Chem.*, Vol. 67, Springer, Heidelberg 1996.

134 E. Ruiz, P. Alemany, *J. Phys. Chem.* **1995**, *99*, 3114.

135 E. Ruiz, S. Alvarez, P. Alemany, R. A. Evarestov, *Phys. Rev. B* **1997**, *56*, 7189.

136 R. A. Evarestov, A. I. Panin, *Int. J. Quant. Chem.* **2002**, *88*, 472.

137 L. H. Thomas, *Proc. Cambr. Phil. Soc.* **1927**, *23*, 542.

138 E. Fermi, *Atti Accad. Naz. Lincei, Rend. Cl. Sci. Fis. Mat.* **1927**, *6*, 602; *ibid.* **1928**, *7*, 342; *Z. Phys.* **1928**, *48*, 73.

139 E. Teller, *Rev. Mod. Phys.* **1962**, *34*, 627; N. L. Balàzs, *Phys. Rev.* **1967**, *156*, 42; E. H. Lieb, B. Simon, *Phys. Rev. Lett.* **1973**, *31*, 681; E. H. Lieb, B. Simon, *Adv. Math.* **1977**, *23*, 22.

140 J. C. Slater, *Phys. Rev.* **1951**, *81*, 385.

141 K. Schwarz, *Phys. Rev. B* **1972**, *5*, 2466; K. Schwarz, *Theor. Chim. Acta* **1974**, *34*, 225.

142 There are numerous reviews about density-functional theory, and I have found the following two contributions, targeted at chemists and physicists, an enjoyable read: R. G. Parr, W. Yang, *Density Functional Theory of Atoms and Molecules*, Clarendon Press, New York 1989; R. O. Jones, O. Gunnarsson, *Rev. Mod. Phys.* **1989**, *61*, 689.

143 P. Hohenberg, W. Kohn, *Phys. Rev.* **1964**, *136*, 864.

144 W. Kohn, L. J. Sham, *Phys. Rev.* **1965**, *140*, 1133.

145 R. Stowasser, R. Hoffmann, *J. Am. Chem. Soc.* **1999**, *121*, 3414.

146 D. P. Chong, O. V. Gritsenko, E. J. Baerends, *J. Chem. Phys.* **2002**, *116*, 1760.

147 M. E. Casida, *Recent Advances in Density Functional Methods* (D. P. Chong, Ed.), in *Recent Advances in Computational Chemistry*, World Scientific, Singapore 1995; K. Burke, E. K. U. Gross, *Density Functionals; Theory and Applications*, Lecture Notes in Physics, Vol. 500 (D. Joubert, Ed.), Springer, Berlin 1998; M. Petersilka, U. J. Gossmann, E. K. U. Gross, *Electronic Density Functional Theory. Recent Progress and New Directions* (J. F. Dobson, G. Vignale, M. P. Das, Eds.), Plenum Press, New York 1998.

148 F. Aryasetiawan, O. Gunnarsson, *Rep. Prog. Phys.* **1998**, *61*, 237.

149 K. Capelle, G. Vignale, *Phys. Rev. Lett.* **2001**, *86*, 5546.

150 H. Eschrig, W. E. Pickett, *Solid State Commun.* **2001**, *118*, 123.

151 O. Gunnarsson, M. Jonson, B. I. Lundqvist, *Phys. Rev. B* **1979**, *20*, 3136.

152 J. Harris, *Phys. Rev. B* **1985**, *31*, 1770.

153 T. Ziegler, *Chem. Rev.* **1991**, *91*, 651.

154 U. von Barth, L. Hedin, *J. Phys. C* **1972**, *5*, 1629.

155 S. H. Vosko, L. Wilk, M. Nusair, *Can. J. Phys.* **1980**, *58*, 1200.

156 D. M. Ceperley, B. J. Alder, *Phys. Rev. Lett.* **1980**, *45*, 566.

157 J. P. Perdew, Y. Wang, *Phys. Rev. B* **1992**, *45*, 13244.

158 W. Kohn, *Rev. Mod. Phys.* **1999**, *71*, 1253.

159 J. P. Perdew, A. Zunger, *Phys. Rev. B* **1981**, *23*, 5048.

160 A. D. Becke, *J. Chem. Phys.* **1998**, *109*, 2092; H. L. Schmider, A. D. Becke, *ibid.* **1998**, *108*, 9624.

161 J. P. Perdew, Y. Wang, *Phys. Rev. B* **1986**, *33*, 8800.
162 A. D. Becke, *Phys. Rev. A* **1988**, *38*, 3098.
163 J. P. Perdew, *Phys. Rev. B* **1986**, *33*, 8822; J. P. Perdew, *ibid.* **1986**, *34*, 7406.
164 J. P. Perdew, in *Electronic Structure of Solids* (P. Ziesche, H. Eschrig, Eds.), Akademie Verlag, Berlin 1991; J. P. Perdew, J. A. Chevary, S. H. Vosko, K. A. Jackson, M. R. Pederson, D. J. Singh, C. Fiolhais, *Phys. Rev. B* **1992**, *46*, 6671.
165 C. Lee, W. Yang, R. G. Parr, *Phys. Rev. B* **1988**, *37*, 785.
166 P. J. Stephens, J. F. Devlin, D. F. Chabalowski, M. J. Frisch, *J. Phys. Chem.* **1994**, *98*, 11623.
167 W. Koch, M. C. Holthausen, *A Chemist's Guide to Density Functional Theory*, 2nd ed., Wiley-VCH, Weinheim, New York 2001.
168 M. Springborg (Ed.), *Density Functional Methods in Chemistry and Materials Science*, Wiley, New York 1997.
169 G. Csonka, N. A. Nguyen, I. A. Kolossváry, *J. Comput. Chem.* **1997**, *18*, 1534.
170 J. P. Perdew, S. Burke, M. Ernzerhof, *Phys. Rev. Lett.* **1996**, *77*, 3865; J. P. Perdew, K. Burke, Y. Wang, *Phys. Rev. B* **1996**, *54*, 16533.
171 J. Kübler, *Phys. Lett. A* **1981**, *81*, 81; C. S. Wang, B. M. Klein, H. Krakauer, *Phys. Rev. Lett.* **1985**, *54*, 1852; P. Bagno, O. Jepsen, O. Gunnarsson, *Phys. Rev. B* **1989**, *40*, 1997.
172 P. W. Anderson, *Phys. Rev.* **1961**, *124*, 41.
173 L. Fritsche, J. Koller, *J. Solid State Chem.* **2003**, *176*, 652.
174 H. Eschrig, K. Koepernik, I. Chaplygin, *J. Solid State Chem.* **2003**, *176*, 482.
175 O. Gunnarsson, O. K. Andersen, O. Jepsen, J. Zaanen, *Phys. Rev. B* **1989**, *39*, 1708.
176 V. I. Anisimov, J. Zaanen, O. K. Andersen, *Phys. Rev. B* **1991**, *44*, 943; V. I. Anisimov, F. Aryasetiawan, A. I. Lichtenstein, *J. Phys.: Condens. Matter* **1997**, *9*, 767.
177 V. I. Anisimov, A. I. Poteryaev, M. A. Korotin, A. O. Anokhin, G. Kotliar, *J. Phys.: Condens. Matter* **1997**, *9*, 7359; A. I. Lichtenstein, M. I. Katsnelson, *Phys. Rev. B* **1998**, *57*, 6884; A. Georges, G. Kotliar, W. Krauth, M. J. Rozenberg, *Rev. Mod. Phys.* **1996**, *68*, 13.
178 S. Y. Savrasov, G. Kotliar, *Phys. Rev. Lett.* **2000**, *84*, 3670; S. Y. Savrasov, G. Kotliar, E. Abrahams, *Nature* **2001**, *410*, 793.
179 B. L. Hammond, W. A. Lester, P. J. Reynolds, *Monte Carlo Methods in Ab Initio Quantum Chemistry*, World Scientific, Singapore 1994.
180 S. Fahy, X. W. Wang, S. G. Louie, *Phys. Rev. Lett.* **1988**, *61*, 1631; A. J. Williamson, S. D. Kenny, G. Rajagopal, A. J. James, R. J. Needs, L. M. Fraser, W. M. C. Foulkes, P. Maccullum, *Phys. Rev. B* **1996**, *53*, 9640.
181 W. M. C. Foulkes, L. Mitas, R. J. Needs, G. Rajagopal, *Rev. Mod. Phys.* **2001**, *73*, 33; R. J. Needs, P. R. C. Kent, A. R. Porter, M. D. Towler, G. Rajagopal, *Int. J. Quant. Chem.* **2002**, *86*, 218.
182 J. A. Pople, *Rev. Mod. Phys.* **1999**, *71*, 1267.
183 J. Čižek, *J. Chem. Phys.* **1966**, *45*, 4256; J. Čižek, J. Paldus, *Phys. Scr.* **1980**, *21*, 251.
184 J. A. Pople, M. Head-Gordon, K. Raghavachari, *J. Chem. Phys.* **1987**, *87*, 5968.
185 G. Hetzer, P. Pulay, H.-J. Werner, *Chem. Phys. Lett.* **1998**, *290*, 143; M. Schütz, G. Hetzer, H.-J. Werner, *J. Chem. Phys.* **1999**, *111*, 5691; G. Hetzer, M. Schütz, H. Stoll, H.-J. Werner, *ibid.* **2000**, *113*, 9443.
186 C. Hampel, H.-J. Werner, *J. Chem. Phys.* **1996**, *104*, 6286; M. Schütz, H.-J. Werner, *Chem. Phys. Lett.* **2000**, *318*, 370; M. Schütz, *J. Chem. Phys.* **2000**, *113*, 9986; M. Schütz, H.-J. Werner, *ibid.* **2001**, *114*, 661.
187 R. McWeeny, *Proc. Roy. Soc. London A* **1959**, *253*, 242.
188 V. A. Fock, *Dokl. Akad. Nauk SSSR* **1950**, *73*, 735; P. R. Surján, *Top. Curr. Chem.* **1999**, *203*, 63.
189 A. M. Tokmachev, A. L. Tchougréeff, *J. Comput. Chem.* **2001**, *22*, 752; A. M. Tokmachev, A. L. Tchougréeff, *J. Phys. Chem. A* **2003**, *107*, 358.
190 W. Kutzelnigg, in *Methods of Electronic Structure Theory* (H. F. Schaefer III, Ed.), Plenum Press, New York 1977.

191 O. Sinanoğlu, *J. Chem. Phys.* **1962**, *36*, 706.
192 A. Tokmachev, R. Dronskowski, *J. Comput. Chem.*, in press.
193 K. N. Kudin, G. E. Scuseria, *Chem. Phys. Lett.* **1998**, *289*, 611.
194 P. Reinhardt, J.-P. Malrieu, *J. Chem. Phys.* **1998**, *109*, 7632.
195 A. Shukla, M. Dolg, P. Fulde, H. Stoll, *Phys. Rev. B* **1998**, *57*, 1471; M. Albrecht, A. Shukla, M. Dolg, P. Fulde, H. Stoll, *Chem. Phys. Lett.* **1998**, *285*, 174.
196 T. Bredow, R. A. Evarestov, K. Jug, *Phys. Status Solidi B* **2000**, *222*, 495; T. Bredow, G. Geudtner, K. Jug, *J. Comput. Chem.* **2001**, *22*, 89; R. A. Evarestov, T. Bredow, K. Jug, *Phys. Sol. State* **2001**, *43*, 1774.
197 I. I. Ukrainskii, *Theor. Math. Phys.* **1977**, *32*, 816; V. A. Kuprievich, *Phys. Rev. B* **1989**, *40*, 3882.
198 L. Kantorovich, O. Danyliv, *J. Phys.: Condens. Matter* **2004**, *16*, 2575.
199 A. Tokmachev, R. Dronskowski, *Phys. Rev. B* **2005**, *71*, 195202; A. Tokmachev, R. Dronskowski, *Chem. Phys.*, in press (DOI: 10.1016/j.chemphys.2005.09.012).
200 R. G. Pearson, *J. Am. Chem. Soc.* **1963**, *85*, 3533; R. G. Pearson, *Science* **1966**, *151*, 172; R. G. Pearson, J. Songstad, *J. Am. Chem. Soc.* **1967**, *89*, 1827; R. G. Pearson, *J. Chem. Ed.* **1968**, *45*, 581; *ibid.* **1968**, *45*, 643.
201 R. G. Pearson, *Chemical Hardness*, Wiley-VCH, Weinheim, New York 1997.
202 R. G. Parr, R. G. Pearson, *J. Am. Chem. Soc.* **1983**, *105*, 7512; see also related earlier work by: J. Hinze, H. H. Jaffé, *ibid.* **1962**, *84*, 540.
203 J. P. Perdew, R. G. Parr, M. Levy, J. L. Balduz Jr., *Phys. Rev. Lett.* **1982**, *49*, 1691.
204 R. G. Pearson, *Inorg. Chem.* **1988**, *27*, 734.
205 R. Dronskowski, *J. Am. Chem. Soc.* **1992**, *114*, 7230.
206 R. Dronskowski, R. Hoffmann, *Adv. Mater.* **1992**, *4*, 514.
207 R. Dronskowski, *Inorg. Chem.* **1993**, *32*, 1.
208 A. E. van Arkel, *Molecules and Crystals in Inorganic Chemistry*, 2nd ed., Butterworth, London 1956.
209 J. A. A. Ketelaar, *Chemical Constitution: an Introduction to the Theory of the Chemical Bond*, 2nd rev. ed., Elsevier, Amsterdam 1958.
210 For a mathematically exhaustive overview of the different methods, see: J. Callaway, *Quantum Theory of the Solid State*, 2nd ed., Academic Press, Boston 1991.
211 J. Bullett, *Solid State Phys.* **1980**, *35*, 215.
212 B. Ahlswede, K. Jug, *J. Comput. Chem.* **1999**, *20*, 563; *ibid.* **1999**, *20*, 572; K. Jug, G. Geudtner, T. Homann, *J. Comput. Chem.* **2000**, *21*, 974; T. Bredow, G. Geudtner, K. Jug, *ibid.* **2001**, *22*, 861.
213 K. Koepernik, H. Eschrig, *Phys. Rev. B* **1999**, *59*, 1743; I. Opahle, K. Koepernik, H. Eschrig, *Phys. Rev. B* **1999**, *60*, 14035.
214 C. Herring, *Phys. Rev.* **1940**, *57*, 1169.
215 H. Hellmann, *J. Chem. Phys.* **1935**, *3*, 61; *Acta Physicochim. URSS* **1934**, *1*, 913; *ibid.* **1936**, *4*, 225.
216 P. Gombás, *Z. Phys.* **1935**, *94*, 473; *ibid.* **1941/42**, *118*, 164; *ibid.* **1942**, *119*, 318.
217 E. Antončik, *J. Phys. Chem. Solids* **1959**, *10*, 314.
218 J. C. Phillipps, L. Kleinman, *Phys. Rev.* **1959**, *116*, 287.
219 P. Gombás, *Pseudopotentiale*, Springer, Wien 1967; L. Szasz, *Pseudopotential Theory of Atoms and Molecules*, Wiley, New York 1985.
220 K. Balasubramanian, in *Encyclopedia of Computational Chemistry*, Vol. 4 (P. v. R. Schleyer et al., Eds.), Wiley, New York 1998; L. Seijo, Z. Barandiarán, in *Computational Chemistry, Reviews of Current Trends*, Vol. 4 (J. Leszczynski, Ed.), World Scientific, Singapore 1999; M. Dolg, in *Modern Methods and Algorithms of Quantum Chemistry*, NIC Series, Vol. 3 (J. Grotendorst, Ed.), John von Neumann Institute for Computing, Jülich 2000; M. Dolg, in *Relativistic Electronic Structure Theory, Part I: Fundamentals, Theoretical and Computational Chemistry*, Vol. 11 (P. Schwerdtfeger, Ed.), Elsevier, Amsterdam 2002.
221 W. A. Harrison, *Pseudopotentials in the Theory of Metals*, Benjamin, New York 1966.
222 A. Zunger, M. Cohen, *Phys. Rev. B* **1978**, *18*, 5449; D. R. Hamann, M. Schlüter, C. Chiang, *Phys. Rev. Lett.*

1979, *43*, 1494; L. Kleinman, D. M. Bylander, *ibid.* **1982**, *48*, 1425; G. B. Bachelet, D. R. Hamann, M. Schlüter, *Phys. Rev. B* **1982**, *26*, 4199; D. Vanderbilt, *ibid.* **1985**, *32*, 8412; N. Troullier, J. L. Martins, *Solid State Commun.* **1990**, *74*, 613; X. Gonze, R. Stumpf, M. Scheffler, *Phys. Rev. B* **1991**, *44*, 8503; M. Fuchs, M. Scheffler, *Comput. Phys. Comm.* **1999**, *119*, 67; S. Goedecker, M. Teter, J. Hutter, *Phys. Rev. B* **1996**, *54*, 1703; C. Hartwigsen, S. Goedecker, J. Hutter, *ibid.* **1998**, *58*, 3641.

223 D. Vanderbilt, *Phys. Rev. B* **1990**, *41*, 7892.

224 G. Kresse, J. Hafner, *J. Phys.: Condens. Matter* **1994**, *6*, 8245; G. Kresse, J. Furthmüller, *Comput. Mater. Sci.* **1996**, *6*, 15; G. Kresse, J. Furthmüller, *Phys. Rev. B* **1996**, *54*, 11169.

225 E. Wigner, F. Seitz, *Phys. Rev.* **1933**, *43*, 804; *ibid.* **1934**, *46*, 509.

226 J. C. Slater, *Phys. Rev.* **1937**, *51*, 846.

227 T. L. Loucks, *Augmented Plane Wave Method: a Guide to Performing Electronic Structure Calculations*, Benjamin, New York 1967.

228 J. Korringa, *Physica* **1947**, *13*, 392.

229 W. Kohn, N. Rostoker, *Phys. Rev.* **1954**, *94*, 1111.

230 O. K. Andersen, *Phys. Rev. B* **1975**, *12*, 3060.

231 D. Singh, *Phys. Rev. B* **1991**, *43*, 6388.

232 E. Sjöstedt, L. Nordström, D. J. Singh, *Solid State Commun.* **2000**, *114*, 15.

233 K. Schwarz, P. Blaha, G. K. H. Madsen, *Comp. Phys. Comm.* **2002**, *147*, 71.

234 O. K. Andersen, O. Jepsen, *Phys. Rev. Lett.* **1984**, *53*, 2571.

235 O. K. Andersen, C. Arcangeli, R. W. Tank, T. Saha-Dasgupta, G. Krier, O. Jepsen, I. Dasgupta, *Mat. Res. Soc. Symp. Proc.* **1998**, *491*, 3.

236 A. R. Williams, J. Kübler, C. D. Gelatt Jr., *Phys. Rev. B* **1979**, *19*, 6094.

237 P. E. Blöchl, *Phys. Rev. B* **1994**, *50*, 17953.

238 S. Goedecker, *Rev. Mod. Phys.* **1999**, *71*, 1085.

239 S. Y. Wu, C. S. Jayanthi, *Phys. Report* **2002**, *358-1*, 1.

240 P. Ordejón, E. Artacho, J. M. Soler, *Phys. Rev. B* **1996**, *53*, 10441; E. Artacho, D. Sánchez-Portal, P. Ordejón, A. García, J. M. Soler, *Phys. Status Solidi B* **1999**, *215*, 809.

241 R. P. Feynman, *Phys. Rev.* **1939**, *56*, 340.

242 P. Pulay, *Adv. Chem. Phys.* **1987**, *69*, 241.

243 H. B. Schlegel, *Adv. Chem. Phys.* **1987**, *67*, 241.

244 P. Pulay, *Mol. Phys.* **1969**, *17*, 197.

245 M. Scheffler, J. P. Vigneron, G. B. Bachelet, *Phys. Rev. B* **1985**, *31*, 6541.

246 D. M. Ceperley, *Rev. Mod. Phys.* **1999**, *71*, S438.

247 R. Haberlandt, S. Fritzsche, G. Peinel, K. Heinzinger, *Molekulardynamik*, Friedr. Vieweg & Sohn, Braunschweig, Wiesbaden 1995.

248 N. Metropolis, A. W. Rosenbluth, M. N. Rosenbluth, A. H. Teller, E. Teller, *J. Chem. Phys.* **1953**, *21*, 1087.

249 B. J. Alder, T. E. Wainwright, *J. Chem. Phys.* **1957**, *27*, 1208.

250 W. C. Swope, H. C. Andersen, P. H. Berens, K. R. Wilson, *J. Chem. Phys.* **1982**, *76*, 637.

251 *Computer Modelling in Inorganic Crystallography* (C. R. A. Catlow, Ed.), Academic Press, San Diego 1997.

252 J. D. Gale, *J. Chem. Soc., Faraday Trans.* **1997**, *93*, 629; J. D. Gale, A. L. Rohl, *Molec. Simul.* **2003**, *29*, 291.

253 J. D. Gale, *Philos. Magaz. B* **1996**, *73*, 3.

254 B. Eck, R. Dronskowski, *J. Alloys Compd.* **2002**, *338*, 136.

255 R. Car, M. Parrinello, *Phys. Rev. Lett.* **1985**, *55*, 2471; see also: M. P. Allen, D. J. Tildesley, *Computer Simulation of Liquids*, Oxford University Press 1989; R. O. Jones, *Angew. Chem. Int. Ed. Engl.* **1991**, *30*, 630.

256 M. C. Payne, M. P. Teter, D. C. Allan, T. A. Arias, J. D. Joannopoulos, *Rev. Mod. Phys.* **1992**, *64*, 1045.

257 M. E. Tuckerman, D. Marx, M. L. Klein, M. Parrinello, *Science* **1997**, *275*, 817.

258 N. L. Doltsinis, D. Marx, *Phys. Rev. Lett.* **2002**, *88*, 166402.

259 P. W. Atkins, J. de Paula, *Atkins' Physical Chemistry*, 7th ed., Oxford University Press 2001; G. Wedler, *Lehrbuch der Physikalischen Chemie*, 4th ed., Wiley-VCH, Weinheim, New York 1997.

260 F. D. Murnaghan, *Am. J. Math.* **1937**, *49*, 235.

261 F. Birch, *Phys. Rev.* **1947**, *71*, 809.

262 W. C. Mackrodt, N. M. Harrison, V. R. Saunders, N. L. Allan, M. D. Towler, E. Aprà, R. Dovesi, *Philos. Magaz. A* **1993**, *68*, 653.

263 K. H. Jack, *Proc. Roy. Soc. London A* **1948**, *195*, 34.

264 See, for example: W. Schnick, *Angew. Chem. Int. Ed. Engl.* **1993**, *32*, 806; J. Jäger, D. Stahl, P. C. Schmidt, R. Kniep, *ibid.* **1993**, *32*, 709; F. J. DiSalvo, S. J. Clarke, *Curr. Opin. Solid State Mater. Sci.* **1996**, *1*, 241; U. Steinbrenner, A. Simon, *Angew. Chem. Int. Ed. Engl.* **1996**, *35*, 552; N. Scotti, W. Kockelmann, J. Senker, St. Traßel, H. Jacobs, *Z. Anorg. Allg. Chem.* **1996**, *625*, 1435; G. Auffermann, Y. Prots, R. Kniep, *Angew. Chem. Int. Ed.* **2001**, *40*, 547.

265 K. Becker, F. Ebert, *Z. Phys.* **1925**, *31*, 268; R. Blix, *Z. Phys. Chem.* **1929**, *3*, 229; A. N. Christensen, *Acta Chem. Scand. A* **1978**, *32*, 89; J. C. Fitzmaurice, A. L. Hector, I. P. Parkin, *J. Chem. Soc., Dalton Trans.* **1993**, 2435; K. Suzuki, T. Kaneko, H. Yoshida, Y. Obi, H. Fujimori, H. Morita, *J. Alloys Compd.* **2000**, *306*, 66.

266 M. Takahashi, H. Fujii, H. Nakagawa, S. Nasu, F. Kanamaru, *Proceedings of The Sixth International Conference on Ferrites*, The Japan Society of Powder and Powder Metallurgy, p. 508, Tokyo, Japan 1992; K. Suzuki, H. Morita, T. Kaneko, H. Yoshida, H. Fujimori, *J. Alloys Compd.* **1993**, *201*, 11; K. Suzuki, T. Kaneko, H. Yoshida, H. Morita, H. Fujimori, *ibid.* **1995**, *224*, 232.

267 B. Eck, R. Dronskowski, M. Takahashi, S. Kikkawa, *J. Mater. Chem.* **1999**, *9*, 1527; G. A. Landrum, B. Eck, R. Dronskowski, *Mater. Sci. Forum* **2000**, *325/6*, 105; P. Kroll, B. Eck, R. Dronskowski, *Adv. Mater.* **2000**, *12*, 307.

268 A. Svane, Z. Szotek, W. M. Temmerman, H. Winter, *Solid State Commun.* **1997**, *102*, 473.

269 G. L. Olcese, *J. Phys. F* **1979**, *9*, 569; H. Bartholin, F. Florence, G. Parisot, J. Paureau, O. Vogt, *Phys. Lett. A* **1977**, *60*, 47.

270 A. Delin, P. M. Oppeneer, M. S. S. Brooks, T. Kraft, J. M. Wills, B. Johansson, O. Eriksson, *Phys. Rev. B* **1997**, *55*, 10173.

271 G. A. Landrum, R. Dronskowski, R. Niewa, F. J. DiSalvo, *Chem. Eur. J.* **1999**, *5*, 515.

272 There is an incredibly rich solid-state chemistry involving metal–metal bonded species, and it covers transition as well as rare-earth metals. See, for example: A. Simon, *Angew. Chem. Int. Ed. Engl.* **1981**, *20*, 1; A. Simon, *ibid.* **1988**, *27*, 159.

273 S. R. Ovshinsky, *Phys. Rev. Lett.* **1968**, *21*, 1450; A. V. Kolobov, P. Fons, A. I. Frenkel, A. L. Ankudinov, J. Tominaga, T. Uruga, *Nature Mater.* **2004**, *3*, 703; M. Luo, M. Wuttig, *Adv. Mater.* **2004**, *16*, 439.

274 J. K. Burdett, T. J. McLarnan, *J. Chem. Phys.* **1981**, *75*, 5764; J. K. Burdett, P. Haaland, T. J. McLarnan, *ibid.* **1981**, *75*, 5774; J. K. Burdett, S. Lee, *J. Am. Chem. Soc.* **1983**, *105*, 1079.

275 M.-H. Whangbo, R. Hoffmann, R. B. Woodward, *Proc. Roy. Soc. London A* **1979**, *366*, 23; M.-H. Whangbo, in *Crystal Chemistry and Properties of Materials with Quasi-One-Dimensional Structures* (J. Rouxel, Ed.), Reidel, Dordrecht 1986.

276 A. K. Bandyopadhyay, D. B. Singh, *Pramana* **1999**, *52*, 303.

277 A. Decker, G. A. Landrum, R. Dronskowski, *Z. Anorg. Allg. Chem.* **2002**, *628*, 295.

278 G. Kresse, J. Furthmüller, J. Hafner, *Phys. Rev. B* **1994**, *50*, 13181.

279 C. Soulard, X. Rocquefelte, M. Evain, S. Jobic, H.-J. Koo, M.-H. Whangbo, *J. Solid State Chem.* **2004**, *177*, 4724.

280 U. Häussermann, S. I. Simak, R. Ahuja, B. Johansson, S. Lidin, *Angew. Chem. Int. Ed.* **1999**, *38*, 2017.

281 R. Nesper, *Angew. Chem. Int. Ed. Engl.* **1991**, *30*, 789.

282 V. Sliwko, P. Mohn, K. Schwarz, *J. Phys.: Condens. Matter* **1994**, *6*, 6557.

283 G. A. Landrum, R. Hoffmann, J. Evers, H. Boysen, *Inorg. Chem.* **1998**, *37*, 5754.

284 S. Chikazumi, *Physics of Ferromagnetism*, 2nd. ed., Clarendon Press, Oxford 1997.

285 H. Lueken, *Magnetochemie – eine Einführung in Theorie und Anwendung*, Teubner, Stuttgart 1999.

286 M. P. Langevin, *Ann. Chim. et Phys., 8th Series* **1905**, *5*, 70.

287 P. Weiss, *J. de Phys., 4th Series* **1907**, *6*, 661; *C. R. Hebd. Séances Acad. Sci.* **1913**, *157*, 1405.

288 W. Heisenberg, *Z. Phys.* **1928**, *49*, 619.

289 P. A. M. Dirac, *Proc. Roy. Soc. London A* **1929**, *123*, 714.

290 E. C. Stoner, *Proc. Roy. Soc. London A* **1938**, *165*, 372; *ibid.* **1939**, *169*, 339.

291 J. Kübler, *Theory of Itinerant Electron Magnetism*, Clarendon Press, Oxford 2000.

292 J. Hubbard, *Proc. Roy. Soc. London A* **1963**, *276*, 238.

293 R. Strack, D. Vollhardt, *Phys. Rev. Lett.* **1994**, *72*, 3425; M. Kollar, R. Strack, D. Vollhardt, *Phys. Rev. B* **1996**, *53*, 9225.

294 J. E. Hirsch, *Phys. Rev. B* **1989**, *40*, 2354; J. E. Hirsch, *ibid.* **1989**, *40*, 9061; S. Tang, J. E. Hirsch, *ibid.* **1990**, *42*, 771; J. E. Hirsch, *ibid.* **1991**, *43*, 705; J. C. Amadon, J. E. Hirsch, *ibid.* **1996**, *54*, 6364; J. E. Hirsch, *Phys. Rev. B* **1997**, *56*, 11022.

295 D. Vollhardt, N. Blümer, K. Held, M. Kollar, J. Schlipf, M. Ulmke, J. Wahle, *Adv. Solid State Phys.* **1999**, *38*, 383.

296 O. K. Andersen, J. Madsen, U. K. Poulsen, O. Jepsen, J. Kollár, *Physica B* **1977**, *86–88*, 249.

297 J. F. Janak, *Phys. Rev. B* **1977**, *16*, 255.

298 V. L. Moruzzi, J. F. Janak, A. R. Williams, *Calculated Electronic Properties of Metals*, Pergamon Press, New York 1978.

299 L. Fritsche, B. Weimert, *Phys. Stat. Sol. B* **1998**, *208*, 287.

300 G. A. Landrum, R. Dronskowski, *Angew. Chem. Int. Ed.* **1999**, *38*, 1389.

301 G. A. Landrum, R. Dronskowski, *Angew. Chem. Int. Ed.* **2000**, *39*, 1560.

302 Z. S. Basinski, J. W. Christian, *Proc. Roy. Soc. London A* **1954**, *223*, 554.

303 S. C. Abrahams, L. Guttman, J. S. Kasper, *Phys. Rev.* **1962**, *127*, 2052; A. Onodera, Y. Tsunoda, N. Kunitomi, O. A. Pringle, R. M. Nicklow, R. M. Moon, *Phys. Rev. B* **1994**, *50*, 3532.

304 M. Asdente, J. Friedel, *Phys. Rev.* **1961**, *124*, 384.

305 E. Canadell, M.-H. Whangbo, *Chem. Rev.* **1991**, *91*, 965; E. Canadell, *Chem. Mater.* **1998**, *10*, 2770.

306 W. H. Lomer, *Proc. Phys. Soc.* **1962**, *80*, 489.

307 S. Asano, J. Yamashita, *J. Phys. Soc. Jpn.* **1967**, *23*, 714.

308 J. Kübler, *J. Magn. Magn. Mater.* **1980**, *20*, 277.

309 A. Decker, G. A. Landrum, R. Dronskowski, *Z. Anorg. Allg. Chem.* **2002**, *628*, 303.

310 A. J. Cox, J. G. Louderback, S. E. Apsel, L. A. Bloomfield, *Phys. Rev. B* **1994**, *49*, 12295.

311 G. Stollhoff, *Angew. Chem. Int. Ed.* **2000**, *39*, 4471; G. A. Landrum, R. Dronskowski, *Angew. Chem. Int. Ed.* **2000**, *39*, 4475.

312 P. Villars, L. D. Calvert, *Pearson's Handbook of Crystallographic Data for Intermetallic Phases*, 2nd ed., ASM International, Ohio 1991.

313 R. Smoluchowski, *J. Phys. Radium* **1951**, *12*, 389.

314 Y. V. Palguyev, A. A. Kuranov, P. N. Syutkin, F. A. Didorenko, *Phys. Met. Metallogr.* **1976**, *42*, 46.

315 E. F. Wassermann, in *Ferromagnetic Materials* (K. H. J. Buschow, E. P. Wohlfarth, Eds.), Vol. 5, North Holland, Amsterdam 1990.

316 P. Mohn, E. Supanetz, K. Schwarz, *Aust. J. Phys.* **1993**, *46*, 651. This paper includes energy-surface calculations for all three alloys, assuming perfectly ordered crystal structures.

317 Fr. Heusler, *Verh. Dt. Phys. Gesell.* **1903**, *5*, 219.

318 Y. Kurtulus, R. Dronskowski, G. D. Samolyuk, V. P. Antropov, *Phys. Rev. B* **2005**, *71*, 14425; Y. Kurtulus, M. Gilleßen, R. Dronskowski, *J. Comput. Chem.*, in press.

319 W. J. Takei, D. E. Cox, G. Shirane, *Phys. Rev.* **1963**, *129*, 2008.

320 H. Nagasaki, I. Wakabayashi, S. Minomura, *J. Phys. Chem. Solids* **1969**, *30*, 329.

321 W. Reimers, E. Hellner, W. Treutmann, P. J. Brown, *J. Phys. Chem. Solids* **1983**, *44*, 195.

322 E. A. Nagelschmitz, W. Jung, R. Feiten, P. Müller, H. Lueken, *Z. Anorg. Allg. Chem.* **2001**, *627*, 523.

323 R. Dronskowski, K. Korczak, H. Lueken, W. Jung, *Angew. Chem. Int. Ed.* **2002**, *41*, 2528.

324 B. P. T. Fokwa, R. Dronskowski, unpublished.
325 O. Gourdon, S. L. Bud'ko, D. Williams, G. J. Miller, *Inorg. Chem.* **2004**, *43*, 3210.
326 S. Kikkawa, *Mater. Trans.* **1996**, *37*, 420.
327 S. Kikkawa, M. Fujiki, M. Takahashi, F. Kamamaru, H. Yoshioka, T. Hinomura, S. Nasu, I. Watanabe, *Appl. Phys. Lett.* **1996**, *68*, 2756.
328 S. H. Elder, F. J. DiSalvo, L. Topor, A. Navrotsky, *Chem. Mater.* **1993**, *5*, 1545.
329 J. Ferrante, J. R. Smith, J. H. Rose, *Phys. Rev. Lett.* **1983**, *50*, 1385.
330 J. H. Rose, J. R. Smith, J. Ferrante, *Phys. Rev. B* **1983**, *28*, 1835.
331 B. Eck, Y. Kurtulus, W. Offermans, R. Dronskowski, *J. Alloys Compd.* **2002**, *338*, 142.
332 H. Rücker, M. Methfessel, *Phys. Rev. B* **1995**, *52*, 11059.
333 R. Dronskowski, B. Eck, S. Kikkawa, *Jpn. J. Appl. Phys.* **2000**, *39*, 3326.
334 S. Lee, R. Hoffmann, *J. Am. Chem. Soc.* **2002**, *124*, 4811; see also: S. Lee, *Acc. Chem. Res.* **1991**, *24*, 249.
335 A. J. McAlister, J. L. Murray, in *Binary Alloy Phase Diagrams*, 2nd ed. (T. B. Massalski, H. Okamoto, P. R. Subramanian, L. Kacprzak, Eds.), American Soc. for Metals, Metals Park, Ohio 1990.
336 H. Kono, *J. Phys. Soc. Jpn.* **1958**, *13*, 1444; A. J. J. Koch, P. Hokkeling, M. G. van den Steeg, K. J. De Vos, *J. Appl. Phys.* **1960**, *31*, S75.
337 J. J. van den Broek, H. Donkersloot, G. van Tendeloo, J. van Landuyt, *Acta Metall.* **1979**, *27*, 1497; S. Kojima, T. Ohtani, N. Kato, K. Kojima, Y. Sakamoto, I. Konno, M. Tsukahara, T. Kubo, *AIP Conf. Proc.* **1974**, *24*, 768.
338 A. Sakuma, *J. Phys. Soc. Jpn.* **1994**, *63*, 1422.
339 Y. Kurtulus, R. Dronskowski, *J. Solid State Chem.* **2003**, *176*, 390.
340 T. Hughbanks, Y. C. Tian, *Solid State Commun.* **1995**, *96*, 321; D. M. Teter, R. J. Hemley, *Science* **1996**, *271*, 53; S.-D. Mo, L. Ouyang, W.-Y. Ching, I. Tanaka, Y. Koyama, R. Riedel, *Phys. Rev. Lett.* **1999**, *83*, 5046; P. Kroll, *J. Solid State Chem.* **2003**, *176*, 530.
341 E. Kroke, M. Schwarz, *Coord. Chem. Rev.* **2004**, *248*, 493.
342 C. Guimon, S. Khayar, F. Gracian, M. Begtrup, G. Pfister-Guillouzo, *Chem. Phys.* **1989**, *138*, 157; see also: F. Tordini, A. Bencini, M. Bruschi, L. De Gioia, G. Zampella, P. Fantucci, *J. Phys. Chem. A* **2003**, *107*, 1188.
343 N.-G. Vannerberg, *Acta Chem. Scand.* **1962**, *16*, 2263.
344 M. G. Down, M. J. Haley, P. Hubberstey, R. J. Pulham, A. E. Thunder, *J. Chem. Soc., Dalton Trans.* **1978**, 1407; A. Harper, P. Hubberstey, *J. Chem. Res. (S)* **1989**, *7*, 194; M. Becker, J. Nuss, M. Jansen, *Z. Anorg. Allg. Chem.* **2000**, *626*, 2505.
345 U. Berger, W. Schnick, *J. Alloys Compd.* **1994**, *206*, 179; O. Reckeweg, F. J. DiSalvo, *Angew. Chem. Int. Ed.* **2000**, *39*, 412; W. Liao, R. Dronskowski, *Acta Cryst. E* **2004**, *60*, i124.
346 R. Dronskowski, *Z. Naturforsch. B* **1995**, *50*, 1245; R. Riedel, A. Greiner, G. Miehe, W. Dressler, H. Fueß, J. Bill, F. Aldinger, *Angew. Chem. Int. Ed. Engl.* **1997**, *36*, 603; M. J. Cooper, *Acta Crystallogr.* **1964**, *17*, 1452; X. Liu, A. Decker, D. Schmitz, R. Dronskowski, *Z. Anorg. Allg. Chem.* **2000**, *626*, 103.
347 F. P. Bowden, H. M. Montagu-Pollock, *Nature* **1961**, *191*, 556; M. Becker, J. Nuss, M. Jansen, *Z. Naturforsch. B* **2000**, *55*, 383.
348 O. Reckeweg, F. J. DiSalvo, *Z. Anorg. Allg. Chem.* **2003**, *629*, 177; R. Srinivasan, M. Ströbele, H.-J. Meyer, *Inorg. Chem.* **2003**, *42*, 3406; W. Liao, C. Hu, R. K. Kremer, R. Dronskowski, *ibid.* **2004**, *43*, 5884.
349 M. Becker, M. Jansen, *Z. Anorg. Allg. Chem.* **2000**, *626*, 1639.
350 X. Liu, P. Müller, P. Kroll, R. Dronskowski, *Inorg. Chem.* **2002**, *41*, 4259.
351 I. Ruppert, *Angew. Chem. Int. Ed. Engl.* **1977**, *16*, 311.
352 X. Liu, P. Müller, P. Kroll, R. Dronskowski, W. Wilsmann, R. Conradt, *ChemPhysChem* **2003**, *4*, 725.
353 A. M. Oleś, F. Pfirsch, P. Fulde, M. Böhm, *Z. Phys. B* **1987**, *66*, 359.
354 N. Rösch, S. B. Trickey, *J. Chem. Phys.* **1997**, *106*, 8940.
355 I. Grinberg, Y. Yourdshahyan, A. M. Rappe, *J. Chem. Phys.* **2002**, *117*, 2264.

356 H. L. Woodcock, H. F. Schaefer III, P. R. Schreiner, *J. Phys. Chem. A* **2002**, *106*, 11923.

357 R. Metselaar, *Pure Appl. Chem.* **1994**, *66*, 1815; K. Miga, K. Stanczyk, C. Sayag, D. Brodzki, G. Djéga-Mariadassou, *J. Catal.* **1999**, *183*, 63; G. Hitoki, T. Takata, J. N. Kondo, M. Hara, H. Kobayashi, K. Domen, *Chem. Comm.* **2002**, 1698; G. Hitoki, T. Takata, J. N. Kondo, M. Hara, H. Kobayashi, K. Domen, *J. Electrochem. Soc. Jpn.* **2002**, *70*, 463.

358 M. Jansen, H. P. Letschert, *Nature* **2000**, *404*, 980.

359 R. Marchand, Y. Laurent, J. Guyader, P. l'Haridon, P. Verdier, *J. Eur. Ceram. Soc.* **1991**, *8*, 197; E. Guenther, M. Jansen, *Mat. Res. Bull.* **2001**, *36*, 1399.

360 G. Brauer, J. R. Weidlein, *Angew. Chem. Int. Ed. Engl.* **1965**, *4*, 241; G. Brauer, J. R. Weidlein, J. Strähle, *Z. Anorg. Allg. Chem.* **1966**, *348*, 298.

361 Yu. A. Buslaev, M. A. Glushova, M. M. Ershova, E. M. Shustorovich, *Neorg. Mater.* **1966**, *2*, 2120; Yu. A. Buslaev, G. M. Safronov, V. I. Pakhomov, M. A. Glushova, V. P. Repko, M. M. Ershova, A. N. Zhukov, T. A. Zhdanova, *Neorg. Mater.* **1969**, *5*, 45.

362 D. Armytage, B. E. F. Fender, *Acta Cryst. B* **1974**, *30*, 809; C. M. Fang, E. Orhan, G. A. de Wijs, H. T. Hintzen, R. A. de Groot, R. Marchand, J.-Y. Saillard, G. de With, *J. Mater. Chem.* **2001**, *11*, 1248.

363 E. Orhan, F. Tessier, R. Marchand *Sol. Stat. Sci.* **2002**, *4*, 1071.

364 M. Weishaupt, J. Strähle, *Z. Anorg. Allg. Chem.* **1977**, *429*, 261.

365 M.-W. Lumey, R. Dronskowski, *Z. Anorg. Allg. Chem.* **2003**, *629*, 2173.

366 N. R. Crawford, J. S. Knutsen, K.-A. Yang, G. Haugstad, S. McKernan, F. B. McCormick, W. L. Gladfelter, *Chem. Vap. Dep.* **1998**, *4*, 181.

367 M. J. Redman, E. G. Steward, *Nature* **1962**, *198*, 867; R. W. Grimes, K. P. D. Lagerlöf, *J. Am. Ceram. Soc.* **1991**, *74*, 270; T. B. Joyner, F. H. Verhoek, *J. Am. Chem. Soc.* **1961**, *83*, 1069; F. Lihl, P. Ettmayer, A. Kutzelnigg, *Z. Metallk.* **1962**, *53*, 715.

368 M. R. Daniel, A. P. Cracknell, *Phys. Rev.* **1969**, *177*, 932.

369 K. Suzuki, T. Shinohara, F. Wagatsuma, T. Kaneko, H. Yoshida, Y. Obi, S. Tomiyoshi, *J. Phys. Soc. Jpn.* **2003**, *72*, 1175.

370 J. Dicarlo, A. Navrotsky, *J. Am. Ceram. Soc.* **1993**, *76*, 2467.

371 M.-W. Lumey, R. Dronskowski, *Adv. Funct. Mat.* **2004**, *14*, 371.

372 F. Rudberg, *Pogg. Ann.* **1830**, *18*, 240; H. Johnen, *Dissertation*, Universität Münster, 1952.

373 A. T. Dinsdale, *Calphad* **1991**, *15*, 317; K. Hack (Ed.), *The SGTE Casebook, Thermodynamics at Work*, The Institute of Materials, London 1996.

374 S. Fries, H. L. Lukas, *COST 507*, G. Jaroma-Weiland, IKE, Stuttgart 1998.

375 Y. Fujinaga, Y. Syono, *Mater. Transact.* **1997**, *38*, 1063.

376 P. Xiao, L. Song, H. Kang, M. Zhao, *J. Less-Common Met.* **1988**, *144*, 1; H. J. Bray, *J. Inst. Met.* **1958-59**, *87*, 49.

377 R. H. Kane, B. C. Giessen, N. J. Grant, *Acta Metallurg.* **1966**, *14*, 605.

378 R. Blachnik, *D'Ans & Lax: Taschenbuch für Chemiker und Physiker, Teil 3*, Springer, Berlin, Heidelberg, New York 1998.

379 J. von Appen, K. Hack, R. Dronskowski, *J. Alloys Compd.* **2004**, *379*, 110.

380 I. N. Nikolaev, V. P. Marin, V. W. Panyushkin, L. S. Pavlyukov, *Sov. Phys. Solid State* **1973**, *14*, 2022.

381 H. T. Hall, in *Accurate Characterization of the High-pressure Environment* (E. C. Lloyd, Ed.), NBS Spec. Publ. 326, p. 313, Washington 1971.

382 M.-W. Lumey, R. Dronskowski, *Z. Anorg. Allg. Chem.* **2005**, *631*, 887.

383 H. G. Drickamer, C. W. Frank, *Electronic Transitions and the High Pressure Chemistry and Physics of Solids*, Chapman and Hall, London 1973.

384 G. Brauer, H. Reuther, *Z. Anorg. Allg. Chem.* **1973**, *395*, 151.

385 B. Wang, B. C. Chakoumakos, B. C. Sales, J. B. Bates, *J. Solid State Chem.* **1996**, *122*, 376.

386 J. M. Cocciantelli, P. Gravereau, J. P. Doumerc, M. Pouchard, P. Hagenmuller, *J. Solid State Chem.* **1991**, *93*, 497.

387 P. Kroll, private communication, Aachen 2004.

388 G. Auffermann, Y. Prots, R. Kniep, *Angew. Chem. Int. Ed.* **2001**, *40*, 547.

389 G. V. Vajenine, G. Auffermann, Y. Prots, W. Schnelle, R. K. Kremer, A. Simon, R. Kniep, *Inorg. Chem.* **2001**, *40*, 4866.

390 X. Liu, R. Dronskowski, *Z. Naturforsch. B* **2002**, *57*, 1108.

391 G. O. Brunner, D. Schwarzenbach, *Z. Kristallogr.* **1971**, *133*, 127.

392 J. Pflüger, J. Fink, W. Weber, K.-P. Bohnen, G. Crecelius, *Phys. Rev. B* **1984**, *30*, 1155.

393 S. Nagakura, *J. Phys. Soc. Jpn.* **1961**, *16*, 1213; Y. Hirotsu, S. Nagakura, *Acta Metall.* **1972**, *20*, 645.

394 M. Launay, R. Dronskowski, *Z. Naturforsch. B* **2005**, *60*, 437.

395 X. Liu, M. Krott, P. Müller, C. Hu, H. Lueken, R. Dronskowski, *Inorg. Chem.* **2005**, *44*, 3001.

396 X. Liu, M. A. Wankeu, H. Lueken, R. Dronskowski, *Z. Naturforsch. B* **2005**, *60*, 593.

397 G. W. Wiener, J. A. Berger, *J. Met.* **1955**, *7*, 360; G. W. Wiener, J. A. Berger, *Ann. Chim.* **1983**, *8*, 533; S. K. Chen, S. Jin, T. H. Tiefel, Y. F. Hsieh, E. M. Gyorgy, D. W. Johnson Jr., *J. Appl. Phys.* **1991**, *70*, 6247; H. Jacobs, D. Rechenbach, U. Zachwieja, *J. Alloys Compd.* **1995**, *227*, 10.

398 S. Matar, P. Mohn, G. Demazeau, B. Siberchicot, *J. Phys. (France)* **1988**, *49*, 1761; C. A. Kuhnen, R. S. de Figueiredo, V. Drago, E. Z. da Silva, *J. Magn. Magn. Mater.* **1992**, *111*, 95.

399 H. H. Stadelmaier, A. C. Fraker, *Trans. Metall. Soc. Aime* **1960**, *218*, 571; B. Siberchicot, S. F. Matar, L. Fournès, G. Demazeau, P. Hagenmuller, *J. Solid State Chem.* **1990**, *84*, 10; S. Matar, L. Fournès, S. Chérubin-Jeanette, G. Demazeau, *Eur. J. Solid State Inorg. Chem.* **1993**, *30*, 871; D. Andriamandroso, S. Matar, G. Demazeau, L. Fournès, *IEEE Trans. Magn.* **1993**, *29*, 2; C. A. Kuhnen, R. S. de Figueiredo, A. V. dos Santos, *J. Magn. Magn. Mat.* **2000**, *219*, 58.

400 P. Mohn, K. Schwarz, S. Matar, G. Demazeau, *Phys. Rev. B* **1992**, *45*, 4000; C. A. Kuhnen, A. V. dos Santos, *J. Alloys Compd.* **2000**, *297*, 68; S. Matar, P. Mohn, J. Kübler, *J. Magn. Magn. Mat.* **1992**, *104–107*, 1927; R. S. de Figueiredo, C. A. Kuhnen, A. V. dos Santos, *ibid.* **1997**, *173*, 141; C. A. Kuhnen, R. S. de Figueiredo, A. V. dos Santos, *ibid.* **2000**, *219*, 58; R. S. de Figueiredo, J. Foct, A. V. dos Santos, C. A. Kuhnen, *J. Alloys Compd.* **2001**, *315*, 42; A. V. dos Santos, C. A. Kuhnen, *ibid.* **2001**, *321*, 60; C. Paduani, *J. Magn. Magn. Mat.* **2004**, *278*, 231.

401 C. Cordier-Robert, J. Foct, *Eur. J. Solid State Inorg. Chem.* **1992**, *29*, 39.

402 F. Tessier, A. Navrotsky, R. Niewa, A. Leineweber, H. Jacobs, S. Kikkawa, M. Takahashi, F. Kanamaru, F. J. DiSalvo, *Solid State Sci.* **2000**, *2*, 457.

403 J. von Appen, R. Dronskowski, *Angew. Chem. Int. Ed.* **2005**, *44*, 1205.

404 A. Houben, P. Müller, J. von Appen, H. Lueken, R. Niewa, R. Dronskowski, *Angew. Chem. Int. Ed.*, in press (DOI: 10.1002/anie.200502579).

405 M. Jansen, *Angew. Chem. Int. Ed.* **2002**, *41*, 3746.

406 J. Hulliger, M. A. Awan, *Chem. Eur. J.* **2004**, *10*, 4694; J. Hulliger, M. A. Awan, *J. Comb. Chem.* **2005**, *7*, 73.

407 K. Jug, G. Geudtner, *J. Solid State Chem.* **2003**, *176*, 575.

408 J. A. Hedvall, *Reaktionsfähigkeit fester Stoffe*, Johann Ambrosius Barth, Leipzig 1938; K. Hauffe, *Reaktionen in und an festen Stoffen*, 2nd ed., Springer, Berlin 1966; H. Schmalzried, *Solid State Reactions*, 2nd ed., Verlag Chemie, Weinheim 1981.

409 A. Simon, *Angew. Chem. Int. Ed.* **1997**, *36*, 1788; S. Deng, A. Simon, J. Köhler, *ibid.* **1998**, *37*, 640; S. Deng, A. Simon, J. Köhler, *J. Am. Chem. Soc.* **2002**, *124*, 10712; S. Deng, A. Simon, J. Köhler, *J. Solid State Chem.* **2003**, *176*, 412.

Index

Computational Chemistry of Solid State Materials. Richard Dronskowski

ISBN: 3-527-31410-5

d

p

q

Acknowledgments

A couple of figures have been adapted from previously published materials, and I want to thank the copyright holders for granting the permission to include them in this book.

Fig. 1.3 adapted from J. K. Burdett et al., *Solid State Commun.* **1981**, *40*, 923, by permission of Elsevier. Fig. 2.5 adapted from M. Boggio-Pasqua et al., *J. Molec. Struct. (Theochem)* **2000**, *531*, 159, by permission of Elsevier. Fig. 2.6 adapted from W. Kutzelnigg, *Einführung in die Theoretische Chemie*, 1993, by permission of Wiley-VCH. Figs. 2.14 and 3.12–3.17 adapted from A. Decker et al., *Z. Anorg. Allg. Chem.* **2002**, *628*, 295, by permission of Wiley-VCH. Figs. 2.23–2.27, 3.18–3.22 and 3.25–3.27 adapted from G. A. Landrum and R. Dronskowski, *Angew. Chem. Int. Ed.* **2000**, *39*, 1560, by permission of Wiley-VCH. Fig. 2.29 adapted from M. Kohout and A. Savin, *J. Comput. Chem.* **1997**, *18*, 1431, by permission of John Wiley & Sons. Figs. 3.5 and 3.6 adapted from B. Eck et al., *J. Mater. Chem.* **1999**, *9*, 1527, by permission of The Royal Society of Chemistry. Figs. 3.9–3.11 adapted from G. A. Landrum et al., *Chem. Eur. J.* **1999**, *5*, 515, by permission of Wiley-VCH. Figs. 3.23–3.24 adapted from A. Decker et al., *Z. Anorg. Allg. Chem.* **2002**, *628*, 303, by permission of Wiley-VCH. Figs. 3.28 and 3.29 adapted from R. Dronskowski et al., *Angew. Chem. Int. Ed.* **2002**, *41*, 2528, by permission of Wiley-VCH. Figs. 3.30 and 3.31 adapted from B. Eck et al, *J. Alloys Compd.* **2002**, *338*, 142, by permission of Elsevier. Fig. 3.33 adapted from R. Dronskowski et al., *Jpn. J. Appl. Phys.* **2000**, *39*, 3326, by permission of the Institute of Pure and Applied Physics. Figs. 3.34–3.39 adapted from Y. Kurtulus and R. Dronskowski, *J. Solid State Chem.* **2003**, *176*, 390, by permission of Elsevier. Figs. 3.41 and 3.57–3.59 adapted from M. Launay and R. Dronskowski, *Z. Naturforsch. B* **2005**, *60*, 437, by permission of Verlag der Zeitschrift für Naturforschung. Figs. 3.42–3.44 adapted from X. Liu et al., *ChemPhysChem* **2003**, *4*, 725, by permission of Wiley-VCH. Figs. 3.45 and 3.46 adapted from M.-W. Lumey and R. Dronskowski, *Z. Anorg. Allg. Chem.* **2003**, *629*, 2173, by permission of Wiley-VCH. Figs. 3.47 and 3.48 adapted from M.-W. Lumey and R. Dronskowski, *Adv. Funct. Mater.* **2004**, *14*, 371, by permission of Wiley-VCH.

Figs. 3.49–3.52 adapted from J. von Appen et al., *J. Alloys Compd.* **2004**, *379*, 110, by permission of Elsevier. Figs. 3.53–3.56 adapted from M.-W. Lumey and R. Dronskowski, *Z. Anorg. Allg. Chem.* **2005**, *631*, 887, by permission of Wiley-VCH. Figs. 3.60–3.63 adapted from J. von Appen and R. Dronskowski, *Angew. Chem. Int. Ed.* **2005**, *44*, 1205, by permission of Wiley-VCH.